Africa Now!

"The editors have to be congratulated for assembling an impressive range of continental-based African scholars and researchers proximate to the current status of African and some diasporic social research. This compendious, timely handbook's chapters represent the best in distilling the topical, theoretical, and case illustrative scholarship centering on displacement and migration, while chapters variously synthesize scholarship on, amongst other thing, land and food security, indigenous technology, securitization, governance, the sex trade, and Chinese investment. Highly recommended".

—Pablo Idahosa, Professor, *African and Development Studies, York University*

Adebusuyi Adeniran • Lanre Ikuteyijo
Editors

Africa Now!

Emerging Issues and Alternative Perspectives

palgrave
macmillan

Editors
Adebusuyi Adeniran
Department of Sociology and Anthropology
Obafemi Awolowo University
Ile-Ife, Nigeria

Lanre Ikuteyijo
Department of Sociology and Anthropology
Obafemi Awolowo University
Ile-Ife, Nigeria

ISBN 978-3-319-62442-6 ISBN 978-3-319-62443-3 (eBook)
https://doi.org/10.1007/978-3-319-62443-3

Library of Congress Control Number: 2017954333

Cover illustration: kkong / Alamy Stock Photo
Cover design by Fatima Jamadar

Printed on acid-free paper

This Palgrave Macmillan imprint is published by Springer Nature
The registered company is Springer International Publishing AG
The registered company address is: Gewerbestrasse 11, 6330 Cham, Switzerland

Contents

About the Editors

Adebusuyi Isaac Adeniran holds a PhD in Development Sociology. He is presently a Senior Lecturer/Researcher at Obafemi Awolowo University, Ile-Ife, Nigeria and a Visiting Researcher at The Harriet Tubman Institute for Research on Global Migrations of African Peoples, York University, Canada and The Nordic Africa Institute, Uppsala University, Sweden. He has been a recipient of an International Sociological Association doctoral scholarship (2011), an Africa Initiative doctoral research grant (2011) and a Comparative Research Network grant from the Council for the Development of Social Science Research in Africa (2012). He has published widely in renowned local and international outlets. Among his recent publications are 'Engendering an Endogenous Framework for Socio-economic Development in the ECOWAS Sub-region' (*Critical Sociology*, Vol. 40(2), March 2014) and *Regional Economic Communities: Exploring the Process of Socio-economic Integration in Africa* (a co-edited book) (CODESRIA, 2014). He specialises in ECOWAS (the Economic Community of West African States) and international development, Chinese–African relations, transnationalism, migration, integration and identity studies.

Lanre Olusegun Ikuteyijo is a lecturer and researcher in the Department of Sociology and Anthropology at Obafemi Awolowo University, Nigeria. He obtained his PhD in Sociology and Anthropology from the same university, specialising in criminology, migration and urbanisation. His research interests include youth migration, urban crime and community policing. He has published in reputable journals, contributed to a number of books and written a number of encyclopaedia entries published by SAGE and Blackwell Publishing.

He was a visiting researcher at the Department of History, McMaster University, Canada between 2011 and 2012 under the auspices of the Centre for International Governance Innovation, Waterloo, Canada. Ikuteyijo has received a number of academic grants and fellowships, including: a Qualitative Research Network Africa Grant for his PhD thesis; a Council for Development of Social Research in Africa research grant; and an Africa Initiative research grant for work at the Brown International Advanced Research Institute (Population and Development). His latest publications include: 'Illegal Migration and Policy Challenges in Nigeria' (*African Portal*); 'The Image of Nigeria Police: Lessons from History' (*Journal of Applied Security Research*, Volume 9, Taylor and Francis Group, 2014); and 'EU Migration Policies and the Criminalisation of the Senegalese Irregular Migration Flows' (in *Between Migrations to Europe and Returns: International Commentary*, edited by Michelle Gonneli: Vol. 10, no. 35, Scuola Superiore Sant'Anna, Pisa, Italy).

List of Tables

1

Introduction: Africa in the Twenty-First Century

Adebusuyi Isaac Adeniran
and Lanre Olusegun Ikuteyijo

While a plethora of scholarly studies have been presented on issues related to African development, this book aims to provide alternative frameworks that will not just be participatory in conception and practice, but will present a contextual study of varying aspects of the developmental enterprise in 21st Century African society. Changing connotations of the practice of international development have meant that aside from the traditional multilateral players, other important individual and group participants, such as China and even India, have become involved. In addition, annual remittances to Africa are now becoming central to the development process. In 2014 alone, accrued remittances to Africa from its diaspora communities surpassed \$60 billion, which significantly outweighed all funds from the multilateral funding framework within the same period. Many African economies have steadily remained among the fastest growing globally. The case of Ethiopia, which has recorded success in converting raw materials, especially leather and coffee, into finished

A.I. Adeniran (✉) • L.O. Ikuteyijo
Department of Sociology and Anthropology, Obafemi Awolowo University, Ile-Ife, Osun State, Nigeria

A. Adeniran, L. Ikuteyijo (eds.), *Africa Now!*,
https://doi.org/10.1007/978-3-319-62443-3_1

products for export, has been quite remarkable. Ironically, a paradox still exists. Despite having alternative and less cumbersome sources of funding, with commendable economic growth indices, African countries have been unable to translate the related opportunities into sustainable human development outcomes for the majority of their citizens. Poverty continues to rage, unemployment continues to soar, electoral violence and insurgency are rife, and the state of social services and infrastructures continues to worsen. Democratic structures in most African countries are still in their nascent forms; some countries (for example, Burundi, Rwanda, Zimbabwe and Egypt) struggle with teething problems in organising free, fair and credible elections, which should ordinarily serve as the basis for engendering a sustainable process of democratisation. Similarly, the threat arising from the proliferation of small arms and light weapons, with an attendant rise in the rate of armed robbery, insurgencies and terrorism, all account for the abysmal level of human security on the continent, for instance in Sudan, Somalia, South Africa and Nigeria.

In an age of rapid mobility, Africans are moving. But, in contrast to what some literature would have us believe, larger migrant flows occur within the continent than from African to non-African countries. This, of course, comes with its attendant challenges—irregular migration has become a front-burner issue on the African Union's agenda. Displacement due to natural causes, conflicts and man-made disasters have resulted in an increase in the number of refugees and internally displaced persons.

It is obvious that Africa has played a springboard role in the emergence of contemporary civilisation. Developments such as the evolution of modern hydraulics from rudimentary water technology along the Nile, the advanced level of brass carving technology among the Nupes in Central Nigeria, the sophistication of democratic governance in the old Oyo Empire (Western Nigeria) before contact with the West, and the social organisation among the Pygmies of Central Africa all add credence to the idea that Africa provided the initial lead to the modern period.

In spite of this conceptual advantage, African leaders, especially in the immediate post-colonial era, have pursued strongly misguided policies in attempting to promote development in human and resource management, science and technology, and the economy and society. Fanon (1961: 2) observed that nationalism within the African context has often failed to achieve 'liberation across class boundaries because its aspirations are primarily those of the colonised bourgeoise—a privileged middle class who perhaps seeks to defeat the prevailing colonial rule only to usurp its place of dominance and surveillance over the working class lumpenproletariat'. As noted by Ake (1996: 10), 'the problem with Africa has not in any way been a failure of development per se for it has never even been on the agenda of the political leadership who were mainly engrossed with the utilisation of the statist apparatus to augment their individual resource base which had been vastly devastated by the stifling colonial rule'. Since the post-colonial era, African political elites have unarguably remained the problem of Africa. In actual fact, development as a concept has been left—unpatriotically—in the hands of the international community. The result has been that issues relating to dependency, and hitherto assumed to have been resolved in the political arena, have resurfaced on the socio-economic, educational and technological planes, largely to the detriment of Africa's citizens.

However, matters of leadership do generate debates and altercations globally and will continue to do so because of the significance of leadership to national development. The relationship between leadership and development is so intertwined that the nature of one affects the other. The level to which the available resources are programmatically and judiciously applied for the enhancement of the socio-economic and technological well-being of the citizenry is usually a function of leadership, just as the extent of development often influences leadership functionality.

As surmised by Adeniran (2006: 45), 'unlike their counterparts in Malaysia and other South African nations in the 1960s and 1970s, African leaders have often found it difficult to ensure the sustainability of a reform agenda once adopted'. Of course, this has been an offshoot of political bickering, epitomised by poor leadership that has been glaringly defective in vision. This book, therefore, interrogates the menace of inept leadership, which is conspicuous in all spheres of African society in relation to the high levels of contemporary societal dislocation. Can Africa possibly develop with its prevailing leadership structures? Can the continent's acclaimed initial advantage actually be resuscitated? These and related issues are addressed in this book, from an African perspective in particular.

Contextualising Development Crises in Africa

Various factors have been offered in the literature to explain the apparent failure of the development enterprise in parts of Africa: the colonial legacy; social pluralism and its centrifugal tendencies; the corruption of the continent's leaders; poor labour discipline; the lack of entrepreneurial skills; weak planning and incompetent management; inappropriate policies; the stifling of market mechanisms; low levels of technical assistance; the limited in-flows of foreign capital; falling commodity prices; unfavourable terms of trade; and the low levels of saving and investment. While these factors are not irrelevant to the inherent developmental crises, they are formidable impediments—alone or in combination—to the process of development. However, the assumption so readily made that there has been a failure of development is misleading. The problem is not so much that it has failed as that it never really existed. By all indications, retrogressive political dynamics in Africa have comprised the greatest impediment to the development process. By the time independence was being attained, the centrifugal tendencies had become increasingly complex as a result of a needless retention by the emergent independence leaders of the typologies that structured colonial politics.

In Africa, a conspicuous foundational flaw in tackling developmental challenges could be attributed to the historical absence of an endogenous emancipating policy framework that would have recognised the historical specificities of the people in conception and in implementation. Working within the framework of EAGER/PSGE, it has been argued that no African country (with the probable exception of Mauritius) has ever been able to sustain a socio-economic reform agenda. This has had a negative impact on the state of human development and level of technological re-appraisal in Africa (Adeniran 2005). It has been observed that every government has engaged in some form of recidivism; many have abandoned reform efforts altogether, only to revive them in the face of threatened or actual cuts in donor support. This is in contrast to their Asian counterparts in the 1960s and 1970s, who ensured that transformational reform, once adopted, was sustained.

The book covers issues that relate to or touch on African development (from both endogenous and exogenous cleavages): governance and models of development; natural resources and environmental management; agricultural practice; afforestation/reforestation; regional integration projects; migration and cross-border mobility; land grabbing; the China–Africa partnership; global North–South relations; electoral violence; and ethnic agitation.

The book is comprised of five parts with 19 chapters: Part I: Conceptualising the Development Process in Twenty-First Century Africa; Part II: Environment, Natural Resources Management and Development; Part III: Governance, Economy and Sustainable Development; Part IV: Transnationalism, Education and Human Development; and Part V: Crime, Conflicts and Regional Dynamics.

In Chap. 1, Adeniran and Ikuteyijo present a conceptual understanding of changing dynamics of development as a human process, from an African perspective especially. The role of Africa in the evolving globalised world is explored in the chapter.

Links interrogates the right to development in a reconfigured global order in Chap. 2. She presents in particular a theoretical and conceptual interpretation of changing connotations of international development from an emergent South–South partnering perspective; in this, the budding China–Africa relationship signifies an alternative framework for enabling societal development.

The ethnicisation of the oil industry in Nigeria is explored in Chap. 3. Babatunde suggests that this development has made the continued existence of Nigeria as a single socio-political entity highly improbable. Rather than facilitating development, the oil industry in Nigeria has served as a significant impediment to engendering national integration. Undoubtedly, however, continuing social injustice in the country has played a part in this regard.

Chapter 4 utilises a phenomenological approach to study the governance of the extractive industries in Côte d'Ivoire, with a focus on the development outcomes of inherent governance processes. As noted by Aristide, various regulatory agencies of the government in charge of the extractive industries have routinely jettisoned established procedures.

Equally, the functions of these agencies have visibly overlapped. Therefore, rather than helping human and national development, the existence and governance of the extractive industries in Côte d'Ivoire have been detrimental to it.

In Chap. 5, Fajobi studies the association between the practice of land degradation and the unsustainability of the human development process in Nigeria. In his view, the management of land resources in Nigeria has been poorly approached. If the present practice is permitted to continue without intervention, Fajobi believes that the challenge of food insecurity in Nigeria could worsen significantly within the foreseeable future. Additionally, rural–urban migration in the country would worsen as a result of the unabated destruction of fertile land through excavation and related retrogressive purposes.

Cavanagh explores in Chap. 6 how land and natural resources impact on the process of governance in contemporary Kenya, employing both historical and evaluative approaches. According to Cavanagh, while the 2010 Kenyan constitution certainly presents a new institutional framework for governing access to state rents, so too does it promise to shift the terms of access and ownership of land—a crucial aspect of Kenya's political economy since at least the nineteenth century. Here, at least in *de jure* terms, the land reforms enshrined in both the 2010 Constitution and the accompanying National Land Policy of 2009 seem promising.

Chapter 7 focuses on the connections among urban agriculture, food security and poverty alleviation in post-apartheid metropolitan Durban, South Africa. Khan affirms that in Durban, the boundaries of the former apartheid city have been extended by 68%; this represents 900 square kilometres of the entire land space. Regardless of persisting poverty, many of these areas are endowed with natural resources that are conducive to agriculture and, of course, food security for the urban poor. Buoyed by this possibility, the Durban metropolitan government, in partnership with a range of stakeholders, has pioneered urban agriculture projects across the city. In this chapter, Khan evaluates the significance of these projects in providing food security for the urban poor in South Africa.

In Chap. 8, Gidado seeks to establish a relationship between the notion of internal democracy within major political parties and the intrinsic crisis of leadership in the Nigerian legislature. In Gidado's view, since the return of Nigeria to civil rule in 1999, the crisis of leadership that usually heralds the birth of each legislative assembly has been spawned by the spoils of office. He contends that the impunity that is largely associated with the primaries of political parties reflects the character of politics in a rentier state; public office is merely exploited for personal enrichment and the structure of the state routinely promotes impunity in governance. He surmises that the crisis of leadership is unconnected with the desire to serve, but is largely driven by the opportunities that such exalted office provides for the distribution of patronage.

Xiao explores in Chap. 9 observable social interactive tendencies between Chinese immigrants and their Nigerian hosts, especially as these pertain to unethical transnational practices. Using an ethnographic design, Xiao unearths the socio-cultural implications of Chinese–Nigerian interactions in the context of corruption. As he notes, transnational mobility has intensified the normalisation of corruption in the everyday life of Chinese immigrants in Nigeria. When the tact implicit in Chinese *guanxi* (interpersonal relationship) practices encounters explicit Nigerian corruption, Chinese immigrants attempt to adapt strategies to the Nigerian context by employing *guanxi* practices. Nevertheless, incompatible affectivities in interpersonal relationships between Chinese and Nigerians lead to misunderstandings. According to Xiao, intrinsic trans-border peculiarities have allowed Chinese immigrants in Nigeria to become embedded in the Nigerian patron-clientism, in which ordinary Nigerians also function as agents.

In Chap. 10, Agunyai identifies emerging patterns of governance crises in twenty-first century Nigerian society. Agunyai observes that even with a change of government in 2015, indicators of good governance, such as the rule of law, service delivery, transparent electoral processes, the competency of appointed officials and so forth, have continued to be elusive. In Nigeria, good governance has been repeatedly threatened by disrespect for the rule of law, non-adherence to court judgements and incompetent political appointees, who do nothing except accuse past governments. Agunyai surmises that corruption and the inordinate interests of the political leaders over national interests have continued to undermine good governance in contemporary Nigeria.

Adeyanju in Chap. 11 examines the short- and long-term outcomes of migration of Africans to Canada for undergraduate education. Using as case studies Nigerian undergraduate immigrants in a group of Canadian universities, selected Nigerian parents with wards in undergraduate programmes in Canada, international student advisers, a high school teacher in Nigeria, and international student recruitment agents, Adeyanju discovers that Nigerians are attracted to Canadian higher institutions of learning because Canada offers a quality of education that is unavailable in Nigerian public universities. The findings also show that most of the students intend to take up permanent Canadian residence upon graduation, based on the understanding that the status will enhance their life chances globally. He equally notes that the students' decision to remain in Canada after graduation has implications for development, especially as it pertains to issues relating to 'brain drain'.

The deconstruction of the significance of imperialism as a barrier to the growth and development of science and technology in Africa is the focus of Chap. 12. Kolawole analyses essentially point to the veracity of the intrinsic digital divide between the Global North and the Global South.

In Chap. 13, Tawfik investigates how innate trans-border security challenges in North Africa can be explored within the process of enabling its regional integration agenda. Tawfik affirms that North African countries have faced ever-increasing security challenges, especially after the popular uprisings of 2010. She argues that in spite of such challenges, the region still lacks the necessary conditions to strengthen security and developmental interdependence when juxtaposed with other parts of the continent. In terms of conflict, various countries in the region are cooperating more with international powers that are eager to intervene directly in the region, rather than delegating the responsibility of this management to regional African powers. For instance, in order to resolve the historical tensions between Morocco and Algeria, Tawfik suggests investing in the creation and maintenance of a sub-regional organisation. Such an organisation would provide a framework for security and economic cooperation, and for supporting the building of state institutions in countries such as Libya. Indeed, taking a step in this direction is key to fostering security and developmental integration within the region.

With the application of ethnographic techniques, Osezua, in Chap. 14, presents an alternative perspective of survivors of human trafficking from Nigeria to Europe. She takes the Bini sub-ethnic group in Southern Nigeria as the study's unit of analysis. Osezua confirms that there is a different side to the story, which previously obscured the agency of the trafficked female victim—their lived realities have been routinely ignored. According to her, many of the survivors are still eager to go back to Europe if the opportunity presents itself, despite the ongoing campaign against human trafficking, due to the prevalent poverty in Nigeria. It is worth noting that the proceeds of international sex trade are perceived as the quickest way out of poverty among the study's subjects.

In Chap. 15, Singh analyses the concept of human security as an important component of individuals' well-being. She employs South Africa as her unit of analysis, suggesting that 22 years of transition of the country from apartheid to democracy has not marked a commensurate change in the subsisting criminal justice system of the country. In the chapter, she assesses the roles of the law enforcement components of South Africa, that is, the police as well as other functionaries of the criminal justice system. She suggests how they should be sensitised towards a civilised relationship with the citizens, especially in terms of observing human rights.

The implications of environmental conflicts for forced migration in the Nigerian Niger Delta is the focus of Chap. 16. Okwechime highlights the causal relationship between environmentally induced conflicts and forced migration in Nigeria. Focusing on the Niger Delta region, he draws attention to the role of the subsisting capitalist mode of production and affirms that its patterns of social relations have generated scarcities, which have invariably resulted in environmental conflicts on one hand and forced migration on the other hand in the oil-producing region. He presents a criticism of the standpoints of Homer-Dixon's population-conflict thesis and the Nigerian state's alliance with the global capitalist system. Okwechime outlines the inherent implications of such a cleavage for environmental justice, security and sustainability in the oil-producing communities.

By exploring the conditions of informal immigrant traders from member countries of the Southern Africa Development Commission (SADC) in Johannesburg, South Africa, Nshimbe and Moyo in Chap. 17 present inherent challenges to the regional integration project within the region.

They utilise three levels of analysis—regional, national and local—with discussions restricted to economically active individuals. Taking a qualitative approach to informal cross-border traders in Johannesburg from the SADC countries, Nshimbe and Moyo illustrate the absence of any supportive legal or institutional mechanism that would promote and protect the subsistence of immigrant traders. In their view, this would imply that an important element in the SADC integration project is unwelcome or ignored. Nshimbe and Moyo suggest that SADC states should establish localised SADC institutions in order to ensure appropriate domestication, implementation and monitoring of regional integration mechanisms.

In Chap. 18, Atoyebi analyses emergent dynamics in female migration patterns in West Africa. As he notes, in recent times, women are increasingly engaging in migration, in spite of the inherent risks and challenges; this is coupled with the physical, environmental and other human-induced threats that continue to hinder cross-border, socio-economic interactions within the sub-region. In order to improve their living conditions, Ejigbo-Yoruba women from Nigeria have used trans-border migration as a coping mechanism. Atoyebi concludes that the functionality of inter-regional mobility, as practised by the Ejigbo-Yoruba women migrants, should be utilised as a platform for reducing poverty and for fostering growth and development within the sub-region.

Chapter 19 presents an overview of the entire book. According Adeniran and Ikuteyijo, this book, unlike other preceding scholarly efforts, contains relevant endogenous procedures that could be used to address the challenge of transforming ideas into sustainable opportunities in Africa. They suggest tenable procedures for understanding Africa within the context of emerging global realities. The book will be useful for researchers and policy-makers focusing on Africa-related issues and subjects across the world.

Bibliography

Adeniran, A. (2005, June). *Poverty and political stability in Nigeria*. Unpublished seminar paper, Department of Sociology, University of Lagos, Lagos.

Ake, C. (1996). *Democracy and development in Africa*. Washington, DC: The Brookings Institution.

Fanon, F. (1961). *The wretched of the Earth*. Benjamin Grave 98, Brown University.

Internet Source

EAGER. (2001). Past reform growth policy: The role of leadership. Retrieved from www.eagerproject.com

Newspaper Article

Adeniran, A. (2006, August 23). A non-dependent framework for development. *Thisday*, p. 45.

Adebusuyi Isaac Adeniran holds a PhD in Development Sociology. He is presently a Senior Lecturer/Researcher at Obafemi Awolowo University, Ile-Ife, Nigeria and a Visiting Researcher at The Harriet Tubman Institute for Research on Global Migrations of African Peoples, York University, Canada and The Nordic African Institute, Uppsala University, Sweden. He has been a recipient of an International Sociological Association doctoral scholarship (2011), an Africa Initiative doctoral research grant (2011) and a Comparative Research Network grant from the Council for the Development of Social Science Research in Africa (2012). He has published widely in renowned local and international outlets. Among his recent publications are 'Engendering an Endogenous Framework for Socio-economic Development in the ECOWAS Sub-region' (*Critical Sociology*, Vol. 40(2), March 2014) and *Regional Economic Communities: Exploring the Process of Socio-economic Integration in Africa* (a co-edited book) (CODESRIA, 2014). He specialises in ECOWAS (the Economic Community of West African States) and international development, Chinese–African relations, transnationalism, migration, integration and identity studies.

Lanre Olusegun Ikuteyijo is a lecturer and researcher in the Department of Sociology and Anthropology at Obafemi Awolowo University, Nigeria. He obtained his PhD in Sociology and Anthropology from the same university, specialising in criminology, migration and urbanisation. His research interests include youth migration, urban crime and community policing. He has published in reputable journals, contributed to a number of books and written a number of encyclopaedia entries published by SAGE and Blackwell Publishing. He was a visiting researcher at the Department of History, McMaster University, Canada between 2011 and 2012 under the auspices of the Centre for International Governance Innovation, Waterloo, Canada. Ikuteyijo has received a number of academic grants and fellowships, including: a Qualitative Research Network Africa Grant for his PhD thesis; a Council for Development of Social Research in Africa research grant; and an Africa Initiative research grant for work at the Brown International Advanced Research Institute (Population and Development). His latest publications include: 'Illegal Migration and Policy Challenges in Nigeria' (*African Portal*); 'The Image of Nigeria Police: Lessons from History' (*Journal of Applied Security Research*, Volume 9, Taylor and Francis Group, 2014); and 'EU Migration Policies and the Criminalisation of the Senegalese Irregular Migration Flows' (in *Between Migrations to Europe and Returns: International Commentary*, edited by Michelle Gonneli: Vol. 10, no. 35, Scuola Superiore Sant'Anna, Pisa, Italy).

Part I

Conceptualising the Development Process in Twenty-First Century Africa

2

The Right to Development in a Reconfigured Global Order: Conceptual Dynamism and South–South Cooperation

Stacey Links

Introduction

Ensuring the equitable distribution as well as the sustainability of international development remains a critical challenge and quest, one yet to be fulfilled in contemporary international society. Development, a term that once denoted the goals of pure economic progress, has in past decades morphed into a far broader concept, encompassing the goals of equality, equity, fairness and sustainability. Despite the conceptual evolution of development as a process integral to international society, it has long been recognised as a fundamental and inalienable right, thereby forming part of the human rights corpus. Development, therefore, no longer solely describes a process of social, economic and political change that lubricates the cogs enabling society to function optimally. Rather, it has broadened and deepened to include the RTD of both individuals and collectivities in the international system. The point of departure of this chapter is that the notion of development as a right reflects the international recognition

S. Links (✉)
Rotterdam, Netherlands

A. Adeniran, L. Ikuteyijo (eds.), *Africa Now!*,
https://doi.org/10.1007/978-3-319-62443-3_2

that development is integral to achieving an equitable, just and fair international order. The right is enshrined in the United Nations Declaration on the Right to Development (DRTD), one of the primary international manifestations recognising the necessity of development in the international system. Although other global instruments related to development exist (for example, international development agencies, various United Nations development mechanisms, and so on), it is the declaration on the right itself that has elevated development as a concept and process to the status of an international right, entrenching its global importance and priority. The RTD is thus an instrumental, conceptual prism through which development can be understood in an ever-changing international arena. The RTD, however, is not only instrumental in how we understand development as a concept and phenomenon, but additionally frames how we embark on securing this process and goal as a right.

This chapter is focused on development as a phenomenon and the challenges in securing what I term the 'international dimension' of the RTD. It departs from the premise that our conceptual understanding of this right necessarily affects how we pursue securing such development (through policy, both national and international). Put simply, our understanding of the 'content' of the RTD (how we conceptualise the substance of the right) directly affects the avenues through which international society seeks to achieve such development. Similarly, the way in which we conceptualise the right determines how and what we perceive the challenges to the RTD to be.

In this quest to understand the challenges to development in contemporary international society, recognising the dynamism of international relations, as evidenced by the reconfiguration of global power structures, is central to contextualising current challenges. These reconfigurations are not only integral to how we conceptualise development but also affect the broader discourse of international development specifically. The gradual transition to an increasingly multipolar world has witnessed the emergence of 'new' international players, signifying a shift in the traditional 'gatekeepers' of knowledge and centres of power to include often excluded voices, both nationally and internationally. This, it is argued, is already moving power from those who have enjoyed a monopoly over development discourse and subsequent policy to emerging global players.

These changes will exert, and are already in the process of exerting, a significant impact on how development is conceptualised and, more importantly, how we seek to secure the RTD. These 'new' voices stand as equal stakeholders in the discourse of development and are essentially comprised of equal rights holders. They are, moreover, reconfiguring conventional ideas on development both at national and international levels, all of which will have a significant impact on how we seek to secure the RTD and thus substantively realise development internationally. To reiterate, these shifts will impact *what* we identify as the fundamental challenges to development of our time, as well as *how* we seek to address these challenges.

As a starting point, it is perhaps necessary to clarify what is meant by development within the scope of this analysis. Development in the context of this chapter looks at what is traditionally termed 'international development'. Thus, although concepts of development exist in an array of disciplines, development within the scope of this chapter refers to international and/or global development as framed in the post-Second World War era. The concept of international development that followed categorised the world in terms of those nations that were least developed to those that were developing and, finally, those that were deemed developed. International development thus encompasses what is referred to in the DRTD as:

> [A] comprehensive economic, social, cultural and political process, which aims at the constant improvement of the well-being of the entire population and of all individuals on the basis of their active, free and meaningful participation in development and in the fair distribution of benefits resulting therefrom. (UN DRTD 1986: Preamble)

This kind of development has historically been largely synonymous with the idea of economic development within the international economic system. However, in taking the RTD as a starting point, the conception of development addressed in this chapter stretches further to include a notion of *fair* and *equitable* development. It is this aspect of equitable, fair and just development that distinguishes the RTD as a fundamental inalienable human right from development as an end in itself.

While recognising the integral nature of the national dimension of the RTD, this chapter, due to space and scope restraints, is concerned with the international dimension of the RTD—it focuses on the amenability of the international system (of nation states, institutions and policies that shape and guide development globally) to bring about equitable, fair and just development.

In order to understand the international dimension of the RTD, understanding its initial raison d'être is critical. The paper contextualises the emergence of the right as emerging with the contemporary idea of international or global development, particularly in the aftermath of the Second World War and the subsequent period of decolonisation and the emergence of newly independent states. This historical contextualisation is essential if we are to understand the conceptualisation of the right as well as its goals and vision. As will be subsequently highlighted, the international financial and development institutions that emerged focused 'on alleviating poverty and improving the living conditions in previously colonised countries' (Gregory 2009: 155–156). The question, however, remains as to whether or not these international institutions and actors are able to achieve the poverty alleviation and basic living conditions promised, given the current conceptualisations of development?

In spite of the vast array of international human rights currently enshrined in international law, the RTD remains uniquely positioned in its capacity for transformation and emancipation of international society. The aspects of transformation[1] and emancipation in particular, though often overlooked, situate the right as intimately connected to ideas of global and social justice, which prioritises an equitable and just distribution of power, access, wealth and opportunity. The right has frequently been cited as being far from fulfilled (United Nations Office of the High Commissioner 2013; Oduwole 2014), resulting in the need to investigate factors that potentially contribute to these seemingly perpetual shortfalls. This chapter holds that current dominant conceptualisations of the right, in which national dimensions frequently override international ones, have in fact masked an essential aspect of the right itself, namely equal, equitable and fair development globally.

What has become apparent is that the goals of equitable and just development are central to the idea of development and the RTD itself.

However, in a reconfigured international system, particularly with a more vocal and prominent Global South, conventional ideas on development and how to achieve development have been interrogated and renegotiated. The rise of China, for example, to the status of international economic heavyweight has altered traditional modes of viewing the international economic order. As will be explored later, the case of China in Africa and Sino-African relations more broadly speaking has sparked a significant amount of speculation regarding the merits of this relationship, particularly insofar as human rights are concerned. Illustrating the exploration of alternative conceptualisations of the RTD, this chapter also seeks to challenge common accusations levelled against China's involvement in Africa, namely, that this relationship necessarily hinders equitable, just and fair development on the continent. Instead, it suggests that a holistic and dialogical understanding of the RTD, one which includes its international dimension as indispensable, is required. Including the international dimension allows for a re-conceptualisation of both the Sino-African relationship as well as an expanded understanding of chronic underdevelopment on the African continent.

This chapter proposes steering away from purely economic- and individual-oriented models in order to 'reclaim' the international dimension of the right as integral to any development initiative. As such, the Sino-African relationship is taken as indicative of the conceptual dynamism of the RTD as well as indicative of how challenges to securing the right are framed. Tackling the often criticised example of China's involvement in Africa additionally highlights the limitations of dominant conceptualisations, which frequently overlook the emancipatory and more collectivist aspects of global justice inherent in the international dimension of the right. In the light of this analysis, it will be argued that as far as challenges to securing the RTD are concerned, traditional power configurations and structures present a persistent challenge to securing the right itself. Moreover, these seemingly immutable power structures are argued to hinder the much-needed development of the developing world. Instead, a dismantling of these traditional centres of power through alternative avenues to the dominant institutions of the Global North are viewed as a possible path to securing the right in its international dimension. It is argued that alternative institutions, such as the New Development Bank (NDP), formerly referred to as the BRICS Development Bank, and the Asian Infrastructure Investment Bank (AIIB), stand out as possible alternatives to the traditional powers that have simply perpetuated the status quo and are essentially complicit in reproducing unequal power relations in international development. Viewing these emerging institutions as possible complementary avenues for securing the RTD opens up the space for a more equitable interpretation of the right in its international dimension.

The RTD in Context

The Context and Emergence of the RTD

The RTD within the United Nations framework emerged as a third-generation right subsequent to the establishment of the International Covenant on Civil and Political Rights and the International Covenant on Economic, Social and Cultural Rights. In the categorisation of rights by way of 'generations', this right is widely considered to form part of the group belonging to 'solidarity rights', which include among others the right to peace, the right to the environment, the right to ownership of the common heritage of mankind, and the right to communication (Udombana 2000: 761). Although many, particularly from the developing world, have described the RTD as a 'core right' in that 'all other rights [are seen to] stem from, or point to, this right' (Udombana 2000: 770), it has received a significant amount of attention and debate. Despite debate regarding the justiciability of the right,[2] and consensus support for the right, it is irrefutable that the issue of development has become more central than ever to the maintenance and progression of a just and peaceful international economic and political system—particularly in the current era of increased interconnectedness and globalisation. The RTD is thus a right which, by its very nature, calls for the insights of a multidisciplinary perspective if it is to be understood, consolidated and safeguarded. In a disciplinary sense, the RTD is most notably related to the fields of international political economy, development and international law.

As already mentioned, for current purposes the RTD will refer to the international right as articulated in article 1(1) of the UN DRTD. This states: 'The Right to Development is an inalienable human right by virtue of which every human person and all peoples are entitled to participate in, contribute to, and enjoy economic, social, cultural and political development, in which all human rights and fundamental freedoms can be fully realized' (1986). Articles 2(3) and 3(3) go further, emphasising the duty on states to ensure appropriate national development policies and urging international cooperation in order to ensure development as well as the elimination of potential obstacles to development (UN RTD 1986). The right is therefore simultaneously individual and collective, meaning that both individuals as well as collectivities, such as groups of peoples, have both the legal and moral right and corresponding responsibility to ensure development. As is encompassed in the DRTD, and noted by the Office of the United Nations High Commissioner for Human Rights (OHCHR), 'The human person is identified as the beneficiary of the Right to Development, as of all human rights. The Right to Development can be invoked both by individuals and by peoples. It imposes obligations both on individual states—to ensure equal and adequate access to essential resources—and on the international community—to promote fair development policies and effective international cooperation' (OHCHR 1996–2016). Thus, in as far as the international dimension of the right is concerned, obligations to ensure such development rest both on individual states and the international community more broadly speaking. This was similarly recognised in the International Conference on Human Rights (1968), which concludes that the RTD encompasses a 'collective responsibility of the international community to ensure the attainment of the minimum standard of living necessary for the enjoyment of human rights and fundamental freedoms by all persons throughout the world' (Proclamation of Tehran 1968). It is this international aspect of the RTD that is the focus of this analysis. It is critical to address how this international dimension has been conceptualised and what effects this conceptualisation has had on what we perceive to be some of the fundamental challenges in achieving international development.

In order to grasp the content of the RTD, an important starting point is to understand the historical context of its emergence. In doing so, we better positioned to engage with its current international application. Further, understanding its contextual and historical raison d'être allows for insight into its fundamental aims and goals in contemporary international society. This in turn enables the identification of some of the most pre-eminent challenges presented in securing this right in its international dimension.

Many studies have focused on the national conditions and challenges faced in securing the RTD and thus development itself. These have often been identified as weak and poor governance, a lack of state legitimacy and a lack of the institutional frameworks necessary to carry out just, equitable and fair development. In these dominant narratives, government elites are often denoted the gatekeepers of the 'fruits' of development and are thus at the heart of the problem.[3] While this is at once a critical truth and a serious challenge, the excessive focus on national elites has led to the neglect of the international dimension and the context of development as an arena for violation and potential change. The international factors that contribute to chronic underdevelopment and/or present challenges to the development agenda are often downplayed in favour of laying the sole onus on the lack of the adoption of liberal democratic values within a national context. However, as is pointed out by General Comment No. 3 of the Nature of States Parties' Obligations of the UN Committee on Economic, Social and Cultural Rights, economic and/or political systems are irrelevant to realising the RTD. Thus, at a national level, 'the Covenant is neutral and its principles cannot accurately be described as being predicated exclusively upon the need for, or the desirability of a socialist or a capitalist system, or a mixed, centrally planned, or laissez-faire economy, or upon any other particular approach' (UN Committee on Economic, Social and Cultural Rights 1990). This is reiterated in the United Nations 'Commemoration of 25 years of the United Nations Declarations on the Right to Development', which states:

> What constitutes 'development' is largely subjective, and in this respect development strategies must be determined by the people themselves and adapted to their particular conditions and needs. No one model of

> development is universally applicable to all cultures and peoples. All development models, however, must conform to international human rights standards. (United Nations Office of the High Commissioner 2013: 60)

Despite the policy prescriptions of reform frameworks such as those of the Washington Consensus and the accompanying failures of strucutral adjustment policies (SAPs), for example, the national adoption of and adherence to a narrow set of liberal democratic values on a national level do not ensure a just, fair and equitable development trajectory, nor is this a mandatory prescription. With regards to development internationally then, the international system in itself has remained a largely undemocratic and asymmetrical space—with few alternative avenues. Thus, while it seems that international institutions function 'democratically' in principle, the owners of production, knowledge and thus power remain largely in the hands of the developed world.[4] This imbalance has created a fundamental challenge to the aspiration of equitable, fair and just global development.

Added to this, there is the commonly cited realist position that no overarching governing body exists or can exist in international society—in short, the international system is deemed anarchical. If this is our starting point then democratisation in the form of equality and equity (inclusive ownership, distribution of ownership and opportunity, and so on) becomes essential to the establishment of just, fair and equitable development. Similarly critical is the opening up or perhaps the claiming of 'space' (in the form of proactive democratisation) for the developing world to shape, in as far is possible, their own course of development within the framework of international human rights. In creating such a space, the traditionally marginalised voices of the Global South are able to join the conversation and become equal contributors to the terms of development.

Due to limits of scope, this chapter is unable to delve into questions related to the most optimal international arrangement; however, it analyses the situation as it currently stands and in subsequent sections look at current global reconfigurations, such as the Sino-African relationship, as potential opportunities to be harnessed vis-à-vis the RTD. Thus, while these configurations are riddled with challenges, it is argued that they in fact present new opportunities for an international developmental framework that can in part secure the RTD. The characterisation of this relationship as necessarily inimical to human rights-centred development is therefore on the one hand partially false, while on the other neglects to understand the relationship vis-à-vis the international dimension of the RTD. While concerns raised regarding the 'national' dimension of the right are warranted—in particular regarding the fair, just and equitable development implications of these relations—the wholesale characterisation of Sino-African relations, for example, as detrimental to the protection of the RTD neglects another important, yet overlooked international dimension of the right. It is argued that these often-neglected dimensions require equal consideration and attention if we are to fully secure the RTD.

It is important to bear in mind, however, that one of the main driving forces that led to the right's emergence was the historical process of decolonisation. The rationale and implications of colonisation as well as the subsequent process of decolonisation is integral to understanding the sentiment that drove the demand for a RTD. The wave of independence and decolonisation that developed in the late 1960s, particularly out of Africa, saw the 'elevation of economic development goals to the top of the international agenda' (Udombana 2000: 763) and sparked the beginnings of the conceptualisation of the RTD. Although the RTD was not the focal point or primary motive behind the colonial independence movements (as was the right to self-determination), one could argue that it emerged as the natural result of the integration of the former colonies as independent sovereign states into what was an already established international market economy. The introduction of a host of newly independent and autonomous states into the fold of the international market economy meant that the power dynamics integral to the functioning of this system would soon be exposed. The idea, therefore, that newly independent states could enter into the international system as equal stakeholders in spite of a history of formal colonisation and dependence, one could argue, was both naive and short-sighted. Shortly after the euphoria of self-determination and self-rule had subsided, new challenges were ready to present themselves to these new members of the international system. The chance for these newly emerged states to catch their breath before encountering the next challenge, namely equitable and fair integration into the world market economy, was underestimated by many in both the developed and developing world.

The swift realisation of the challenges faced by the newly independent states emerged largely endogenously, and resulted in a strong desire for former colonial powers to be held to account for the damage that they had caused during the colonial era. This was envisioned to be optimally secured in the form of reparations. Despite the reticent attitude of the former colonial powers regarding reparations, the newly independent states were adamant about the former colonial powers' continued obligations—particularly in as far as social, economic and cultural rights had been violated through colonial exploitation and subsequently recognised as gross violations of international law. These were seen as being owed to the former colonies, and as not a question of choice, welfare or charity, but rather as entitlements (Udombana 2000: 763).

A string of international developments and pivotal moments, such as the 1973 Arab oil embargo and the accompanying search for a New International Economic Order, followed by the United Nations Development Decade starting in 1961, culminated in the integration of the RTD as a fundamental international human right. The obstacles faced by the developing world were becoming increasingly apparent and, as such, the United Nations General Assembly adopted the declaration on the RTD in 1986, which marked a significant international development.[5] It was subsequently internationally reaffirmed and legitimised through the 1993 Vienna Declaration and Programme of Action, which set out the conditions (article 10) of the right as an integral part of fundamental human rights (1993).

A further set of factors that prompted the international recognition of the RTD was the recognised failure of national and international development policies vis-à-vis newly independent states. As will be discussed later, the predominant policies of the 1980s placed newly independent states under the administration of the World Bank and focused on structural adjustment, economic efficiency and macro-regulation as the cornerstones of policy (Sano 2000: 740). However, the failures of structural adjustment policies in Africa and the continued concentration of wealth and power in the Global North signalled a need for redress.

Rights of the Third World

Within the discourse of international human rights, the Third World—or rather the Global South—are often characterised as the 'recipients' or consumers of rights as well as the occupants of the geographic space of the site of human rights violations. This common characterisation, evident in the narratives of Africa as the 'dark continent' or even the notion of 'harmful traditional practices', has raised serious concerns with regards to the reproduction of problematic narratives. These narratives not only relate to agency and autonomy but also include the power dynamics of knowledge production and contribution when it comes to international human rights. Makau Mutua succinctly frames and problematises the narrative espoused by the international human rights regime as one characterised by 'victims, savages and saviours' (2001). In this narrative, the Global South is frequently characterised as the victims and/or savages, while the North maintains the status of saviour. This narrative is deeply problematic for a number of reasons (though the scope of this chapter does not allow for a delineation of these here). However, one important aspect of this narrative is the static power dynamic that is reproduced. This has effects on what is seen as legitimate knowledge, who is taken seriously, and who dictates the boundaries, course and story of international human rights.

One aspect related to the issue of power in international human rights is the fact that many international human rights have indeed had their origins in the Global North.[6] As a consequence, many discussions on international human rights have been centred on Euro- American experiences, histories and perspectives. But while much of the international corpus of formalised human rights can be traced back to the European experience, I contend that this does not signify that the idea of human rights is not to be found across other societies. Similarly, the contextual setting of the 'origins' of human rights should not lead to ownership of the discourse. However, because the formalisation of human rights largely took place in the Global North, the dominant story of human rights has been told through a Eurocentric lens (Paupp 2014; Mutua 2001; Anghie 2005).

Nevertheless, in spite of 'origin' debates, the significance of the RTD has been instrumental in reversing the traditional narratives that place the Global North at the centre of knowledge production and the Global South as the site where this knowledge is consumed. Instead, the historical emergence of the RTD has, if anything, shifted the emphasis onto the 'self'-emancipation of the countries of the Global South, as active contributors to the body of international human rights. Roland Rich echoes this sentiment by stating: 'it provided a means of "ownership" by the "South" of a significant part of the human rights discourse that had previously been monopolised by the "North"' (2002: 26). Notably, the RTD was *first* recognised in the African Charter of Peoples Rights of 1981, under article 22, both as an individual and collective right. It was only subsequent to its incorporation into the African Charter that the United Nations adopted and recognised the right internationally. This is of significance not only in terms of recognising the founding role and agency of the African continent in this now internationally recognised right, but moreover in terms of recognising the African, and more broadly speaking the developing world's, experience in 'creating' this right.

As with the right to self-determination, the RTD created a distinct category of nations outside the Global North within international law (Rich 2002: 26). In particular, and as will be elaborated upon later, the 'collective' nature of the RTD was spurred on by the globalised international economic order into which newly independent states were incorporated. It would subsequently be recognised that national governments were limited in 'unilaterally meet[ing] their peoples' needs', (Rich 2002: 26) thus requiring concerted collective effort. This is similarly true for the contemporary international landscape, in which 'problems confronting any contemporary society can no longer be met by even the most resolute action of any single state' (Wellman 2000: 24) This speaks to the 'solidarity' characterisation of the RTD, as Karel Vasak famously termed it (1977). It is precisely the international dimension of the right that demands concerted collective effort from across the international community. Just as the right to self-determination rests on collective claims, so too does the international dimension of the RTD, without which the fulfilment of this right will remain illusory.

As already mentioned, the plight of the former colonies in substantiating the right to self- determination was swiftly followed by the international formalisation of the RTD and as such is often viewed as the extension of the right to self-determination, or put it differently, the economic dimension of the right to self-determination (Udombana 2000: 769). These 'twin' rights are additionally viewed largely as collective rights in that they are peoples' rights, and thus the sum of their individual components. This is also supported in the UN development document, which states that 'the Right to Development is related to the right to self- determination, which has many aspects, both individual and collective' (United Nations Office of the High Commissioner 2013: 60). More still, the right was in its inception both promoted and critiqued as a collective right of states in particular (Sengupta 2013: 76). This emphasis on collectivism stands in contrast to the individual emphasis often advocated by proponents of rights in the Global North. Thus, while there has been an insistence on universalism, particularly in the Western discourse of human rights, one of the biggest challenges regarding the RTD has come from the lack of support from the West itself (Isa 2005). As a result, group rights remain a category and topic of much debate within human rights circles, particularly within classical liberal frameworks, where rights are first and foremost seen as protection for the individual against the state and where civil and political rights have been the primary focus. But as Felipe Gomez Isa rightly points out, 'This [has been] one of the main flaws in Western discourse on the universality of human rights. While the majority of these countries stand up for universality, they are thinking exclusively in the universality of civil and political rights, forgetting entirely that the dignity of human beings also demands the universality of economic, social and cultural rights' (Isa 2005: 28). This similarly applies to the contention of individual versus collective rights, and while the merits of this kind of individual protection are evident, there is also an equally valid and legitimate claim to group rights.[7] In particular, such rights have been deemed instrumental in their capacity to facilitate the realisation of the principles of equity and equality in societies.[8] Thus, group rights as held by entire groups rather than by its individual members carry with them the ability to address issues of systemic disadvantage or redistribution. Because such injustices affect entire groups, the action required is generally far more powerful and significant in a collective context than in the case of multiple individual claims. With regards to the international dimension of development, therefore, the developing world as a collective brought this right to the attention of the international community. This, coupled with the historical context of its emergence and its raison d'être, can characterise it as a right of the 'Third World'. It is within this context that the RTD has subsequently been conceptualised as the 'precondition of liberty, progress, justice and creativity. It is the alpha and omega of human rights, the first and last human right, the beginning and the end, the means and the goal of human rights' (Bedjaoui 1991: 1177).

One of the current challenges to development and an aspect that the right advocates for is fair, equitable and just development, nationally and internationally. In the context of this aim, however, equality of opportunity is not sufficient. Here, Amartya Sen's capabilities approach to human rights has been instrumental in the subsequent conceptualisation of development as led by the United Nations Development Programme (Sen 1999). Sen emphasises the importance and centrality of the capability to enjoy the same rights. In this sense, Sen is not solely concerned with equality in its formal or absolute sense, but rather integrates principles of justness and fairness into his approach. An emphasis on equity is required. Regarding development in the international context, the question can be asked whether all states or societies are equally able (in terms of opportunity and capability) to achieve development. Ensuring this kind of equity on an international scale, I argue, is vital in securing both the international and national dimensions of the RTD—ensuring an international economic system in which all nations are capable of achieving a minimum level of development. This includes removing potential barriers that would hinder the ability to achieve this, and nationally in the sense that national development is necessarily hindered by blockages encountered on the international level. An example of such challenges or international barriers would be unequal trade agreements that disadvantage developing states, or the conditioning of much-needed loans that inevitably circumscribe the possibility for growth and development.[9] While a one-size-fits-all solution is not possible, a re-evaluation of the merits of traditional modes of engagement is a starting point in finding a viable solution.

In considering the historical and contextual emergence of the RTD, the political dimensions of this right stand out prominently. The politically contested nature of the RTD therefore should come as no surprise: it emerged out of a need to address past injustices that primarily affected the previously colonised regions of the world. Implicit in the right has been the demand that the former colonial powers are held accountable for the maintenance or creation of a system of underdevelopment which has plagued the Global South and continues to perpetrate a system of inequality. Therefore, despite the demise of colonial political control, economic control was and is seen as an issue that remains to be addressed. The world capitalist economic system in which certain powers have inherited positions of privilege while others have inherited continued underdevelopment maintains a stark North–South divide. The RTD was and is therefore often seen as the continued struggle for the full independence of the colonial world. As a result, the interpretation and application of the RTD speaks to many underlying and deep-seated international political tensions (Oduwole 2014: 5).

The Continued Third World Predicament: A Cautionary Note

Often, when discussing development and the RTD in contemporary international relations, the example of the 'Asian Tigers' or the successes of the 'Flying Geese Model' are mentioned. These models are often compared and/or even suggested as the way forward for, for example, African states. Although I am unable to go into extensive detail here, it is both important and necessary to highlight briefly a number of key factors that are often ignored in critiques of African development vis-à-vis the Asian experience when discussing possible development trajectories.

The array of challenges faced both on a national and international level regarding the RTD can be summed up in what I term the 'Third World Development Predicament'—inspired by Mohammed Ayoob's work the *Third World Security Predicament* (1995). It is apparent that many of the criticisms, challenges and hurdles in the conceptualisation of the RTD are challenges and particularities shared by much of the developing world. In his work, Ayoob investigates security as it concerns the Third World and highlights the limitation of state-centric explanatory frameworks, instead advocating a more complex interconnected set of explanatory factors. Because of considerations of history and capacity, he argues that the Third World finds itself in a unique predicament. If we turn to the issue of development and look at it from the perspective of a type of the 'Third World Development Predicament,' we can state a case largely in the same vein, namely that the Third World predicament is attributable to the the push and pull forces of the local vs. the global, as well as to the 'catch-up' hamster-wheel within which the Third World finds itself trapped. Central to this predicament is trying to balance local realities with global demands, while having little weight in terms of voice on an international scale. This complex reality limits the ability to change and places the Third World in a quandary when it comes to development. Moreover, most assessments of Third World development or chronic underdevelopment eschew the all too apparent evidence that the nations of the developed world managed to develop in eras of absolute sovereignty, no global regulatory bodies and virtual 'freedom' to develop under limited restrictions. These levels of development have been characterised by self-sufficiency as well as surpluses of both capital and trained people, all of which were accumulated over historical periods of Western dominance of the international economic system (Standing 2014).

The examples of the high performing Asian economies of the 1970s are often referenced as being the prime example of development possibilities in spite of the 'curse' of colonialism and international power. Though these arguments hold some truth, they fail to recognise the very particular and at times 'non-human rights'-centred strategies that such newly industrialised economies pursued.[10] Further, such analyses often oversimplify the pathways and characteristics which such states adopted in order to industrialise and develop. The adoption of a 'developmentalist' state,[11] as based on a strong central government, is a frequently overlooked aspect of Asian development (Fritz and Menocal 2007: 536–537). These developmental states did not face restraints with regard to development with regards for the principles of human rights. Much of this development therefore took place at the expense of human rights. This is largely as a result of the East Asian dismissal of the neoliberal Washington Consensus as a guiding framework for development (Stiglitz 2001).

Moreover, despite the often-cited issue of corruption as necessarily impeding development in its purest sense, many have overlooked the role of political elites in stimulating and catalysing economic growth and broad-based development in East Asia. Thus, while scholars point to the importance of domestic accountability and governance as the source of *chronic underdevelopment* in Africa, the comparison of heterogeneous historical and cultural experiences (such as Africa, Asia and Latin America) greatly oversimplifies the complexities of differing development trajectories (Osinbajo and Ajayi 1994). Notably, East Asia began its developmental transition more than half a century ago, with a distinct set of historical factors, institutions and circumstances. The reference to the Asian development experience when analysing the RTD therefore does not hold significant enough weight in terms of understanding the right in its international and/or national dimensions. Moreover, placing the Asian experience under the rubric of human rights-centred development, as opposed to pure economic development, routinely ignores Asia's path to development. Drawing on comparisons for the assessment of development paths thus requires caution and context. This brief overview simply seeks to highlight the vast complexity of the RTD and what remain some of the fundamental challenges to securing this right.

Conceptual Dynamism

The RTD, like any other right, has navigated a host of conceptual changes since its inception. As is seen in the above section(s), the right emerged within a very particular context and with very specific aims. This undoubtedly had a significant bearing on how the right was conceptualised. Since then, however, this conceptualisation has changed with the times. On the one hand, this has been a welcome development that has enabled the incorporation of equally valid principles, such as equity and social justice, into our conceptualisations. On the other hand, however, as a result of international power configurations, the right in its international dimension has undergone a significant amount of transformation and has simultaneously emphasised certain dimensions over others. This section briefly journeys through the right's conceptual dynamism, highlighting how conceptualisation affects the formulation of challenges and subsequent strategies aimed at safeguarding the right.

To briefly illustrate the importance of conceptualisation, it is useful to look at how the various conceptualisations across time have resulted in the adoption of specific approaches. This has implications for how we currently conceptualise the RTD, what we identify as the primary challenges, and how we seek to tackle these.

The RTD and its Conceptual Evolution

Despite the initial rationale and intent behind the RTD, development discourse has evolved over time. Particularly concerning, however, has been the dominance of economic frameworks advocated by international financial institutions. These frameworks, largely dominated by the developed world, have faltered on a number of counts and have perpetuated unequal terms of trade, the strain of external debt repayment and the concentration of economic power (and thus political power) in the hands of a few (United Nations Office of the High Commissioner 2013: 61; Stiglitz 2015). Although it is not the aim of this chapter to tackle the motives and merits of structural adjustment programmes, for example, it is important to underscore their role in creating and sustaining a particular power dynamic within the international system. Rather, it is argued that by understanding the failure of such programmes we can gain insight into the limitations of earlier understandings of 'development'. If we look at the centrality of neoliberal ideas (as embodied by structural adjustment policies) in the 1970s and 1980s, we are able to recognise how a particular conceptualisation of development led to the adoption of particular policies. Thus, during this period, a purely economic and linear conception of development was predominantly concerned with bringing the newly independent states into the fold of the international market economy. There was, however, little regard for the sustainability of these programmes, or for their legitimacy or appropriateness in the societies concerned. Issues of a structural nature that affected equality of opportunity or access were not of concern.

The failure of the structural adjustment programmes of the 1980s and the early 1990s, however, saw a shift of emphasis towards fairer terms of trade. While these adaptations were welcome, they largely maintained the structures of the international economic system, but sought to protect less developed nations from the detrimental effects of unequal and asymmetrical trade arrangements. Similarly, aid as development in the form of official development assistance took off as a central concept in international development circles. However, these too would later on be deemed as essentially detrimental to the economics of the Global South as it built a system premised on the dependency of the developing world on the developed world. This would further perpetuate an unequal international system that left the developing world largely powerless against the often-harsh restrictions imposed through tied aid. These conditions frequently neglected to understand the local realities and contexts of receiving developing states.

Despite changes in the approach of development agencies, mismatches persisted. The mid-1990s saw yet another change to development conceptualisation, being largely driven this time by the goals of absolute poverty reduction. This vision of development was embodied by the UN Millennium Development Goals and included human rights in development frameworks. What emerged were development programmes that sought to undo the paternalistic patterns of engagement that had emerged in the aid discourse. Instead, the focus was placed on, for example, the empowerment of entrepreneurs in the developing world. This involved programmes of microcredit and finance for local businesses, seeking to stimulate domestic production. With regards to human rights-centred development, the fundamental tenets of international human rights would be the focus and foundation of any development initiative. Thus, local initiatives would be supported only if these fostered the protection of human rights. Development for its own sake was widely criticised and renounced as illegitimate and unacceptable. This era of human rights-centred development still remains the dominant development framework, albeit one adjusted with some additional conceptual considerations.[12]

Non-governmental organisations focusing on human rights, in particular, have heralded human rights-centred development—that is, placing individuals and vulnerable groups at the centre of development agendas—as the way forward. In their focus on mainly individual rights, these groups have largely emphasised civil and political rights, often to the neglect of socio-economic and cultural rights (Bell & Carens 2004). Such programmes largely function at the national level and emphasise the individual conceptualisation of the RTD. At an international level, however, these approaches have had little to say regarding perpetuated power dynamics and the inherent or apparent contradictions of human rights principles of equality, equity and fairness vis-à-vis the international economic system. In fact, human rights-centred development as a particular conceptualisation of the RTD has dominated the development discourse but has frequently overlooked the international dimension of the right. Calls to address the international dimension regarding state equality have often been dismissed as being state-centric and thus too vulnerable to the whims of dominant groups, to the peril of the rights of individuals. This mutually exclusive take on the individual and collective aspects of the right, as mentioned, is neither productive nor reflective of the right and its goals in its entirety (United Nations Office of the High Commissioner 2013: 11). Instead, as is reiterated in the UN report *Realizing the right to development*, the international dimensions of the right 'are of major and increasing significance'. The report goes on to say that 'it is impossible to consider development without regard for the international context in which it takes place' (United Nations Office of the High Commissioner 2013: 8). Thus, while human rights-centred development has largely focused on the national, one dimension cannot be considered to the exclusion of the other.[13]

The international dimension of the RTD is comprised of a number of issues. One of these issues relates to threats to development that are of a universal nature, i.e. that affect all societies. Environmental degradation and climate change are examples of universal threats. This has led to the embrace of sustainable development as a central notion in development circles. There are, however, other international challenges that remain to be addressed.

An increasingly pressing concern is the essential compatibility of human rights development in an international capitalist economic system which has remained largely unchanged since the 1970s. Some argue that the two in themselves thrive on similar standards of inequality and that neither are concerned with issues of equal distribution per se; they do not provide the necessary 'floor' for populations (Paupp 2014; Moyn 2015). Similarly, it appears that neither seeks to redress structural imbalances that perpetuate such inequality. Thus, despite an increased emphasis on corporate social responsibility with regards to multinational corporations, for example, the ability of the international economic system to engender an equitable and just distribution of the 'fruits of development' is increasingly questioned. Though it is not the aim of this chapter to evaluate the merits or legitimacy of the international economic order, the example of the international economic order and its inherent ability to bring about equitable, just and fair development underscores the point that our conceptualisation(s) (of the RTD) necessarily implicate what we identify as the fundamental challenges to securing the right. In conceptualising the RTD as necessarily involving the idea of international or global justice, the interrogation of the fundamental structures of the contemporary international economic system is unavoidable. Questions that relate to who wields power, and who determines the course of development and the parameters of the discourse on development, are all central in problematising the RTD. This brings us to a further aspect of the international dimension of the RTD, one related to power, equality and equity.

What is of interest here is whether the principle of equity as it pertains to the RTD can be incorporated into the current international system, and in particular insofar as states are concerned. It is critical that we understand the power dynamics of the current system in order to uncover avenues for redistribution and the balancing of current power imbalances. The end goal is to create a more equitable international framework in which all states become equal contributors, stakeholders and participants in international development.

As it currently stands, the developed world still wields a significant amount of power over the developing world, creating an unequal playing field in international development. This, I argue, is not only due to the power which is accorded to differing levels of development but also concerns the possibilities for alternative avenues for development.

The International Dimension of the RTD: Power, Participation, Equality and Equity

The rationale behind the emergence and conceptualisation of development as a right and as an extension of the right to self-determination remains an integral component of the international dimension of 'development as a right'. In this conceptualisation, the importance of equality and equity within the international system is crucial. There is an emancipatory aspect to this conceptualisation to which the issue of equal participation and opportunity is central. The initial emancipatory and liberating aspects of the right therefore placed emphasis on the idea of international equality. This idea of international equality in the international arena in particular is ideal in tackling some of the stubborn international challenges relating to power. In this conceptualisation, international equality and equity become prerequisites for securing the RTD. The right, however, is not often framed in this emancipatory light. If the RTD is to retain its applicability and remain true to its original intent, it is necessary that we ensure the consideration of its emancipatory conceptualisation. So far, little attention has been paid to such conceptualisations. Instead, most research and analysis has favoured a focus on the more legally oriented challenges, such as the justiciability, scope and necessity of the right.

In spite of the emancipatory emphasis in the emergence of the RTD, the unequal power dynamic that largely characterised the 1980s and early 1990s remains the bane of the developing world. Among the obstacles highlighted in the United Nations' *Realizing the right to development* report is 'the concentration of economic and political power in the most industrialised countries, the international division of labour and the functioning of the Bretton Woods institutions…and the adverse consumption patterns of the more industrialised countries' (United Nations Office of the High Commissioner 2013: 62). Central to the emergence of the right in the post-colonial global order was the aim of dismantling the hierarchical power relations, not only in light of the economy but also politically. Unfortunately, emancipatory, self-sufficient and equitable development remains an unfulfilled goal of much of the developing world.

What has been problematic in part has been the apparent ideological neutrality of many of the traditional institutions and frameworks of development. As laid out in the 'conceptual dynamism of the RTD', many adopted initiatives followed the directives of the Washington Consensus. Thus, despite calls for the neutrality of development initiatives with regards to social, political and economic directives, 'The World Bank's assistance was sometimes overburdened by prevailing ideology; for example, the free-market Washington Consensus policies foisted on recipients actually led to deindustrialisation and declining income in Sub-Saharan Africa' (Stiglitz 2015, para. 9). Understanding this inherent subjectivity strips away the semblance of objectivity from these initiatives, and instead exposes the more prescriptive ideological undertones attached to international aid and apparent 'universal' development. Such initiatives, while not wrong per se, therefore need to be understood as they relate to the RTD and the parameters set out in the declaration of this right.[14] Thus, while traditional initiatives might secure particular aspects of the right, this may come at a particular cost. As a result of this and in an effort to achieve and secure more broad-based development, South–South institutions such as the newly created BRICS development bank and the AIIB have emerged as alternative initiatives. As will be discussed later, these new initiatives attempt to redress some of the previous imbalances. While by no means perfect, in the light of the abovementioned subjectivity of traditional mechanisms, much of the criticism of these new initiatives has tended to be unreflective of their own frameworks and inherent subjectivities vis-à-vis the parameters laid out in the RTD.

Fundamental Principles: Equality, Equity and Social Justice in the International Arena

Despite the fact that the RTD was initially founded on a principle of equality in which newly independent states sought equal status in relation to their developed colonial counterparts, it remains questionable whether equality has been secured in the international sphere. Looking at the international imbalances of power and who dictates the terms of development, international level inequality is persistently visible. Thus, while domestic factors, as mentioned, are of importance in bringing about development, the international issue of inequality remains equally troublesome. Despite the 'formal' emancipation of the Global South and its legal and political incorporation into the international (economic) system, there has been increasing recognition of the 'failure of an alien growth-centred profit-oriented development model, based on an unequal international division of labour, to eliminate inequalities, achieve social well-being, and to consolidate political independence through economic independence' (United Nations Office of the High Commissioner 2013: 4). Not only has the international economic system itself come under fire for its inability to address inequality but it has also been branded complicit in driving and sustaining international inequality. In recognising this, there has been a demand for increased equality and equity in the international system.

On the other hand, scholars such as Terrence Paupp have advocated the replacement or alteration of international financial institutions under a global human rights regime as well as the complete rejection of a neoliberal model, which—he argues—is at the very heart of continued inequality and underdevelopment (2014). For Paupp, 'neoliberal ideology is incompatible with the project of inclusionary and humane governance because it confines the role of the new multilateralism to the civil and political domain, thereby avoiding accountability to socio-economic claims related to human rights, equity and justice concerns' (2014: 449). Nevertheless, others have pointed out the uncanny similarities of human rights (as conceived of within a liberal paradigm) and market fundamentalism, which both took off in the 1970s as individualistic and anti-statist (Moyn 2015: 17). These fundamental characteristics have resulted in the relegation of human rights to the 'powerless companion of market fundamentalism... [with] nothing to say about inequality, which we now know to be the central achievement, locally and globally, of the new political economy' (Moyn 2015: 17).

This position, however, has been largely driven by unwarranted narrow liberal understandings of human rights that have unnecessarily confined themselves to the dictates of particular political, economic and social ideologies.[15] This is not to say that marketisation in itself as Paupp (2014) would have it has been inevitably disastrous,[16] but rather that ceilings and floors (in the form of equity/equality/distribution) and/or alternative avenues need to be put in place in order to counterbalance these effects. In conceptualising what we understand as the challenges and thus remedies to the RTD, we are able to embark on discovering new, innovative and novel configurations in which the goals of fairness, justness and equality can be achieved. While the likelihood of the complete dismantling of the international economic order remains slim, alternative possibilities exist and should be pushed for increased accountability and democratisation and for the increased participation of the Global South within the international arena and international institutions. This kind of reformist agenda requires prioritisation by political powers and international institutions. However, this too requires political, which is more often than not elusive. As already highlighted, global calls to solidarity, though admirable in theory, go against the anarchist international system driven predominantly by power considerations, wealth accumulation and profit maximisation.

Moreover, the idea of global social justice, with states engaging on an equal footing, remains to be realised and presents itself as a major challenge to securing the RTD. The goals of equality tied to ideas of social justice are what are referred to as equity. The principle of equity in turn is intimately connected to the concept of fairness. Thus, while there have been many calls to equality, the issue of equity in the RTD has not received adequate attention. Whereas equality emphasises the idea of equal or 'the same' treatment, equity refers to the ability to offset disadvantages in order to achieve fairness. The core principles for the realisation of the RTD include equity, equality, non-discrimination and international cooperation, among others (UN Economic and Social Council 2002: 4). This kind of equity on an international level 'would require that the formulation and conduct of economic policies and programmes be geared towards ensuring a fair and just distribution of the "fruits" of development between competing regions, population segments and individuals' (UN Economic and Social Council 2002: 5). These would include structural obstacles and inequalities within the international order and international relations that inevitably affect the achievement of genuine and universal participation (United Nations

Office of the High Commissioner 2013: 59). As stated in the UN report *Realizing the right to development*, 'in order to reverse the situation of growing inequalities in the world, affirmative action in favour of the disadvantaged groups and increased assistance to disadvantaged countries will be required. The removal of barriers to economic activities such as trade liberalisation is not sufficient' (United Nations Office of the High Commissioner 2013: 60). Added to this is the importance of participation of the developing world in international economic decision-making and as a 'central factor' in realising the RTD (UN Economic and Social Council 2002: 10 and United Nations Office of the High Commissioner 2013: 14). The goals of participatory, equitable and therefore just development remain a significant challenge for the international dimension of the right.

Although there exist a number of ethical duties, such as calls to international solidarity and the moral duty of reparations, which are invoked regarding the RTD, these remain highly utopian in an 'anarchical' international system. Instead, I contend that seeking out a viable alternative in the context of the international dimension of the right is a more realistic task. I propose that the reconfiguration of the international political and economic space in itself (which will be described in the following sections) has initiated a process of redistribution that can in fact bring us closer to realising the RTD. This is not to argue that recourse to international legal remedies or the prospects of making the right internationally justiciable are invalid or unnecessary. Rather, it is that these endeavours can be complemented and further bolstered by existing or emerging international developments. Harnessing these reconfigurations to the benefit of 'realising' the right in the case of Africa, for example, requires African agency and assertiveness. In recognising the opportunity represented by what are in fact complementary options (the Western vs. the Southern model), however, the 'space' needs to be given for the existence of such alternative models. Viewing them simply as competitors will not help in the quest for equal, fair and just international development. Again, as emphasised in the UN development report, 'constituent peoples continue to enjoy the right to their own cultural identity and to determine their own economics, social and political system' (United Nations Office of the High Commissioner 2013: 60). How these new configurations are represented within the discourse of international human rights will be instrumental to harnessing this reconfiguration as an opportunity.

Alternative Global Spaces

In the light of the above-mentioned changes, it is evident that not only are alternatives required in order to overcome the challenges to the right, but that such alternatives are perhaps already underway in the 'natural' reconfiguration and diversification of the international economic order, represented by the increased presence of new and emerging powers. This reconfiguration of economic ties and networks—in particular with the increasing prominence of South–South cooperation—has emerged as a kind of 'reconfiguration underway'. Although many of these 'new' networks require increased strengthening, 'maturation' and solidification, the sheer attention they have and continue to receive, as well as the pace at which they have emerged and stabilised, has not only been impressive but indicative of a global, albeit partial, recognition of their increasing power and importance as equal stakeholders and decision-makers in the international arena. The BRICS development bank and the AIIB, for example, though they are yet to reach full development, represent such possible alternatives.[17]

These institutions are not only valuable in terms of their contribution as economic or financial alternatives, but more importantly represent an altogether different intellectual framework that challenge the traditional structures of power and knowledge production that have dominated development discourse.[18] Both these initiatives, among others, have carved out a sizeable and ever-growing space for the ambitions and concerns of the Global South, the implications of which remain to be seen.

Within the international arena, safeguarding the RTD requires such a space, particularly for developing nations, to assert and determine their own course of development. Thus, while all peoples automatically *have* this right, the conditions required to *enjoy* the right are intimately tied to the hierarchical power asymmetries that characterise contemporary international political economy. These power asymmetries dictate whether or not a given state is able to achieve the substance of the right, namely, development itself, depending on where they find themselves in the hierarchical international pyramid. The locus of power, moreover, naturally determines the discourse on development in terms of what constitutes appropriate development and the conditions of development. This externally imposed conditioning of development vis-à-vis the developing world has perpetuated an imbalanced understanding of the RTD that has more often than not underscored national dimension(s), restrictions and requirements over and above the international and emancipatory dimensions of the right.

Here it is necessary to briefly elaborate on what is behind the idea of national and international dimensions of the RTD. Notably, the right, as a human right, has been categorised largely as a collective and/or solidarity right. Human rights, as a normative structure, have, despite the insistence on the equality of individual and collective considerations, often divorced themselves from a state-centric focus. Instead, the focus on individual humans has become the dominant lens through which a variety of rights are understood. Despite the emphasis on the individual or 'human' component of rights, human rights as a normative framework remains limited and itself partakes in an international system that still deems the state as a principle, if not *the* principle, actor in international affairs. With this in mind, while the increasing turn away from the state is evidently underway with the transnational forces of globalisation, the state remains an integral and indispensable part of any human rights framework. This framework thus not only acts as a normative base that limits the power of the state, but is simultaneously deemed 'essential to promote the development of friendly relations between nations' (Preamble to the Universal Declaration of Human Rights).

In the Declaration itself, it is member states that are the primary responsible actors that 'have pledged themselves to achieve, in cooperation with the United Nations, the promotion of universal respect for and observance of human rights and fundamental freedoms'. Such a role, however, is only possible if the state as actor is viewed not only as an equal stakeholder of the human rights discourse central to the protection of human rights, but moreover is given the space and autonomy to exercise its role in ways it deems fit. In this regard, while a cautionary stance is generally taken vis-à-vis the state, states remain pivotal with regards to ensuring development. The state is therefore an inescapable, albeit paradoxical, facet of international human rights that requires incorporation into our conceptualisations of various rights, including the RTD. Particularly for the developing world, emancipation in the form of autonomous freedom to exercise and assert suitable 'paths of development' represents a major element in securing independence and self-determination.[19]

In relation to the developing world in particular, the RTD, however, is limited by systemic forces of neoliberal globalisation that have often proved disadvantageous. The assertion and exercise of power by developed nations over developing nations in uneven and unequal trade agreements, for example, have contributed to development frustrations and cycles of dependency. The exertion of such power, however, should not be thought of purely in terms of economic power alone, but is also connected to knowledge—thus what we know and how we conceive of development. Thus, where alternative modes of development do not fit the traditional moulds, these are treated with high degrees of scepticism (as evidenced in the scepticism around new configurations). This, however, need not be the case, and—as will be elaborated—can be seen as complementary or a process of democratising our conceptualisations of development so as to create more suitable, fair and equitable development agendas.

Notably, one of the first references to the right, in the first UN World Conference on Human Rights in Teheran, 1968, elaborated on the right by stating that the 'widening gap between the economically developed and developing countries impedes the realization of human rights in the international community' (Boutros-Ghali 1995: 247).[20] Here the widening gap is framed as a gap characterised predominantly by economics. While undoubtedly significant, an accompanying political and intellectual and political gap remains unaccounted for. Accordingly, it is highlighted that 'the prevailing terms of trade, monetary policy and certain conditions tied to bilateral and multilateral aid, which are all perpetuated by the non-democratic decision-making processes of international economic, financial and trade institutions, also frustrate the full realization of the Right to Development as a human right' (United Nations Office of the High Commissioner 2013: 61). One of the fundamental priorities of securing the RTD is thus the establishment and adherence to notions of international equity and equality.

However, this equity and equality goes further than the mere democratic representation of the Global South as a sovereign equal on the international stage. Rather, or additionally, it includes opening up the space for alternative and complementary forms of knowledge regarding development, which that are equally legitimate. While the full breadth of alternative forms of knowledge remains to be explored, institutions such as the BRICS development bank have already emphasised alternative and/or complementary priorities, such as its strong focus on infrastructure—a focus area that traditional players have regularly overlooked.[21] This opening up of our political imagination in terms of what development entails fundamentally signifies the creation of spaces that move away from the monopolisation of development discourse by traditional powers to allow for 'other', equally legitimate producers of knowledge to contribute to integrative and holistic approaches to development.

Despite this opportunity, it remains to be seen whether it will be fulfilled—speculation is warranted with regards to who wields power within these new configurations and whether such power remains the exclusive domain of a few. Opportunities presented by these new configurations can, therefore, be viewed as a challenge to the hegemony of traditional powers, which have often been complicit in the reproduction of path-dependent development trajectories and 'glass' development 'ceilings'. They additionally present a challenge to the limited conceptualisation of what the right has come to mean.[22] As mentioned, however, the dominant conceptualisations of development and the right RTD have come at the cost of other equally legitimate claims, which, if considered, could shed light on how we identify and frame challenges to development in contemporary international politics. In the case of increased Chinese international lending, for example, critics have framed these practices as similarly conditional on traditional and/or Western models of lending and development (Mattlin and Nojonen 2015). What is overlooked is precisely the conceptual differentiation of the *nature* of Chinese lending, the reasoning behind it and the fact that it goes into much-needed development projects, such as infrastructure development on the African continent.

Moreover, development analyses are often overshadowed by the narrative of the 'China threat', which automatically frames Chinese lending in a particular light. Similarly, despite clear power asymmetries between China, for example, and smaller developing states, China's own position as a developing state itself is frequently ignored as irrelevant and/or contested by means of narrow gross domestic product analyses.

What remains a challenge will be the reaction of traditional powers to these new initiatives. Thus far, these have been met with caution, scepticism and hesitation. In particular, the United States has seen these developments as a direct challenge to its power and influence. The case of the creation of the AIIB and the scepticism displayed by the United States was evident in its plea to potential members not join the bank as founding members (Kurlantzick 2016; Stiglitz 2015). Though largely ignored, this move signalled an offensive, competitive and combative position on the part of the United States. In this instance, collaboration and cooperation could have been sought as a way out of an apparent impasse; however, this would have marked compromise and weakness on the part of the United States. Instead, an uncompromising hard-line stance was taken. Additionally, and in a similarly speculative vein, the Heritage Foundation—a prominent US-based think tank—warned the United States in a piece on US policy priorities for Africa in 2016 to 'remain watchful toward other countries' growing influence on the continent' (Meservey 2016, para. 3). This stance not only gives us an insight into the self-image of the United States but it additionally speaks to a negative and unwavering outlook on international affairs. Unfortunately, this has offered little by way of progress and/or even discussion on potential ways forward. Instead of viewing these developments as prospective opportunities for international engagement and even cooperation, emergent initiatives emblematic of 'new' global configurations have been thwarted by arguments centred on criticisms related to an apparent lack of standards in these new arrangements. Traditional standards, as advocated by the United States, have, however, mostly ignored or failed to incorporate international dimensions of equality as integral to the RTD. The dictation of these standards by a few has resulted in the obfuscation of the valid perspectives and concerns of the Global South. The result has been the emergence of a narrow and asymmetrical discourse on the RTD.

Positions like that above undoubtedly hinder adequate cooperation and engagement on the international dimensions of the RTD, essentially negating the co-existence of a range of diverse yet potentially complementary systems within the international political and economic arena. Not only are such positions short-sighted with regards to identifying and locating the contemporary challenges specifically regarding the developing world, but they also frame pluralism within the international space as a threat. Furthermore, the complicit role of economic heavyweights in sustaining and perpetuating imbalanced power hierarchies is maintained through the monopolisation of how we conceptualise development and/or brute economic, and thus bargaining, power. Instead, the dictates of traditional powers are maintained through antagonistic positions, essentially perpetuating and in part contributing to systems of inequality and chronic underdevelopment. This is illustrative of the perpetual challenge of political will and points to unwavering allegiances to particular political, economic and social models that in fact hinder the securing of the international dimensions of the RTD.

With regards to the international dimension of the right, some have argued in favour of invoking solidarity as an underpinning ethical principle to guide its fulfilment. The genuine use of this principle, however, remains a precarious ideal at best.[23] Despite the desirability of a notion of solidarity as far as development is concerned, when it is unfounded, recognition should be given to alternative groupings as distinct constellations at minimum. These constellations can prove useful in their acute insight into both local and global dynamics that affect the developing world in particular. Taking advantage of this emerging reconfiguration and dynamism of the international system should be taken as an opportunity to secure the right in its broadest sense, albeit through alternative avenues.

Setting up alternative institutions or configurations can be seen as part of an indispensable process of expanding notions of development to make it more globally inclusive. This process in turn can see to it that the right is, at minimum, secured in its international and emancipatory dimensions. How political actors play these institutions and their own ambitions off against each other, however, will determine whether such institutions can co-exist with existing traditional frameworks in a cooperative as opposed to a necessarily competitive manner. While such a move could see a more holistic fulfilment of the international democratic component of the RTD, how international actors handle and steer these developments will be determined by the willingness and cooperation of both traditional and newly emerged players. This will in turn determine to what degree the envisioned goal of equal development for all can be more meaningfully secure.

South–South Cooperation as a Reflection of the International Dimension of the RTD: Sino-Africa Relations

The dominant discourse on international human rights has deemed Chinese and African engagement as necessarily inimical to the protection of human rights. In looking particularly at the discourse on the RTD, this has similarly been dominated by Northern media and scholarship on the subject. While many of the criticisms levelled against the relationship hold elements of truth, these frameworks tend to be ignorant of their own political, economic and social bias, as highlighted in the preceding sections. Moreover, these dominant conceptualisations worryingly leave out some of the most essential components of the right, which subsequently reinforce a one-dimensional application and understanding of it. Such unitary understandings are problematic in that they hinder our ability to creatively conceptualise innovative ways in which to safeguard the RTD in its entirety. Thus, the position taken here is one of caution and wariness with regards to the uncritical subscription to dominant perspectives—due consideration for alternative conceptualisations and interpretive frameworks is warranted.

If we reflect upon the preceding sections' analyses of the RTD, it becomes evident that this right is a complex and multifaceted one that has been conceptualised in a number of ways. These conceptualisations need not be mutually exclusive, or standalone conceptions for that matter. What is apparent, however, is that dominant frameworks have frequently ignored the international and emancipatory dimensions of the RTD.

The Sino-African engagement is an intensely scrutinised relationship and has come to characterise the contemporary reconfiguration as well as South–South cooperation. Often, however, when it comes to the analysis of Sino-African engagement, critics fall prey to monolithic analysis of China's involvement in Africa. Frequently, specific incidences are extrapolated to characterise and define the entire relationship, as opposed to understanding the heterogeneity of experiences of China–Africa engagement.[24] Furthermore, there are a host of myths that swirl around this engagement (Brautigam 2009, 2015). The assumption, for example, that these relations are grounded in corruption is very weakly and insufficiently backed up in most analyses. As Frank Ching explains, high-level corruption of this kind is particularly difficult within the parameters of China's investments in Africa: 'Chinese aid is often dispensed in such a way that corrupt rulers cannot somehow use it to buy Mercedes Benzes... [It] is often in the form of infrastructure, such as a railroad...or roads... Or in the form of doctors and nurses to provide health care to people who otherwise would not have access. China provides scholarships for African students to study in its universities and increasingly funds to encourage its businessmen to invest in Africa' (Ching 2005). China–Africa expert Deborah Brautigam (2009) goes further and notes that Western aid, in contrast, is in fact far more easily involved in corruption on the continent.

Additionally, many of the obstacles faced by China's engagement are not exceptional, but instead form part of universal global trends. Therefore, Western and other donors/investors similarly face many of the challenges experienced by China regarding African development.[25] Instead, many of the incidences that are cited as human rights concerns are not particular to Chinese relations but are intimately connected to broader international financial trends and forces of a global capitalist market that perpetuates and deepens inequality. These types of challenge are cached in the fact that the emergence of the RTD, as a human right, has significantly outpaced current levels of state responsibility in international relations (van der Have 2014). Many other examples exist; however, these are but a few examples of how this relationship is frequently mischaracterised.

These examples do not begin and end with the mischaracterisation of these relations but also have a direct effect on the perpetuation of a narrow understanding of why these relations are deepening, as well as how they relate to the RTD. Development as an emancipatory tool in creating a more just and equitable international space remains an aspect necessary to secure the RTD holistically. For this to happen, market diversification, competitive advantage and equitable development are all required (United Nations Office of the High Commissioner 2013). The development projects with which China is concerned in Africa, as well as their support of African development within the context of BRICS and multilateral fora such as the Forum on China–Africa Cooperation are crucial in achieving the necessary diversification and long-term infrastructure to ensure development. Particularly in the light of ensuring self-sufficiency and self-sustaining economies, the kind of long-term investment that China frequently provides is key to achieving development across the continent.[26] This kind of development can allow for the developing world to assume its role not only as an equal stakeholder and player in the political sense but also in the economic sense. These kinds of engagement therefore assist in securing the international dimension of the RTD. Instead, a narrow conceptualisation of the right, often focused on its national and individual dimensions, is used to characterise these developments as wholly detrimental. Arguments that are centred on the self-serving nature of China's engagement in Africa are an oversimplification of the nexus of global justice and international trade and development.

An often-overlooked point in the discourse surrounding Sino-African relations is the development of China as a developing country itself. As a result, much scholarship and scepticism is focused on African development, often neglecting the issue of national Chinese development. If not neglected, it is characterised as an indefensible justification for Chinese investment in Africa. The argument goes that China has deepened relations in Africa purely for its own domestic growth, demand and development.[27] Its role in Africa is subsequently framed largely as negative and self-serving. This picture, however, wholly ignores the struggles of China as a developing nation itself and instead frames this development as being exclusively at the cost of African development. This kind of zero-sum outlook delegitimises the need for internal Chinese development, framing the opportunity for development as mutually exclusive to the relationship, i.e. the assumption is that China and Africa cannot both partake of the fruits of development that flow from their relationship.[28] This narrative is embedded in a fixation with exploitive international relations that assumes mutually beneficial relations to be necessarily impossible. Among other aspects,[29] such analyses tend to subscribe to a linear vision of development that leaves little room for innovative alternative paths towards development.

The international dimension of the RTD, and particularly its emancipatory component, has been frequently overlooked by Northern development agencies. China, along with institutions such as the BRICS development bank and the AIIB, on the other hand, has played into this conceptualisation and one could argue has even 'revived it'. The Chinese rhetoric of brotherly camaraderie in which it posits itself as a fellow developing country has played an instrumental role in Sino-African relations and African development more broadly speaking. While most analyses deem such claims to be pure rhetoric,[30] it would seem as though China's engagement in Africa on a government-to-government level speaks directly to the international emancipatory conceptualisation of the RTD. Understanding right as an extension of the right to self-determination is therefore largely central to the relationship and has largely been supported by the Chinese approach to development in Africa with a strong focus on infrastructure development and eventual economic self-sufficiency. Through these areas of focus the China–Africa engagement thus far has sought to engage Africa on an equal footing. Critics, however, have viewed, for example, the frequently invoked notion of win-win cooperation as a facade, arguing that the relationship is necessarily imbalanced because of the economic clout of the Chinese economy vis-à-vis that of the African economies. However, this take focuses exclusively on a realist framework of power (whether political or economic) as the sole driving force behind cooperation. Often neglected is the realisation that China itself remains a developing nation that is similarly entitled to the RTD.

Moreover, China's policy towards Africa has in fact been far more dynamic and adaptable than is often assumed. This is evident in its incremental change in the Sudan issue for example, as well as its shifting stance on issues of environmental concern and non-intervention on the continent. With these adaptations, China's engagement on the continent cannot be deemed static or unchanging. This malleable approach is in part related to the Chinese approach of 'crossing the river by feeling the stones'.[31] This approach encompasses a spirit of trust in charting unchartered territory together as equals, in which both parties are novices entering unchartered territory. It is this kind of approach that places the emphasis on equality that resonates in the developing world and in Africa in particular. This has naturally differentiated it from the more traditional Northern models. What is evident is that Northern scholarship on development and the RTD has narrowly framed the concept and thus analysed South–South cooperation and more specifically Sino-African cooperation through this lens. These conceptualisations, however, stem from a particular social, economic and political paradigm themselves and are often imposed upon the development trajectories of the developing world. Such approaches not only ignore their own subjectivities but are moreover aloof to the social, economic and political neutrality embedded in the RTD. Thus, while it is easy to analyse South–South cooperation through pure economic and realist lenses, these 'variables' do not provide a full understanding of the dynamics of these reconfigured relations.

It is clear that the influence of reconfigurations of the Global South on the conceptualisation of development as a right is reflected in the kind of development and arrangements increasingly sought out by the developing world. Though it may be easy to reduce these to autocratic, corrupt and self-serving variables frequently invoked to analyse the Global South, these arguments are simplistic and exclude a host of equally valid factors and variables that are essential to making sense of these reconfigurations. The complexity of these relations is not only reflected in the kinds of investments that China undertakes in Africa, for example, but is more broadly reflected in the support given to Africa and other developing nations through various multilateral fora. This kind of 'soft power' support in international fora speaks to the desire for a more just and equitable international economic/political system. While the persistence of the 'Dark Continent' narrative of Africa is evident in macro international, political and economic analyses of the continent that focus on gross domestic product indicators, these frequently miss the broader factors relating to ideas of the emancipation, equality and autonomy of the Global South. Expanding explanatory frameworks to take seriously the historical, contextual and ideational aspects that solidify these relations will allow us not only to better understanding these new configurations, but also to understand what they mean for international human rights such as the RTD.

Concluding Remarks

The RTD remains a contested, challenging and complex right in the corpus of international human rights. Particularly in an ever-changing international arena, complexities multiply by the day and frameworks for making sense of these reconfigurations need to keep up. What remains essential, however, is the self-reflection on dominant modes of thought and inquiry. Ensuring that the various dimensions of the RTD are not only understood but also incorporated into related analyses is paramount. This has important ramifications for the discourse as well as subsequent policy prescriptions and potential cooperation in international relations.

In particular, it has been essential to highlight that the dominant modes of thinking about development and the RTD are value-laden and fit within a particular political, social and economic framework. This often-narrow conceptualisation has frustrated efforts to secure the RTD in its entirety. It is hoped that in recognising these limitations, the example of South–South cooperation is not approached as a reconfiguration that is inevitably antagonistic towards existing models. Rather this self-reflection on dominant models and frameworks provides us with a necessary critique in order to make sense of current reconfigurations and shortfalls. South–South cooperation has not only provided alternative pathways with regards to development, but has also challenged us to rethink our conceptualisation of development to include the often-neglected emancipatory and international dimensions of the right. Viewing these developments and international reconfigurations as an opportunity for ingenuity and creativity that help us to face some of the most difficult challenges to development can move us closer to realising the initial as well as the contemporary aims of the RTD in its entirety.

Notes

1. 'Transformation' in this sense refers to social transformation in which there is an equal distribution of opportunity regarding the factors of production (this includes not only traditional economic production but additionally the production of knowledge, and so on).
2. Critique of the RTD has related to its necessity, and thus the potential redundancy of the right vis-à-vis existing first- and second-generation rights, its justiciability, referring to debate over whether or not the right should be regarded as a positive or negative right, and to its scope, and thus the boundaries of the right. In her inaugural lecture, Olajumoke Oduwole highlights these critiques, emphasising the scepticism of both the practical and legal efficacy of the right. See Oduwole 2014.
3. It is my position that the issues of corruption, clientelism and patronage are rife and problematic for development; the persistence of these issues lies in an intricate web of factors inherent in the post-colonial state.
4. This imbalance is not only in the form of political power, but also with regards to the flow of information—usually from North to South—and its monopolisation, which Filipe Gomez Isa describes. Isa warns against such 'cultural and informational monopoly', stating that it 'could end up seriously endangering political, social and ideological pluralism' (Isa 2005: 26).
5. This was adopted despite controversy regarding the scope, limitations and implications of the right, as vocalised by several member states.
6. The Universal Declaration of Human Rights, for example, was a direct result of the Northern experiences of the atrocities of the Second World War and drew extensively on the French Declaration of the Rights of Man and the Citizen (1789) and the Bill of Rights in the United States Constitution (1791). See Rich 2002, 'Solidarity Rights Give Way Solidifying Rights,' *Dialogue*, 21(3): 25.
7. Just as there are pitfalls in absolute individual rights, so too are there cautions against absolute group rights. It is necessary to understand the balance and trade-offs of both sets of rights, as well as the conditions under which they can be legitimately invoked.
8. Examples include the redress of systemic inequality along racial lines through affirmative action, or quota systems as is highlighted in the UN document 'Realizing the Right to Development': 60.

9. Conditionality itself is ambiguous in its merits, both in the positive and the negative. Pape, R.A. (1997), Paupp, T. E. (2014) as well as Chua, A. (2004) speak of the constrictive effects of conditionality as an often-used market reform policy imposed on developing countries.
10. The idea of 'non-human rights' strategies requires further interrogation and is related to a broader problem concerning what a 'human rights'-centred strategy looks like or entails. The characterisation of East Asian models as having eschewed human rights concerns stems from a particular school of thought within Northern academia that places the emphasis on civil and political rights. It can, however, be equally and legitimately argued that the East Asian model in fact considered socio-economic and cultural rights as the driving force behind their development process. But this difference in emphasis on a particular set of human rights does not negate the East Asian experience as being non-compliant with international human rights standards.
11. State-led macroeconomic planning, of which one characteristic is scepticism of neoliberalism and the centrality of 'good governance', as defined by the Washington Consensus.
12. Such as sustainable development, equality and the environment.
13. This chapter does, however, focus on the international dimension, as it is frequently overlooked.
14. The parameters of the RTD are closely connected to the Covenant on Economic, Social and Cultural Rights. In its famous *General Comment no. 3*, the UN Committee on Economic, Social and Cultural Rights states as follows: "The Committee notes that the undertaking…neither requires nor precludes any particular form of government or economic system…in terms of political and economic systems, the Covenant is neutral and its principles cannot accurately be described as being predicated exclusively upon the need for, or the desirability of a socialist or a capitalist system, or a mixed, centrally planned, or laissez-faire economy." UN Committee on Economic, Social and Cultural Rights, *General Comment No. 3: The Nature of States Parties' Obligations (Art. 2, Para. 1, of the Covenant)*, 14 December 1990, E/1991/23, available at: http://www.refworld.org/docid/4538838e10.html [accessed 21 January 2016].
15. As previously mentioned, the report of the United Nations Office of the High Commissioner (2013) echoes that there are no particular dictates on the political, social or economic systems required to achieve development (p. 60).

16. The example of Chinese marketisation of the 1970s allowed for many to be brought out of poverty (Moyn 2015: 17).
17. See, for example, Stiglitz (2015) on the opportunities offered by the AIIB. While relevant, the issue of a different intellectual framework is not addressed in this chapter due to limitations on length.
18. Mushwana, T. 2015 for example, notes the alternative principles upon which South-South cooperation are founded as opposed to traditional North-South development cooperation.
19. Self-determination refers to 'the right of a people to determine its own destiny. In particular, the principle allows a people to choose its own political status and to determine its own form of economic, cultural and social development…the importance lies in the right of choice' (Unrepresented Nations and Peoples Conference 1993: 6). However, it is notable that self-determination 'is recognised in international law as a right of process (and not of outcome) belonging to peoples and not to states or governments' (Unrepresented Nations and Peoples Conference 1993: 6). The line of argumentation of this chapter, however, assumes interdependence between peoples, their chosen governments (in a variety of forms as long as legitimacy is conferred on the process and/or government via popular sovereignty) and thus states.
20. Proclamation of Teheran, Final Act of the International Conference on Human Rights, 22 April–13 May 1968. For the text of the Proclamation, see Boutros-Ghali (1995: 247).
21. Zambian economist Dambisa Moyo has highlighted the focus on trade and investment, and specifically infrastructure, in the China–Africa engagement as a 'golden opportunity' for Africa (Moyo 2009). Also see: Khanna (2014: 46).
22. For an illustration of how international relations theory in particular obfuscates alternate conceptualisations and the necessity to open up scholarship to these alternatives, see for example, Langlois (2015).
23. Principles of solidarity have been used in various arguments particularly vis-à-vis social, cultural and economic rights. Such arguments, however, are underpinned by particular moral or normative convictions that are often illusory in an anarchical international system. Note that 'solidarity' here does not refer to the idea of solidarity rights as group rights, but rather to solidarity in the form of integrated international cooperation.
24. See, for example, the Zambian mining incident on Deborah Brautigam's blog. 'Human Rights Watch Report on Chinese-Owned Mines in Zambia', *China in Africa: The Real Story*. November 11, 2011. Available

online at: http://www.chinaafricarealstory.com/2011/11/human-rights-watch-report-on-chinese.html
25. Ibid.
26. As is noted in Stiglitz's piece on the AIIB, 'China itself is a testament to the extent to which infrastructure investment can contribute to development' (2015).
27. For examples of this line of argumentation, see for example Daly (2008), Leigh and Pallister (2005), Parris (2008).
28. Mutual benefit is frequently invoked in 'South–South' discourses, but is often read purely as political rhetoric in the guise of 'friendship'. There is a general denial of genuine collaboration for optimal gains for all parties within academic discourses. However, such negation ignores shared histories, values and experiences as valid foundations of engagement. Thus, the categories and concepts used in traditional Western scholarship to explain international relations are inadequate and problematic in their representativeness for the host of factors to be considered. This in turn limits our understanding of international relations to the categories given by dominant scholarship.
29. Here, other aspects refer to an assumed lack of African agency in dealings with China. These assumptions are often bolstered by a notion of corrupt and colluding African state elites and a particular understanding of the Chinese state as necessarily 'authoritarian'. Arguments made from this perspective, while in some cases accurate on certain accounts, are generally uninformed as to the nature and character or Chinese society, government and history. Often these analyses are complicit in reproducing problematic narratives of China, Africa and international relations. See, for example, Asumah (2014).
30. While the scope of this chapter does not allow for an in-depth analysis of the issue of rhetoric, the same accusation of 'rhetoric' could be levelled against Northern development initiatives using a particular rhetoric of 'altruism' and 'aid'. Such arguments therefore do not hold much weight or insight into the fundamental issues that drive international cooperation.
31. 'Crossing the river by feeling the stones' is a saying coined by Deng Xiaoping and denotes the experience of pioneering particularly economically in 'unchartered' terrain with no set roadmap. It is an experimental and piecemeal way of developing or moving forward. Thus, one traverses the unknown way of development by a process of gradual and circumstantial

learning, as opposed to a one-size-fits-all linear approach. Crossing the river while feeling the stones thus allows for learning along the way, moving a few steps forward and perhaps a few to the side or back, in order to progressively cross the river in the most 'careful' or deliberate, cautious, diligent and/or prudent way.

Bibliography

Anghie, A. (2005). *Imperialism, sovereignty and the making of international law*. Cambridge: Cambridge University Press.

Asumah, S. N. (2014). China in Africa: Dislocating cultures, re-examining the role of the nation- state, and the China model in the process of development. In S. N. Asumah & N. Mechthild (Eds.), *Diversity, social justice, and inclusive excellence: Transdisciplinary and global perspectives* (pp. 327–346). Albany, NY: Suny Press.

Ayoob, M. (1995). *The third world security predicament: State making, regional conflict and the international system*. Boulder: Lynne Rienner Publishers.

Bedjaoui, M. (Ed.). (1991). The right to development. In *International law: Achievements and prospects*. Paris: UNESCO.

Bell, D. A., & Carens, J. H. (2004). The ethical dilemmas of international human rights and humanitarian NGOs: Reflections on a dialogue between practitioners and theorists. *Human Rights Quarterly, 26*(2), 300–329.

Boutros-Ghali, B., (1995). The United Nations and human rights, 1945–1995. United Nations Pubns.

Brautigam, D. (2009). *The dragon's gift: The real story of China in Africa*. Oxford: Oxford University Press.

Brautigam, D. (2015). *Will Africa feed China?* Oxford: Oxford University Press.

Ching, F. (2005, July 13). Cosy ties, but China needs to do more for Africa. *Business Times*.

Chua, A. (2004). *World on fire*. New York: Random House Inc.

Collier, P. (1997). The failure of conditionality. In C. Gwin & J. Nelson (Eds.), *Perspectives on aid and development*. Washington, DC: Overseas Development Council.

Daly, J. (2008). Freeding the dragon: China's quest for African minerals. *China Brief*, 8, p. 3.

Fritz, V., & Menocal, A. R. (2007). Developmental states in the new millennium: Concepts and challenges for a new aid agenda. *Development Policy Review, 25*(5), 531–552.

Gregory, D. (2009). Development. In Gregory, D., Johnston, R., Pratt, G., Watts, M. and Whatmore, S. eds., The dictionary of human geography. West Sussex, United Kingdon: John Wiley & Sons.

Isa, F. G. (2005). Globalisation, privatisation and human rights. In K. De Feyter & F. G. Isa (Eds.), *Privatisation and human rights in the age of globalisation*. Antwerp–Oxford: Intersentia.

Khanna, P. (2014). New BRICS bank a building block of alternative world order. *New Perspectives Quarterly, 31*(4), 46–48.

Kurlantzick, J. (2016). Let China win. It's good for America. *Washington Post Online*. Retrieved January 15, from <https://www.washingtonpost.com/opinions/let-china-win-its-good-for-america/2016/01/14/bfec4732-b9b6-11e5-829c-26ffb874a18d_story.html>.

Langlois, A. J. (2015). International relations theory and global sexuality politics [Special issue]. *Politics*. doi:10.1111/1467-9256.12108

Leigh, D., & Pallister, D. (2005, June 1). Revealed: The new scramble for Africa. *The Guardian*.

Mattlin, M., & Nojonen, M. (2015). Conditionality and path dependence in Chinese lending. *Journal of Contemporary China, 24*(94), 701–720.

Meservey, J. (2016). Four U.S. policy priorities for Africa in 2016. *The Heritage Foundation Online*. The Heritage Foundation Issue Brief #4509 on Africa. Retrieved January 15, from <http://www.heritage.org/research/reports/2016/01/four-us-policy-priorities-for-africa-in-2016>.

Moyn, S. (2015). Human rights and the age of inequality. In D. Lettinga & L. van Troost (Eds.), *Can human rights bring social justice. Twelve Essays*. Netherlands: Amnesty International.

Moyo, D. (2009). *Dead aid: Why aid is not working and how there is a better way for Africa*. London: Allen Lane.

Mushwana, T. (2015). BRICS and the New Development Bank: An alternative model of development for Africa? Polity. Available Online http://www.polity.org.za/article/brics-and-the-new-development-bank-an-alternative-model-of-development-for-africa-2015-01-07 (Accessed 28 August, 2017).

Mutua, M. (2001). Savages, victims, and saviors: The metaphor of human rights. *Harvard International Law Journal, 42*(Winter), 201–245.

Nickel, J. W. (2008). Rethinking indivisibility: Towards a theory of supporting relations between human rights. *Human Rights Quarterly, 30*, 984–1001.

Oduwole, O. O. (2014). International law and the right to development: A pragmatic approach for Africa. *Inaugural lecture as Professor to the Prince Claus Chair in Development and Equity 2013/2015*. Delivered on 20 May 2014 at the International Institute of Social Studies, The Hague, The Netherlands.

OHCHR. (1996–2016). Background on the right to development. *Office of the High Commissioner United Nations Human Rights Online*. Retrieved December 2015, from http://www.ohchr.org/EN/Issues/Development/Pages/Backgroundrtd.aspx

Osinbajo, Y., & Ajayi, O. (1994). Human rights and economic development in developing countries. *The International Lawyer, 28*(3), 727–742.

Pape, R.A. (1997). Why Economic Sanctions Do Not Work. International Security. 22 (2):90–136.

Parris, M. (2008, April 19). The new scramble for Africa begins. *The Times*.

Paupp, T. E. (2014). *Redefining human rights in the struggle for peace and development*. New York: Cambridge University Press.

Proclamation of Teheran, Final Act of the International Conference on Human Rights, 22 April–13 May 1968. For the text of the Proclamation, see The United Nations and Human Rights, 1945–95, at 247 (1995).

Rich, R. (2002). Solidarity rights give way to solidifying rights. *Dialogue, 21*(3), 25–33.

Sano, H.-O. (2000). Development and human rights: The necessary, but partial integration of human rights and development. *Human Rights Quarterly, 22*, 734–752.

Sautman, B., & Hairong, Y. (2007). Friends and interests: China's distinctive links with Africa. *African Studies Review, 50*(3), 74–114.

Sautman, B., & Hairong, Y. (2008). The forest for the trees: Trade, investment and the China-in-Africa discourse. *Pacific Affairs, 81*(1), 9–29.

Sen, A. (1999). *Development as freedom*. New York: Oxford University Press.

Sengupta, A. (2013). 'Conceptualizing the right to development for the twenty-first century' in Office of the High Commissioner for Human Rights. In *Realizing the right to development: Essays in commemoration of 25 years of the United Nations Declaration on the Right to Development* (HR/PUB/12/4, United Nations, 2013) 67.

Standing, G. (2014). Conditionality and human rights. United Nations Research Institute for Social Development. Retrieved June 2015, from <http://www.unrisd.org/sp-hr-standing>

Stiglitz, E. J. (2001). From miracle to crisis to recovery: Lessons from four decades of East Asian experience. In J. E. Stiglitz & S. Yusuf (Eds.), *Rethinking the East Asian miracle*. Washington, DC: The World Bank, Oxford University Press.

Stiglitz, E. J. (2015). In defence of the Asian Infrastructure Investment Bank. *The Guardian Online*. Retrieved January 2016, from <http://www.theguardian.com/business/2015/apr/14/in-defence-of-the-asian-infrastructure-investment-bank>.

Taylor, I. (2014). *Africa rising?: BRICS—Diversifying dependency* (p. 208). Woodbridge, UK and Rochester, NY: James Currey. ISBN: 97818470 10964.

Udombana, N. J. (2000). The third world and the right to development: Agenda for the next millennium. *Human Rights Quarterly, 22*, 753–787.

UN Committee on Economic, Social and Cultural Rights. (1990). *General Comment No. 3: The nature of states parties' obligations (Art. 2, Para. 1, of the Covenant)*, 14 December, E/1991/23. Retrieved January 14, 2016, from http://www.refworld.org/docid/4538838e10.html.

UN Economic and Social Council. (2002). *The Right to Development: The importance and application of the principle of equity at both the national and international levels*. E/CN.4/2003/25. Retrieved 28 August, 2017.

UN General Assembly. *Universal declaration of human Rights*, 10 December 1948, 217 A (III). Retrieved August 30, 2016, from http://www.refworld.org/docid/3ae6b3712c.html

UN General Assembly. *Declaration on the right to development: Resolution/adopted by the General Assembly*, 4 December 1986, A/RES/41/128. Retrieved January 20, 2016, from http://www.refworld.org/docid/3b00f22544.html

United Nations Office of the High Commissioner. 2013. *Realizing the right to development. Essays in commemoration of 25 years of the United Nations Declarations on the Right to Development*. New York and Geneva: United Nations Publications. ISBN-13:978-92-1-154194-6.

van der Have, N. (2014). The right to development: Can states be held responsible? In D. Foeken et al. (Eds.), *Development and equity: An interdisciplinary exploration by ten scholars from Africa, Asia and Latin America*. Boston: Brill.

Vienna Declaration and Programme of Action, U.N. GAOR, World Conf. on Hum. Rts., 48th Sess., 22nd plen. mtg., U.N. Doc. A/CONF.157/24 (1993), *reprinted in* 32 I.L.M. 1661 (1993).

Wellman, C. (2000). Solidarity, the individual and human rights. *Human Rights Quarterly, 22*(3), 639–657.

Stacey Links is a South African doctoral researcher at the University of Utrecht in the Netherlands. Links completed her BA in International Relations from the University of Stellenbosch, after which she pursued a double Masters in International Relations and Diplomacy at the University of Leiden, the Netherlands, in conjunction with the Clingendael Institute for International Relations. Her area of expertise lies in the nexus of Sino-African relations, human rights and critical international relations theory. Previously,

Links worked for a number of human rights organisations in the Netherlands, Haiti, Senegal and Ghana, as well as in the fields of international development and security. She is also currently a commissioning editor for E-IR, the world's leading open access website for students and scholars of international politics.

Part II

Environment, Natural Resources Management and Development

3

Oil, Ethnicism and Sustainable National Integration in Contemporary Nigeria

Abosede Omowumi Babatunde

Introduction

The crisis of nation-building in Africa has its root in colonialism, which divided the continent into various national segments with scant regard for ethnic configuration or historical cohesion (Afolayan 2002). In Nigeria, the perennial conflicts that pervade the country clearly underscore the assertion by Abdullahi and Saka (2007) that the union of ethnic groups in Nigeria through the 1914 amalgamation was a marriage of convenience: the whole process was conceived and executed without due respect for the social-cultural and political differences embedded in the political, economic and social structures of the partners. This assertion is not far-fetched; it also gives credence to the arguments by Nnoli (1998) and Osaghae and Suberu (2005) that colonialism was the cradle of ethnicity in Nigeria and a key factor in the crystallisation of contemporary

A.O. Babatunde (✉)
Centre for Peace and Strategic Studies, University of Ilorin,
Ilorin, Kwara State, Nigeria

A. Adeniran, L. Ikuteyijo (eds.), *Africa Now!*,
https://doi.org/10.1007/978-3-319-62443-3_3

identities. This set the stage for the emergence of ethnic nationalism and the genesis of the crisis of state-building in Nigeria.

Ethnicism is the employment or mobilisation of ethnic identity and differences to gain an advantage in situations of competition, conflict or cooperation (Osaghae 1995). Ethnic nationalism expresses a condition of heightened self-consciousness and identity of an organised ethnic group, given a national level of competition and conflictual interaction with similarly organised ethnic groups over a quest for power, wealth, security and status for its members (Eteng 2004). Unlike civic nationalism, which demands that an individual should belong to a nation, blood ties or ethnic ties are subsequently lifted to the level of political supremacy; ethnic nationalism does not necessarily follow the lines of the state (Geertz 1994; Joireman 2003).

In the Nigeria context, ethnic nationalism emerged with the politicisation of the different cultural and ethnic elements, and their mobilisation for political objectives, which included regional autonomy in a multinational state, or even the total break-up of Nigeria into one or more independent sovereign nation states (Hansen 1987). This accentuated centrifugal tendencies in the country, and, as Hansen (1987) argued, eventually aborted the birth of a truly independent and unified nation state. The implication is that ethnic nationalism in Nigeria continues to work against the integration of the different ethnic nationalities, with dire consequences for the continued existence of the country.

Apart from the joining together of the heterogenous groups, another policy of the colonials that engendered ethnicity in Nigeria is its policy of segregation, which breeds mutual distrust and acrimony. In this regard, Soyombo and Attoh (2009) observed that in northern Nigeria, the initial harmonious relation between the migrants and natives was disrupted by the British policy of separating the Hausa-Fulani from their southern counterparts living in the region. As a result, the migrants were forced to live in Sabon-Gari, while the natives lived in Tudun-Wada. The culturally artificial political entity called the 'Strangers Quarters' or 'Sabon-Gari' by the British which Coleman (1958) termed new artificial colonial constructions led to the alienation of the diverse ethnic groups from one another. The ethnic alienation became one of the strong bases for conflictual identity formation and discriminatory practice, leading to numerous ethno-religious crisis in places such as Kano in 1953 and 1966, the Maitatsine religious crisis in 1980, the Jos-Pleateau carnage in 2001, and the Sharia crisis in 2000, as well as the reprisal attacks in the south in places like Shagamu in 1999, among others (Osaghae and Suberu 2005).

Since independence, Nigeria has been battling myriad problems that have consistently undermined its national unity and development. Tracing the challenges of political stability in Nigeria, Olukoju (1997) stated that after Nigeria won independence from Britain in 1960, its artificial origin, coupled with other factors, bequeathed it a number of fundamental problems, one of which was the challenge of integrating, into a cohesive socio-political whole, the various entities lumped together by the colonialists. The narratives of the historical antecedents of the crisis of national disintegration in Nigeria clearly alluded to the fact that the diverse ethnic groups in the state were not effectively integrated towards the end of evolving a true sense of national identity and commitment to the survival and development of the nation state.

This paper examines how the intense competition for the control of oil resources has led to the rise and proliferation of ethnic nationalism, and the attendant implications for Nigeria's quest for national integration. Data is derived from studies relevant to the analysis of the issues of oil and ethnic nationalism in Nigeria. The paper argues that the intense competition for the control of oil resources is a causal factor in the rise and proliferation of ethnic nationalism, which is further heightened by the state's coercive conflict management strategies and poses grave implications for Nigeria's quest for national integration.

In spite of the fact that Nigeria transited to a democracy in 1999, which has provided the opportunity for the citizens to imbibe democratic ideals, the inequitable distribution of oil wealth has led to the resurgence of ethnic nationalism, which has seemed to aggravate the disunity and instability in Nigeria rather than serve as a basis for national integration. The question which arises at this point is that of how the competition for oil wealth has led to the rise and proliferation of ethnic nationalism. What are the implications of the proliferation of ethnic nationalism for the quest for national integration in Nigeria?

Studies have identified several factors which includes the colonial legacy, the plural nature of the state, the contentious revenue allocation formula, competition for oil revenue and the on-shore/off-shore dichotomy, the lack of civic nationalism and patriotism of the citizens, ethnically based political parties, the monopoly of power by the major ethnic groups and the consequent marginalisation of the minority groups as factors responsible for the rise of ethnic nationalism in Nigeria (Ake 1978; Nnoli 1978; Olukoju 1997; Afolayan 2002; Osaghae and Suberu 2005; Ekanola 2006; Aluko and Ajani 2009; Attoh and Soyombo 2011). However, in the context of this paper, it is argued that the competition for oil revenue allocation with its links with many of the other highlighted factors is a major crucial factor in the rise and proliferation of ethnic nationalities in Nigeria. The consequence is the prevailing political stability and lack of national integration, manifesting in the increased level of political violence and a fragmented and divided civil society.

At this juncture, an insight into the plural nature of the Nigerian state and its variant compositions, which Abiodun (2012) described as the basis upon which identities thrive, is crucial to this analysis. Nigeria is comprised of over 270 ethnic groups and not just the three ethnic groups—Hausa-Fulani, Ibo (Igbo) and Yoruba. The majority ethnic groups are the three largest groups mentioned above and they comprise 78 % of the entire population (Paden 2008). All other ethnic groups fit into the minority category with varying degrees of political status depending on their numerical size and political influence (Abiodun 2012). Therefore, as a heterogeneous society, Nigeria has continually faced the problem of ethnicity in its national life.

Although it has been argued that the root causes of the problem of ethnicity in Nigeria can be traced to colonialism, Nigerian leaders who gained power after independence and had an opportunity to promote the integration of the diverse ethnic groups failed woefully. This is because the political elites utilised the established colonial administrative structure to achieve their own personal and elitist ends (Ekanola 2006). Nevertheless, colonialism has been credited with certain positive gains. According to Osaghae and Suberu (2005), whether colonialists are to be blamed for identity problems in Nigeria or not, the truth is that the amalgamation of Nigeria promoted modern economic opportunities in an emerging colonial centre, leading to an influx of people from the Yoruba and Igbo into northern cities such as Kano, Jos, Kaduna and Zaria, and also people from the north to southern cities such as Lagos and Ibadan.

This paper is situated within the theoretical domain of the social constructivism, a theoretical perspective that explains ethnicity as a social construction manipulated to serve elite interest and thereby eroding the basis of good governance and legitimacy (Burr 2003). The constructionists believe that the common past and traditional values on which members of ethnic groups anchor their identity may have very little to do with historical realities (Ake 2000). The emphasis of social constructionist on the perspective of history, culture and socio-economic arrangement makes it ideal for understanding and explaining how the competition for oil led to the resurgence of ethnic nationalities and the implications of this for the quest for national integration in Nigeria.

In this sense, the resurgence of ethnic nationalities does not hinge on historical realities; rather, in this present democratic dispensation, the distribution of oil wealth has engendered fierce competition that has led to the rise and proliferation of ethnic nationalities who violently agitate for the control of power so as to have exclusive access to oil, thereby accentuating the divisions in the polity, with dire consequences for national integration. Within the context of the social constructivism theory, it is possible to reconstruct the Nigerian state to accommodate all ethnic groups agitating for a more egalitarian society; the aim is that all citizens will have a sense of belonging and identify more with the Nigerian nation than their ethnic groups, thereby promoting civic and national integration.

In a plural political state such as Nigeria, the nature of ethnicity is conflictual in the sense that people from different ethnic groups see each other as competitors in the quest for resources rather than compatriots. Since independence, the various ethnic elites have promoted the exclusive success of their members to the detriment of others (Salamone 1997). Politics in the post-independence years, during the first republic (1960–1965), was a triangular competition between the regions, championed by their hegemonic groups and carried out through ethnically controlled parties (Mahmudat 2010). The second republic (1979–1983) was not markedly different. This was a continuation of the same crude contest for supremacy by the former monolithic regions, especially as old foes and contests resurfaced in the 1960s. The aborted third republic, in which two parties of Social Democratic Party and National Republican Convention were allowed to co-exist, was created to streamline ethnic politics and promote unity. However, the tendency to hold on to power by the military truncated the democratic experiment of the third republic (Mahmudat 2010).

The ethnic rivalry culminated in the various conflicts that led to the abortion of the first republic and, consequently, the coup and counter-coup which culminated in the civil war in which the Ibo attempted to secede by declaring the sovereign state of Biafra in 1967 (Attoh and Soyombo 2011). The tactical alienation of the Ibo in the federal power on account of the suspicion engendered by the attempt at secession has continued to be a source of Ibo agitation (Abiodun 2012). Another major factor for the secession attempt by the Ibo was oil, which is the backbone of the Nigerian economy and was found in commercial quantities in the Ibo region and in another southern region, the Niger Delta region.

Therefore, the aspiration of the elites of the other ethnic nationalities to have their share of the oil wealth led to the fierce struggle for the control of oil resources. This, perhaps lend credence to Ekanola's (2006) argument that the Ibo and ethnic minorities in the Niger Delta have been aggrieved by the inadequate benefit derived from the enormous profits resulting from the exploitation of the extensive oil wealth lying beneath what they regard as their ancestral lands. The fact that no Ibo person has made it to the presidency since the Biafra war in the 1970s further reinforces the feelings of marginalisation harboured by the Ibos.

The colonialist ethnic segregation strategy was further exacerbated by the establishment of regional or federal structure of three units—the North, West and East regions—constructed to accommodate the identities of the major ethnic formation, Hausa-Fulani (North), Yoruba (West) and Ibo (East), without adequately capturing the ethnic minority components in the major ethnic groups (Abiodun 2012). One of the consequences of the establishment of this federal structure, which did not ensure the protection and security of minority interests, has been the incessant minority agitation for autonomy and the growing feelings of nationalism among the different groups. The attempt of the government to address the imbalanced tripartite ethno-regional structure with the creation of a mid-western state did not yield any positive result; rather, it encouraged ethno-regional polarisation and collapsed into a bipolar north–south confrontation (Osaghae and Suberu 2005).

In this regard, the assertion of Mustapha (2003) is most apt, when he argued that the many tangible and intangible fundamental changes that have taken place in majority–minority relations since 1966 has been constrained by three fundamental problems: the continuing claims of old-style domination by many minority ethnic groups; the specific situation of the ethnic minorities of the Niger Delta, from where the nation's oil revenue is derived; and the problems associated with the nation's language policy, which discriminates against ethnic minority languages to the extent that it confers a special status on the ethnic majority languages (Mustapha 2003).

Empirical evidence has indicated that the various efforts made by the Nigerian government to integrate the ethnic nationalities have achieved minimal success. Indeed, analysing the Nigerian state's lackadaisical quest to tackle the crisis, Osaghae and Suberu (2005) noted that the government eventually dissolved the four regions into 12, then 19 and finally 36 states, rather than restructuring the lopsided structure of the Nigerian federalism by addressing the minority question through the creation of sub-federal regional units. While state creation was envisaged as creating equality between the majority and minority ethnic groups, in practice it has reproduced the inequalities embedded historically in the Nigerian state (Mustapha 2003). Mustapha (2003) argued that since resources such as grants, jobs, scholarships, social infrastructure, public investments, and so on, are shared on the basis of territorially defined states and local governments; the ethnic majority groups who control the preponderant numbers of these units continue to enjoy a preponderant share, thereby further increasing their already accumulated advantages.

It is hardly contentious that the situation of ethnic tensions and conflict between the diverse ethnic nationalities in Nigeria has the potential to lead to the fragmentation of the country. Indeed, the incessant violence along ethnic lines witnessed in many parts of the country is a cause for concern. The intelligence report from the United States which warned that Nigeria might disintegrate within the next 15 years should not be disregarded. According to Cosmas (2005), a study conducted by Air University in the United States ranked Nigeria among the five least peaceful countries in Africa and concluded that Nigeria was likely to be a failed state by 2030. This assertion is not far-fetched given the prevailing security challenges, not least the increase in the rise of ethnic nationalism, religious violence and insurgency, humanitarian disasters, as well as the proliferation of small arms and light weapons in Nigeria. The activities of the youth militias in the Niger Delta and the Boko Haram insurgency in the north are major consequences of the political instability, which continues to pose a major threat to the survival of the Nigerian state.

Oil and Ethnic Nationalism in Nigeria

Most oil-rich African countries have been embroiled in conflict, largely because of the mismanagement of oil resources. In most of these countries, oil revenues are managed by central governments that are often neither democratic nor financially transparent. The revenue generated from oil resources is often diverted into the personal accounts of government officials, misused or wasted on white elephant projects which do not contribute to national economic development. The mismanagement of oil revenues usually fuels and exacerbates political discontent and tension and can provoke internal political violence. The distinctive characteristic of oil relative to most other primary commodities is its high value relative to its costs of production. The rent from oil, which is the surplus over costs and normal profit, is much larger than rents on agricultural commodities, and this may be why oil is differentially important in conflict risk.

Nigeria possesses Africa's largest oil and gas reserves; these account for over 83 % of federal government revenue, more than 95 % of export earnings and approximately 40 % of gross domestic product (Iwayemi 2006). According to Human Rights Watch (2007), when the country gained independence from British colonial rule in 1960, the possession of such natural resources led many Nigerians to be optimistic about the future of their country. However, rather than making concrete improvements in the lives of ordinary Nigerians, oil revenues have often fuelled political violence, fraudulent elections, police abuse and other human rights violations.

The federal grip over access and control of oil, the main source of government revenue, has been at the heart of Nigerian politics, economics and governance in the last five decades (Soremekun and Obi 1993; Obi 1998; Egwaikhide and Aregbeyen 1999; Babawale 2006). Oil provides a corrosive underpinning for official malfeasance in Nigeria—members of the governing elite have squandered and siphoned off public funds to enrich themselves at the expense of basic health and education services for most ordinary citizens (HRW 2007). The state governors in the oil-producing areas who were supposed to use the available oil revenue allocated to the state to alleviate the suffering of the people are also participants in the plundering of state funds. Compounding the situation is the fact that these corrupt politicians used oil wealth to mobilise violence in support of their political aims (HRW 2008).

Nigeria is a rentier state and the ruling class is primarily a state class, based less on the exploitation of wage-labour and more on the exploitation of state power and wealth derived from oil resources (Mustapha 2003). In the heated competition for resources and power, politicians have to resort to radicalising their ethnic base to gain access to the oil wealth. One reason that Mustapha (2003) identified as accounting for the disadvantaged position of the ethnic minorities relates to the distributive logic of the rentier state in Nigeria, particularly after its coffers were filled with oil revenues from the 1970s: most of the governors and administrators of the ethnic minority states did not always reflect the wishes of their constituencies because of what Guyer (1994) succinctly referred to as 'representation without taxation'. The main thrust of argument is that since taxation was not the basis of public finance, most governments in Nigeria, including those in ethnic minority areas, did as they pleased, with little effective oversight from the ethnic constituencies who voted for them.

Since the mid-1960s, oil has played a strategic role in the economics and politics of the country. The sudden and sharp rise in world crude oil prices, combined with a rapid increase in crude oil exports, translated into an unprecedented inflow of foreign exchange for Nigeria and other oil-exporting countries. The competition for oil wealth, Olurode (1999) has argued, set in motion new forms of ethnic rivalry and conflict. For the ethnic minority, oil heightened the fear of domination, which did not abate even with all the measures put in place to overcome such fears. Thus, the politics of oil led to the emergence of ethnic militias in the

Niger Delta who are aggrieved by the monopolising of their oil resources by the central government and the attendant environmental dislocation without any commensurate development of the Niger Delta region.

It is hardly contentious that the vast revenues from oil have not had any appreciable impact on the lives of the ordinary people, particularly the ethnic minorities in the Niger Delta, from whose land the oil is extracted. The marginalisation of the Niger Delta ethnic minorities has led to anger, frustration and violent agitation by the people, particularly young people, through the formation of ethnic militias and the resurgence of ethnic nationalism in different parts of the country. Mustapha (2003) pointed out that the pattern of agitation by the ethnic militias shifted from the more civic to the more parochial following the repressive response of the Nigerian government to the peaceful agitation of the Ogoni. On the other hand, as Okonta and Douglas (2001) have argued, groups such as the Ogoni had to become 'tribesmen' as a necessary condition for having their civic rights respected by the Nigerian state.

Various political and institutional measures were taken by the central government to resolve the lingering conflict in the Niger Delta. Government strategies for tackling the militancy ranged from piecemeal efforts to develop the region such as the adoption of 13 % derivation principle for the oil-producing state of the Niger Delta, the setting up of development commissions and the recent granting of an amnesty to the militias of the region and other coercive measures. All these efforts have failed to make an appreciable impact on the oil-producing areas largely because the development agenda of these authorities have not been people-centred and participatory.

The establishment of development initiatives by the Niger Delta Development Board established in 1961, the Niger Delta Basin Development Authority, the defunct Oil Mineral Producing Areas Development Commission established in 1992, the present Niger Delta Development Commission established by the civilian government in 2001, and the Ministry of the Niger Delta have yielded only a marginal impact in improving the environmental and human conditions of the oil-producing areas, which remain pervasively poor and underdeveloped. Empirical evidence indicates that all these agencies are a further demonstration of skewed revenue allocation and the marginalisation of the ethnic minorities—these agencies have always been deliberately left under-funded and have sometimes been made redundant by both the Nigerian government and the oil companies (Babatunde 2010).

Consequently, there is a cycle of political instability and legitimacy crises in the Niger Delta region. One feature of the resource control conflict has been proliferation of ethnic minority organisations, such as Movement for the Survival of Ogoni People (MOSOP), the Ijaw National Congress, the Itsekiri Nationality Patriots, the Ijaw Youth Council, the Niger Delta Volunteer Force, and the Movement for the Emancipation of the Niger Delta (MEND), among others in the Niger Delta (Obi 2002). The agitation in the Niger Delta for the control of oil resources is premised on the argument that Nigeria is the only country in the world which has so cruelly plundered its oil-producing district without any viable policy of compensation and repatriation.

These ethnic militias have been pursuing violent campaigns of hostage-taking and kidnapping, the destruction of oil facilities, oil bunkering and other criminal activities for resource control and the rejuvenation of their environment. The government's coercive reaction to the violent agitation of the ethnic militias seems to have aggravated the crises. As Abdullahi and Saka (2007) point out, government has sustained a long pattern of repression of local resistance, unleashing the might of the state to suppress these ethno-nationalist manifestations. They further note that the incidence of the Odi massacre, the Zaki-biam attack and the razing of Onitsha by Nigerian troops, contrary to the rule of law, were indicative of the repressive response of the government to ethnic agitation. The militarisation of the region—a joint task force has been permanently stationed in the oil-producing areas to protect oil installations which have been the target of the militia organisations—is another indication of the repressive strategy employed by the government (Suberu 1998; Obi 2002).

The insurgency in the Niger Delta region, which begun in 1998, was ignited by MOSOP, which carried out peaceful action in the early 1990s against the degradation of the Ogoni environment as a result of oil exploration by the multinational oil companies (Obi 2002). The repression of the Ogoni uprising and the hanging of the leader of MOSOP, Ken Saro Wiwa, by General Sani Abacha's military regime sparked off an international uproar that encouraged other Niger Delta ethnic groups, especially the Ijaws, who are the largest minority group in the country, to follow suit (Osaghae 1995; Duruji 2010). The struggle of Ijaw youths, then aimed at achieving self-determination and justice, attracted a brutal response from state security forces. The state's response to such a struggle is the massive deployment of military forces to crush popular pressures. To resist the deadly military might of the state, armed confrontation was adopted by the youths as a defence mechanism (Osaghae *et al.* 2011).

The ethnic militias in the Niger Delta see violence as a strategy to demand a greater share of the oil wealth. The criminal actions of the ethnic militias, such as kidnapping and the vandalisation of oil pipelines, which have resulted in loss of many lives and properties attest to the level of violence in the region. The recent amnesty programme by the government temporarily halted the violence in the Niger Delta. However, at present, there is a resurgence of violence as a result of the unwillingness of the government to genuinely address the challenges facing the oil-producing areas. In Nigeria, politics is conceptualised by various factions of the elites as a competition for the crude accumulation for personal wealth, but which they portray as a means of enhancing ethnic interest. The Buhari-led administration is faced with the daunting challenge of finding a lasting solution to the Niger Delta crises. So far, the government is yet to make any meaningful inroads into addressing the complex and daunting problem bedevilling the region. The implementation of the United Nations environmental report on the Ogoni by the Buhari government is a positive step towards tackling the region's many ills.

The renewal of hostilities by the Niger Delta militants underscored the fact that continued militancy in the Niger Delta could have more serious consequences in the near future for Nigeria. Also, the intensification of the separatist agitation championed by the Movement for the Actualisation of the Sovereign State of Biafra (MASSOB) arising from their long-standing position that Ibos can never realise their political aspirations within the Nigerian federation is another dimension of the threat to Nigeria's survival.

These ethno-nationalist movements which are predominantly in the south and middle-belt regions queried the dominance of the Hausa-Fulani political elites in the affairs of the state and, by implication, the control of the oil resources. Within this context, the emergence of ethnic nationalities in the democratic dispensation, such as MEND, MASSOB and the Niger Delta Avengers, was seen as a way of challenging the unfair distribution of oil resources and the over-concentration of power in the centre, as well as the Hausa-Fulani political dominance of the country.

The various sub-ethnic disturbances plaguing the Nigerian state, such as the renewed hostilities in the Niger Delta, MASSOB's recent agitations for a sovereign state of Biafra, and Boko Haram insurgency, is not only indicative of the ineffectiveness of the state authority but also underscores the ethnic rivalry and agitation for the control of power and oil resources. To cap it all, in Nigeria's political terrain, ethnicity is both a potent factor and a tool used by the dominant elite to secure for themselves strategic advantages in the competition for political power and resources. A case in point is the event that led to the transition to democratic rule in 1999. In this light, it has been argued that the annulment of the 1993 election by the military administration of Ibrahim Babangida, in which Chief M. K. O. Abiola, a Yoruba, emerged the winner, is indicative of the Hausa-Fulani quest for hegemony and the dominance of the northern elites in Nigeria's politics (Adejumobi 1996). However, the heated agitation and outcry over political marginalisation and exclusion by the southern ethnic nationalists, particularly the Odua People's Congress, influence the 'installation' of the Obasanjo-led administration in 1999, thereby ushering in the fourth republic (Babawale 1999; Onuoha 1999).

One of the most contentious issues that make ethnic agitation and the emergence of ethnic nationalities inevitable is the issue of oil revenue allocation. Numerous studies have provided empirical evidence to support the assertion that the agitation of the ethnic minorities in the Niger Delta centres on resource control. In particular, Mustapha (2003) observed that a major demand of the Niger Delta ethnic minorities, enunciated in various documents, including the Ogoni Bill of Rights and the Kaiama Declaration, is that of 'resource control'. Mustapha (2003) further pointed out that from 1954, the principal criterion for the allocation of federally collected revenue to the regions was the principle of derivation. Under this principle, regions were entitled to 50% of revenue collected from their territories.

However, the sharing of oil rent from the exploitation of oil among the three tiers of government—federal, state and local—in the federation became a contentious issue after 1970. Obi (2002) posited that in order to concretise its monopoly of oil revenues, the federal government changed the basis of the revenue allocation formula from derivation (which benefitted the hegemonic nationalities in the old regions) in favour of the principles of equality and the population of states (which again benefitted the big ethnic nationalities), thereby shutting off and alienating the oil-producing ethnic minorities from any direct access to oil. Gradually, derivation, as a principle of revenue allocation, which had been a mainstay of regional fiscal autonomy, declined from 50 % in 1969 to 2 % in 1981 and 1 % in 1989, but improved to 3 % in 1992 and 13 % in 2000 (Suberu 2001).

In their agitation for control over their oil resources, the communities in the Niger Delta pointed out that in the 1950s and 1960s, when agricultural commodities such as cocoa, groundnuts and palm oil were the main export income earners for Nigeria, the various regional governments where these products were produced controlled the income from their exports. But now that the main export income earner is oil, produced in the land of ethnic minorities, the wealth is being controlled by the central government dominated by the majority ethnic groups, yet the people are not allowed e any democratic input into how governments are formed or how the wealth is used (Constitutional Rights Project 1999).

Indeed, the contest for the control of oil and oil rents has been at the core of the political economy of Nigeria since the civil war years of the late 1960s (Onimode 2001). The advent of oil over the years eroded fiscal federalism in favour of fiscal centralism. Onosode (2003) contended that much more significant is the fact that among oil-producing areas, not only has derivation declined but other criteria subsequently introduced for sharing federally collected revenue, such as population, land area, school enrolment and so on, did not favour producing areas. Consequently, non-oil-producing states perceive oil revenues as a windfall that they can spend without the need for serious effort at internal revenue generation. Corruption by state and local government in these states has become rampant. It is in this light that one may be inclined to agree with Ihonvbore (1994) that oil played a pivotal role in the crises that have been rocking post-colonial Nigeria. Thus, rather than strengthen the fiscal muscles of the state to facilitate development, oil wealth led to 'extravagance, mismanagement and misplaced priorities'.

It is thus evident that in the politics of oil revenue in Nigeria, the major ethnic groups in the country strived to gain control of the centre in order to gain access to the oil resources—this was pursued through party politics and military coups (Akinwumi 2004). In the light of intense competition for oil resources, it is argued that one can attribute all the military coups and the civil strife in the country to the struggle to gain control of the centre and the resources it controlled. This perhaps explains the underlying motives of the military rulers who promulgated numerous decrees to ensure unfettered access to the oil resources. Soremekun (1995) pointed out that the Nigerian military government introduced a decree shortly after the civil war, which ceded 100 % duty on oil to the federal government. By Decree No. 9 of 1971, rents and royalties of off-shore petroleum mines were transferred from the various states to the federal government. He further stated that all the key laws governing oil production activities in Nigeria—the Land Use Act, the Petroleum Act, the Oil and Pipelines Act, the Mineral act and Petroleum (Anti-Sabotage) Decree—were brought under military rule without consulting the Nigerian people.

In the democratic era, the outcome of the 2011 election, which ushered in a President from a southern minority ethnic minority group heightened the expectation of the Niger Delta people for improved governance and the development of the region. The anticipated development sadly failed to materialise—the plight of the ethnic minority in the Niger Delta remained unresolved. The Jonathan administration used oil wealth to 'buy' social peace rather than address head-on the plight of the Niger Delta people. Many of the militias were awarded contracts to guide oil facilities in the region as a way of dousing their violent agitations. The failure of President Jonathan to address the developmental challenges in the Niger Delta during his tenure bring to the fore the ulterior motives of political elites who merely hide under the guise of ethnicity to protect their personal interests.

During Jonathan's tenure, the Hausa-Fulani agitation for a power shift to the north became louder, to the extent that there is threat of mass revolt if their quest is ignored. With the shift of power to Buhari, a northerner in the seat of power, the agitation of the ethnic minorities in the Niger Delta intensified with the resurgence of hostilities and the emergence of new militias, notably the Niger Delta Avengers.

It can be argued that the displacement of other natural resources by oil, thereby endangering the over-dependence of the central and states government on oil revenue, is another dimension of the problem. The inability of each state of the federation to produce and control the natural resources in their region, as was the case before the discovery of oil, has led to an over-reliance on oil resources and heated competition for access to the oil wealth. There has been persistent agitation for resources control by the ethnic nationalities in the Niger Delta, who are also demanding for the convening of a national conference of ethnic groups as a path to the restoration of a truly federalist or confederal constitution which will guarantee the autonomy and right to 'self-determination' of the various ethnic minorities of the Niger Delta.

The efforts of President Jonathan to convene a national conference when Nigerian reached 100 years of existence in 2014, however, failed to achieve the desired outcome. The national discourse was widely criticised for its failure to carry along all the ethnic and religious groups, and for the manipulation of the composition of the various groups by the government as well as for the financial inducements given to members. The resolution reached at the national conference, such as the additional state creations, have also been criticised by Nigerians as an unrealistic solution to the daunting identity crisis confronting the country. This assertion is not far-fetched considering that state creations in the past as solutions to ethnic conflict and to create equality between the majority and minority ethnic groups have, in practice, reproduced existing inequalities that have been historically structured into the construction of the Nigerian state (Mustapha 2003).

The short-term political calculations of those in power make it extremely difficult for them to provide the leadership needed, even when they wish to do so. In an attempt to halt the rising tide of opposition, political leaders, both civilian and military, exploit and manipulate the entrenched ethnic divide in Nigeria for political purposes in the struggle to enforce legitimacy. Enforced legitimacy cannot, especially in an ethnically differentiated society, stimulate national development. Power consolidation through the manipulation of the fragile ethnic relationships made the overriding objectives of national integration and development an impossibility. The legitimacy deficit will thus affect the government's moral authority in the negotiations which are urgently needed to resolve its many domestic conflicts. In the prevailing situation of the proliferation of ethnic nationalism and the absence of genuine citizenship, Nigeria continues a forlorn search for the existence of civic nationalism where citizens experience a fulfilling sense of belonging and are included in the country's social, economic and political activities.

Ethnic Nationalism and the Quest for National Integration in Nigeria

Nigeria's transition to democratic rule, which ushered in the fourth republic in 1999, could be perceived as the instrumentality for addressing the monopolisation of power by a 'single ethnic group' or a 'group of ethnic groups' in the country as well as the restoration of political stability in a multi-ethnic society. However, it would seem that democracy has failed to reduce ethnic tension as the resurgence of ethnic militias and ethnic nationalities has now taken a new dimension in the quest for power in the country. The instability in the polity has been perceived as the result of the struggle to impose new power relations and advance the interest of one ethnic group against others (Abiodun 2012). The long period of military rule was able to curtail the aggravated ethno-religious and regional diversities and suppress the manifestation of ethnic nationalism in the country from 1983 to 1999 (Duruji 2010). The Hausa-Fulani hegemony and the marginalisation of other ethnic groups during the military era was the issue that dominated the national political discourse prior to 1999. In this regard, Abiodun (2012) pointed out that the perception of some section of the society is that the military is serving the interest of the Hausa-Fulani major ethnic group, which dominates the military institutions of Nigeria by exercising hegemony over the security apparatus of the state, thereby exacerbating the present ethnic tension and identity relations in Nigeria.

One frightening dimension to these manifestations is the unprecedented rate at which ethno-religious and ethno-regional divisions are used to advance the cause of these groupings, resulting in violent ethnic and religious conflicts across the country (Akinwumi 2004; Adebanwi 2004). This assertion is buttressed by Maier (2000), who says that Nigeria has faced perennial crises of territorial or state legitimacy, as typified by the Biafra secessionist bid, the Niger Delta agitation for resource control, the south-west quest for sovereign national conference, and the sectarian religious crisis, which have often challenged its national cohesion, democratisation, stability and economic transformation (Maier 2000).

To underscore the deadly incidences of violence in the state, Jega (2003) noted that Nigeria witnessed 40 major civil disturbances in the form of ethno-religious and commercial clashes between 1999 and 2002 alone. In mid-2005 alone, it was estimated that at least 50,000 people had been killed in various incidents of ethnic, religious and communal violence and between 2009 and 2012, while Boko Haram has threatened the national coexistence of ethnic groups in Nigeria (HRW 2012). Thousands of lives and properties are lost in the incessant violence across the country, thereby impacting negatively on the security of the people, undermining development and constituting a threat to national security, stability and integration. It has been argued that the issues that generate the fiercest conflict among ethnic diversities are those considered to be fundamental to the existence and legitimacy of the state, over which competing groups tend to adopt exclusionary and winner-takes-all strategies (Osaghae and Suberu 2005).

It is extremely important to underscore the fact that the competition for the control of resources (oil) is the major reason for the conflictual nature of ethnic relation in contemporary Nigeria. The contest for political supremacy has been seen as a means of gaining access to the national wealth, which Joseph (1991) aptly posited Nigeria's brand of 'prebendal' politics—an unremitting and unconstrained struggle for possession and access to state resources. Consequently, the sharing of the federal revenue has been a contentious political issue. With the windfall from oil revenue in the 1970s, this contention has become increasingly marked between the federal, state and local governments, on the one hand, and by inference between the different ethnic groups, on the other.

It is disheartening that a country where the three major regions used to have control of their own resources in the past has descended to a level whereby only one resource—oil—is central to the entire country and is hotly contested along ethnic, political or religious lines. Therefore, the unequal distribution of oil wealth and the attempts by ethnic groups to gain access to power so as to control the enormous revenue from oil is the reason that, in spite of the emphasis on political accommodation and inter-segmental balance, increases in revenue allocation, appeasement of the south-west with the presidential position, the emergence of Jonathan from the Niger Delta ethnic minority and state creation for minority ethnic groups, ethnic tension and agitation is yet to abate.

Indeed, Nigeria is a federation with very strong unitarist elements in its constitution and operations such that the concentration of powers in the centre is cause for both instability and disaffection. Nigeria's federal structure comprises the federal government and the 36 state governments, as well as the 774 local government areas, which serve as the third tier of government. The revenue allocation formula ascribed more than 50 % of monthly revenue to the federal government such it takes 52.8 % of all federally collected revenue; the 36 states share 26.72 % while the 774 local councils share 20 %. The implication is that the federal government control more resources than the state and local government. In addition, the centre has been made too powerful and attractive because the resources lie in the exclusive control of the centre, where it is distributed to the 36 states in the federation.

To tackle the lopsided nature of Nigeria's federalism in the long term, it is vital that each state should develop or initiate strategies of developing their domains without having to depend on the oil revenue allocation from the federal government. There is also a call for the review of Nigeria's constitution to remove provisions which restrict state governments from developing their domains. In this light, the 'anti-federal provisions' in the constitution, such as the overbearing powers vested in the federal government, which also make security the exclusive responsibility of the central government, should be expunged.

Moreover, civil society, which is supposed to serve as the engine room of civic integration, has been divided and fragmented as a result of the interplay of ethnic nationalism in Nigeria. Oyadare (2004) traced the problem of the civil society in Nigeria to its heterogeneity and adversarial nature. As such, civil society in Nigeria is relatively weak, parochial in outlook and divided most especially along ethnic, religious and linguistic lines and this has adverse effects on Nigeria's nascent democracy.

The inability of the government to command the patriotism of its citizens through the provision of basic services is another dimension of the problem. Most importantly, Nigeria's government has abdicated its governmental responsibilities to its citizenry by its inability to adequately provide food, shelter, security and other basic necessities. The centrifugal tendencies of the elites erode the bases of good governance and have led to a crisis of legitimacy, which is an indication of state failure and collapse and poses a threat to national integration (Aluko and Ajani 2009; Mahmudat 2010). The implication is that without a veneer of legitimacy, the government is delinked from society and alienated from it (Ihonvbere (2001). In a situation where there is loss of legitimacy, the state loses the willing allegiance and legitimising support of its population, giving way to alternative centres of power within the territorial space of the nation state (Adejumobi 1996).

It would be naive, however, to assume that ethnicity in itself is a negative, disruptive and conflictual factor in governance as it could also be cooperative, non-conflictual and positive. Osaghae (2003) has argued that ethnicity is not altogether negative but can also be viewed as emancipatory because it offers the weak and oppressed groups a platform to articulate their grievances and seek redress. For instance, agitation of the Ogonis through the activities of MOSOP and the violent agitation of MEND over the neglect and degradation of the Niger Delta environment by the state and oil companies brought them to both national and global attention (Attoh and Soyombo 2010). Equally worthy of note is the emancipatory role played by the Odua People Congress (OPC) by consistently bringing to national and global attention the perceived injustice meted out to the Yoruba by the annulment of the June 1993 election, which was won by the late Chief Abiola. Through its actions, the right to contest the 1999 presidential election was conceded to the Yoruba. Chief Olusegun Obasanjo, a Yoruba, won the election and became the President for two terms (Attoh and Soyombo 2010). The thrust of the argument is that, despite the manipulative efforts of ethnic elites, the same elites find it politically expedient to seek a compromise when the stability of the polity is threatened—such compromises are needed and should be canvassed to build a virile state.

Moreover, the establishment of the Federal Character Commission, as well as the invocation of the federal character principle, the zoning formula adopted by the People's Democratic Party and other such acts are all geared towards political stability and national integration, but with little success. It has also been argued that the emancipatory roles played by the ethnic militias and ethnic nationalities in the Niger Delta earned the region a Vice President in 2007, who became the President following the death of the former President Yar'adua. This gave the region a sense of belonging and led to marginal reductions in the various acts of violence perpetrated by the militants in the region (Attoh and Soyombo 2010).

However, the assertion that such emancipatory roles should be viewed positively because they help other ethnic groups to understand the feelings and aspirations of their fellow citizens and thus help to engender harmony seems far-fetched. This is because the violent dimension of the ethnic agitation is a major security challenge and may lead to reprisals by other ethnic groups and spiral into a crisis that could engulf the whole country. The political recent phenomenon of the Boko Haram terrorist group, which have been targeting both Christians and Muslims through bombings and other violent attacks in the northern region, is instructive. In the case of the ethnic militias in the Niger Delta, this argument could be viewed as too simplistic, given that even though relative peace was achieved in the region after the granting of an amnesty to the militants, the insecurity in the region has not substantially abated. Kidnapping activities and oil bunkering are still rampant, as are sporadic violent attacks—an indication that the peace achieved can be termed the 'peace of the grapevine'. The recent emergence of Buhari, a northerner, as the President of Nigeria has fuelled the resurgence of violence by the ethnic militias in the Niger Delta and the renewed agitation by the Ibos for secession.

Indeed, it is of the utmost importance that one takes into account the fact that all the cited instances of compromises and policies have been inadequate to enhance political stability and national integration; rather, the polity continues to become more hostile with the attendant insecurity arising for the parochial interest of the political elites. Therefore, it is incontrovertible that there is a problem with power relations in Nigeria politics. In this sense, Aluko and Ajani (2009) state that the propaganda of the northern oligarchy is that the south has always been in control of the economic power and that, to balance the equation, the north must hold on to the political power.

The quest for political stability and national integration may remain elusive unless Nigeria has leaders who do not identify as members of a particular ethnic group but as nationalists—the kind of leader who, for instance, might be of Yoruba ethnicity, but who defines or sees himself as a Nigerian first and foremost. Such a leader would be totally devoid of nepotism and ethnicity, and consequently see all the other ethnic groups in Nigeria as partners in the Nigerian project. The problem of ethnicity is so pervasive in Nigeria because the leaders are ethnic leaders who identify with their ethnic groups first and Nigeria second.

The Nigerian government urgently needs to gear itself towards a political restructuring of Nigeria, the type that will ensure the continued existence of Nigeria as a polity. The alternative is a revolution by the ethnic nationalists, the consequences of which may either bring about a truly united Nigeria or a disintegrated Nigeria. The projection by the United States intelligence report that the Nigerian state may disintegrate in the near future is becoming more real than imagined as the current manifestation of insecurity in Nigeria is giving credence to such a prophecy.

Concluding Comments

In Nigeria, the quest for a viable state where there will be peaceful coexistence of the diverse ethnic groups and in which the people will have a sense of belonging and enjoy the dividends of good governance has been jettisoned as a result of the unhealthy competition among the elites for power and resources. As such, this situation continues to threaten the existence and unity of the Nigeria state as an entity. Ethnic nationalism has provided a way of challenging the prevailing inequity, injustice and marginalisation of the various ethnic groups that constitute the Nigerian state. The implications of the proliferation of these ethnic nationalism is the fact that they may draw on a mix of traditional or cultural symbols as well as those of economic inequality to express their deep-seated frustration, coupled with the desire, backed by force, for good governance with improved dividends of democracy for the marginalised.

Indeed, Ethnic nationalist movements were beginning to move beyond ethno-nationalism and religious fundamentalism into a shared sense of under-class alienation from the control of natural resources, deprivation from livelihoods and largesse. Therefore, improvements in the local economy, governance and the delivery of justice as part of planned institution building for socially inclusive growth with equity can contain the rise of ethnic nationalism and the associated insecurity and instability that have permeated the Nigerian state.

References

Abdullahi, A. A., & Saka, L. (2007). Ethno-religious and political conflicts: Threat to Nigeria nascent democracy. *Journal of Sustainable Development in Africa, 9*(3), 21–36.

Abiodun, F. (2012). Democracy and national identities: The travails of national national security in Nigeria. *British Journal of Arts and Social Sciences, 9*(II), 126–140.

Adejumobi, S. (1996). Citizenship, rights, and the problem of conflicts and civil wars in Africa. *Human Rights Quarterly, 23*, 148–170.

Afolayan, A. (2002). *Political concepts and ideologies: Nationalism and the nation state*. Ibadan: Hope Publications.

Ake, C. (1978). *Revolutionary pressures in Africa*. London: Zed Press.

Ake, C. (2000). *The feasibility of democracy in Africa*. Dakar Senegal: Codesria.

Akinwumi, O. (2004). *Crises and conflicts in Nigeria: A political history since 1960s*. Germany: Die Deutsche Bibliothek (e) LITVERLAG Munster.

Aluko, M. A. O., & Ajani, O. A. (2009). Ethnic nationalism and the Nigerian democratic experience in the fourth republic. *Afrian Research Review, 3*(1), 483–499.

Attoh, F., & Soyombo, O. (2011). The politics of ethnic balancing in Nigeria. *International Journal of Sociology and Anthropology, 3*(2), 40–44.

Babatunde, A. O. (2010). Managing environmental conflict in the oil-producing areas of the Niger Delta, Nigeria. In I. O. Albert & I. O. Oloyede (Eds.), *Dynamics of peace processes* (pp. 487–500). Ilorin: Centre for Peace and Strategic Studies University of Ilorin, Ilorin, Nigeria.

Babawale, T. (1999). Military governance as crisis of legitimation of the Nigerian state. In A. Ninalowo (Ed.), *Crises of legitimation* (pp. 99–123). Lagos: Obaroh and Ogbinaka Publishers Ltd.

Babawale, T. (2006). *Nigerian in the crises of governance and development: A retrospective and prospective analysis of selected issues and events*. Lagos: Political and Administrative Resources Center (PARC).

Burr, V. (2003). *Social constructionism* (p. 6). New York: Routledge.

Coleman, J. S. (1958). *Nigeria background to nationalism*. Berkeley and Los Angeles: University of California press.

Constitutional Rights Project. (1999). *Land, oil and human rights in Nigeria's Delta region*. Lagos: CRP.

Cosmas, E. (2005, August 2). Senate, U.S Military Chief Discuss Intelligence Report. *Daily Champion*.

Duruji, M. M. (2010, March). Democracy and the challenge of ethno-nationalism in Nigeria's fourth republic: Interrogating institutional mechanics. *Journal of Peace, Conflict and Development*, (*15*), 92–108.

Egwaikhide, F. O., & Aregbeyen, O. (1999). Oil production externalities in the Niger Delta: Is fiscal solution feasible? In *fiscal federalism and Nigeria's economic development*. Selected papers presented at the Annual Conference. The Nigeria Economic Society, Department of Economics, University of Ibadan, Ibadan, Nigeria.

Ekanola, A. B. (2006, Fall). National integration and the survival of Nigeria in the 21st century. *The Journal of Social, Political and Economic Studies, 31*(*3*), 279–293.

Eteng, I. (2004). Ethnicity and ethno class relationship: Crisis of Nigeria's enduring national question. In D. Oni et al. (Eds.), *Essay on Nigeria and globalization* (p. 46). Lagos: CBAAC.

Geertz, C. (1994). Primordial and civic ties. In Hutchinson & Smith (Eds.), *Nationalism* (p. 31). London: OUPress.

Guyer, J. (1994). Representation without taxation: An essay on Democracy in rural Nigeria: 1952–1990. *CASS Occasional Monograph No. 3*, CASS, Port-Harcourt.

Hansen, E. (1987). *Africa: Perspectives on the peace and development* (pp. 57–58). London: Zed Books.

Human Rights Watch. (2007, October). *Criminal politics: Violence, "godfathers" and corruption in Nigeria*. (Vol. 19, No. 16(A)). New York: HRW.

Human Rights Watch. (2008, March). *Politics as war: The human rights impact and causes of post-election violence in rivers state, Nigeria*. (Vol. 20, No. 3(A)). New York: HRW.

Human Rights Watch. (2012, October). *Spiraling violence: Boko Haram attacks and security force abuses in Nigeria.* New York: Human Rights Watch.

Ihonvbere, J. (2001). *Dismantling the Leviathian: Constitutionalism and the national question in Nigeria.* A Paper delivered at the 12th Convocation Lecture of the Lagos State University.

Ihonvbore, J. O. (1994). The state and environmental degradation in Nigeria: A study of the 1988 toxic dump in Koko. *Journal of Environmental Systems, 23*, 3.

Iwayemi, A. (2006). Nigeria's oil wealth: The challenges of sustainable development in a non-renewable natural resources dependent economy. The Postgraduate School University of Ibadan, 31st Interdisciplinary research discourse.

Jega, A. (2003). Democracy, economic crisis and conflicts: The Nigerian experience. *Keynote Address*, 22nd Nigeria Political Science Association (NPSA) National Conference, Ahmadu Bello University, Zaria, January 13–15.

Joireman, S. F. (2003). *Nationalism and political identity* (p. 12). New York: Continuum Press.

Joseph, R. (1991). *Democracy and prebendal politics in Nigeria: The rise and fall of the second republic.* Ibadan: Spectrum Books.

Mahmudat, M. O. (2010). Intra-class struggle in Nigeria. *Journal of Public Administration and Policy Research, 2*(7), 88–95.

Maier, K. (2000). *This house has fallen: Nigeria in crisis.* London: Penguin Books.

Mustapha, A. R. (2003). Ethnic minority groups in Nigeria: Current situation and major problems. United Nations Commission on Human Rights. Working Group on Minorities, Ninth session, 12–16 May.

Nnoli, O. (Ed.). (1998). *Ethnic conflicts in Africa.* Dakar Senegal: Codesria Book Series.

Obi, C. (1998). The impact of oil on Nigeria's revenue allocation system problems and prospects for national reconstruction. In K. Amuwo, A. A. B. Adigun, R. T. Suberu, & G. Herault (Eds.), *Federalism and political restructuring in Nigeria.* Ibadan: Spectrum Books.

Obi, C. (2002). Oil and the minority question. In M. Abubakah (Ed.), *The national question in Nigeria: Comparative perspectives.* Aldershot: Ashagate.

Okonta, I., & Douglas, O. (2001). *Where cultures feasts: Forty years of shell in the Niger Delta.* Benin: ERA/FOEN.

Olukoju, A. (1997). Nigeria: A historical review. In F. U. Okafor (Ed.), *New strategies for curbing ethnic and religious conflicts in Nigeria* (pp. 12–13). Enugu: Fourth Dimension.

Olurode, L. (Ed.). (1999). *Nigeria: People and culture*. Lagos: Rebonik Publications.
Onimode, B. (2001). Fiscal federalism in Nigeria: Options for the 21st century (Vol. 1). *Research Report*. Ibadan.
Onosode, G. (2003). *Environmental issues and the challenges of the Niger Delta: Perspectives from the Niger Delta*. Environmental Survey Process, Yaba. Lagos: The CIBN Press.
Onuoha, B. (1999). Political power blocs and legitimation crisis. In A. Ninalowo (Ed.), *Crisis of legitimation* (pp. 35–59). Lagos: Obaroh and Ogbinaka Publishers Ltd.
Osaghae, E. E. (1995). *Structural adjustment and ethnicity in Nigeria*. Upsalla: Nordiska Afrikan Institute.
Osaghae, E. E., Ikelegbe, A., Olarinmoye, O. O., & Okhomina, S. I. (2011). *Youth militias, self determination and resource control struggles in the Niger-delta region of Nigeria*. Council for the Development of Social Science Research in Africa (CODESRIA). Avenue Dakar, Senegal. Retrieved from www.codesria.org
Osaghae, E. E., & Suberu, R. T. (2005). A history of identities, violence and stability in Nigeria. CRISE. *Working Paper*. University of Oxford.
Paden, J. (2008). *Faith and politics in Nigeria: Nigeria as pivotal state in the muslim world*. Washington, DC: United State Institute of Peace Press.
Salamone, F. A. (1997). *Ethnicity and Nigeria since the end of the civil war*. Netherlands: Kluwer Academic Publishers.
Soremekun, K. (1995). Oil and the democratic imperative in Nigeria. In D. Olowu (Ed.), *Governance and democratization in Nigeria*. Ibadan: Spectrum Books.
Soremekun, K., & Obi, C. (1993). Oil and the national question. In NES (Ed.), *The national question and economic development in Nigeria* (pp. 209–231). Nigerian Economic Society Annual Conference.
Soyombo, O., & Attoh, F. (2009). Trend of political crime and violence in Nigeria. In R. I. Salawu, A. Akiade, & S. O. Adetona (Eds.), *Curbing political violence in Nigeria: The role of security profession*. Lagos: Institute of Security Nigeria.
Suberu, R. T. (2001). *Federalism and ethnic conflict in Nigeria*. Washington, DC: United States Institute of Peace Press.

Abosede Omowumi Babatunde teaches at the Centre for Peace and Strategic Studies, the University of Ilorin, Nigeria. She holds a PhD degree in Peace and Conflict Studies from the University of Ibadan, Nigeria. Dr Babatunde is a member of the Society for Peace Studies and Practice, Nigeria; the International Peace Research Association; and the Council for the Development of Social Science Research in Africa. She has been awarded several distinguished academic fellowships, including an African Humanities Postdoctoral Fellowship (2010), and an American Council of Learned Societies Research Fellowship in Peace Studies (2013) at the Consortium for Peace Studies, the University of Calgary, Alberta, Canada. She was a 2016 Individual Research Grantee of the African Peacebuilding Network of the Social Science Research Council.

4

A Phenomenological Study of the Development Outcomes of Extractive Sector Governance in Côte d'Ivoire

Kabran Aristide Djane

Introduction

Environmental concerns appear as a new frontier of the social question: public policies for equity or the reduction of social inequalities that do not take into account the environmental dimension ignore an essential aspect of this. More clearly, social question refers to the state of deep interrogation in which politicians and citizens of all stripes are confronted with the problems associated with the radical transformation of labor to the benefit of wage-earning, and the models of governance which modify the relationship between social classes and the production process. Conversely, the evolution of inequality may reinforce certain environmental imbalances, and represents a real challenge for greening growth. The environment includes the social future of children growing up there (the subject of my doctoral thesis). In addition, some studies have shown how a vicious spiral could reduce the social-environmental opportunities

K.A. Djane (✉)
Abidjan, Côte d'Ivoire

A. Adeniran, L. Ikuteyijo (eds.), *Africa Now!*,
https://doi.org/10.1007/978-3-319-62443-3_4

of unborn children due to the environmental pollution suffered by the mother during pregnancy (Olukoshi 2008). Less tragically, the quality of daily life is determined in part by access to what we call 'environmental amenities'. Therefore, this study aims at interrogating the social and environmental inequalities of local residents, emerging diamond production zones and challenges due to poor governance of the extractive resources sub-sector in Côte d'Ivoire, particularly in the northern part of the country.

Extractive resources have significant implications for both governance and the life qualities of the surrounding population. As noted by Olukoshi (2008), the nature of this social and environmental crisis '*over the years* [has been] *related to the process of exploiting natural resources, the nature of the concessions granted by governments to mining companies and controversies about the distribution of income derived from of natural resources*' (2008: 27) must be investigated. In order to address these difficulties, which apply in other countries that seek popular participation and transparency in the management of their extractive resources, Côte d'Ivoire has adhered to the principles and criteria of the Extractive Industries Transparency Initiative (EITI), which was launched during the 2002 Summit on Sustainable Development in South Africa. Côte d'Ivoire has met the requirements of the EITI and its National Council has become 'EITI Compliant' since 22 May 2013. Measuring this transparency requirement, both at an internal and an international level, the NC EITI decided, with the support of the Government of Côte d'Ivoire, a history and diagnostic analysis of the EITI to identify the prospect for equities in 2015. From this perspective, the evaluation of the EITI implementation workshop and the development of the EITI work plan 2013–2014 was held from 22 to 23 August 2013 with a framework to reflect on the applicability of the dictates of laws and good governance practices linked to extractive resources. The willingness of the National EITI Council reflects observation and persistent social concern on the efficient use of extractive resources in the socio-economic development of Côte d'Ivoire via executive fiat. Also, in order to reduce economic and social crises in the country, the Ivorian government has embarked on the implementation of a large-scale campaign series involving public investment and private participation. These actions have been funded mainly by external financing and borrowings from the regional financial market with the budget deficit at 3.5% of its gross domestic product in 2012. The fiscal resources have been concentrated more on inland customs revenues, property and other tax. Although many investors have denounced a major tax burden in Côte d'Ivoire, the governments at various levels in the country have pegged investment tax at 17%, which is clearly below the 75% benchmark of West African Economic and Monetary Union countries.

During last decades, the lack of action to fight against social and environmental injustice linked to non-renewable resources extraction has increased a growing interest in research by researchers (Le Billon 2007); Other authors (Evans 2010) go further by questioning the social conflicts linked to the governance of extractive resources (AIDS 1981–1982). Hendrix & Saleyman (2012) consider, for their part, that climate change also has a link in the emergence of these environmental conflicts. United Nations Group of Experts on Natural Resources and Conflict in Africa has also concluded to this result in their scientific work (Witness 2007). So they suggest that extractive resources have two-dimensions: non-renewable ones (Diamond, oil...) and renewable ones (agricultural products); as the working group mentioned it, the related conflicts non-renewable extractive resources are much more intense; but those related to renewable extractive resources are larger in Africa than the first one. In any case, several African researchers (Le Billon 2007) have lean on the issue by constantly linking the problematic of extractive resources with conflicting issues. All these, based on the fundamentals of the Brundtland Report (Guesnet & Al. 2009), indicate that in these terms *"disruption of the environment is both a cause and an effect of political tension and military conflicts"*, so *"there is no [...] military solutions to the insecurity of the environment"* and it is necessary to combine environmental sustainability with the reduction of poverty and development Economic *"by common management"*. Thus, Benjaminsen et al. (Muhammed 2003) reinforce the political governance of the environment by demonstrating that agropastoral conflicts are more determined by three structural factors such as (1) the expansion of cultivated areas, (2) the effects of a weak political authority at the local level and (3) the annuity behaviour of government officials (Hyden, O., & Hastings 2000). Therewith the variable *"governance"* is the cause of all this antinomy which falls under the central authority. But CODESRIA Studies in 2006 on Governance Trends in West Africa goes further by stating that *"consciously or unconsciously the term governance has become a code to describe the corruption of the elite and / or reprehensible practices of an official. It has drastically limited its scope, reducing involuntarily to a repetitive speech of one of its pathologies. ....* " (Olukoshi 2008, p. 6). The governance remains a concept so badly mastered even if major scientific outputs of African researchers are, in most cases, brought into connection with democracy and citizenship. Indeed, as he says *"... governance increase poverty and social inequality in contemporary and measurement, environmental inequality."* (Olukoshi 2008, p. 27).

Based on these scientific concerns, we raise up those following questions: What limits the Ivorian executive in its deployment of a transparent policy to guide its extractive sub-sector? What is the level of impact of this governance and what effect do the opaque governance practices have on the environment? How are social and environmental conflicts arising from this development managed by governance at both local and national level?

This study, therefore, aims to present a phenomenological analysis of the governance of the extractive sub-sector, and the impact of the exploitation of natural resources in Côte d'Ivoire vis-à-vis the country's commitment to the EITI initiative.

Methodology

Participants in the study were selected from various public stakeholders: the Ministry of Mines, the Ministry of Economy, the National Council for Initiative Extractive Industries Transparency (NC EITI), mining companies, local communities, and civil society organisations. Data was sourced through intensive interviewing conducted with individuals and organisations involved in EITI implementation. The study took place in three phases: focus groups were organised mainly with civil society organisations; this was followed by individual interviews and then structured surveys. A total of 137 individuals were involved.

The analysis combines two methods: first, the analysis of texts, including legal texts (the Mining Code) and various reports, in particular the

EITI reports; and, second, analysis using SPSS software of the results of interviews, the focus groups and the structured surveys.

The main difficulty in the collection of data was the unavailability of some keys players. This somewhat limited the participatory principle behind the execution of the research. This unavailability has inevitably negatively impacted the data collection process, limiting the representativeness of the data collected.

Synergy Between Stakeholders in the EITI and the Government

NC EITI is composed of 26 members grouped into three entities: the state; the extractive sector; and the civil societies. CN is the EITI multi-stakeholder, though still tripartite, group.

Among the representatives of the state are: a representative of the National Assembly, representatives of the Government and a representative of the Chamber of Accounts (the judiciary). They collaborate with representatives of the private sector and those of civil society organisations. NC EITI, in its composition, presents the synergies among key players in the extractive industries. The different structures represented in the NC EITI are the focal points—their role is to provide information to NC EITI in elaborating its annual reports.

Article 11 establishes the Committee that serves as the interface between the government and the NC EITI through the improvement of communication between the two entities. In other words, the establishment of this Committee should assist the government in receiving prompt updates on all actions undertaken within the framework of EITI implementation. Unfortunately, up till the point of concluding this study, this Committee remained elusive.

The compositions of the Committee can be analysed on two fronts. First, the Chamber of Accounts has been put in a problematic position as both judge and jury. Indeed, how could independence have been enabled if the Chamber is to consider the sub-sector report that it was expected to participate in compiling? Second, the NC EITI is placed under the authority of the Prime Minister. But the Constitution guarantees the independence of judges, especially those of the Supreme Court. The Chamber of

Auditors, as the Supreme Court, cannot be subjected to the Prime Minister.

The absence of representative(s) of the National Secretariat for Governance and Strengthening Capacities, and representatives of artisan miners and cooperative societies operating in the extractive sector, have limited the functionality of existing frameworks for administering the industries.

Findings from the Study

The governance issue was examined through the lens of accountability, transparency, participation and the level of corruption.

a. Accountability

Substantially, all of those interviewed in the study believed that the government and its agencies have not adequately matched words with actions in respect of activities in the extractive sub-sector. The structures for accountability exist only in documentation (such as the National Assembly, the General State Inspection, the Court of Auditors, etc.) but in practice, there has been an obvious lack of culture of accountability. The majority of the research participants believed that the Ivorian government has been lackadaisical in the management of the extractive industries, especially in respect of monitoring and evaluation. According to one of the interviewees, passivity on the parts of the population, the political parties and civil society organisations has permitted the government to mismanage the sub-sector.

b. Transparency and openness

Data sourced from the study did reveal that the Ivorian government has not been sufficiently open and transparent in respect of the extractive industries. This could be seen in the difficulty associated with accessing information on the industries. However, the research subjects were unanimous on the fact that they would trust the government if it were to be transparent in terms of the management of the industries.

c. Participation

Civil societies in Côte d'Ivoire have remained relatively inactive in monitoring the government's operations in the extractive industries. They have working poorly in matters concerning governance and the economic sector. Nevertheless, they are seen to be working at three levels: activism, independence and influence. In politics, they have been very active, especially in the field of human rights. Civil societies in this country are theoretically independent, but in practice they are not, largely due to financial dependency.

d. The level of corruption and impunity

The research participants were unanimous in decrying the high level of corruption in the industries despite the existence of various laws against corruption. Lack of publication of contract details, unverifiable financial reporting mechanisms, and lack of investment control in various project host communities have remained fertile grounds for corruption. It is worth noting that this high level of corruption has been recognised officially.[1]

Concluding Comments

Transparency in the extractive industries in Côte d'Ivoire has been predicated on the social construction of a governmental perspective on the sub-sector, especially as it pertains to the establishment of the extractive resources initiative (NC EITI). The indicators for examining the industries as set out in this study have assisted in devising a reliable assessment for determining the level of confidence that the government itself grants to its agencies. It is noted that while these agencies have been saddled with the responsibility of ensuring equity and clarity in the social redistribution of extractive resources, they have been manipulated for other purposes by the same government and its agencies. Hence, this study affirms that the development has further widened the gulf between the governance process and the livelihoods of the citizenry, especially those living around various mining areas in Côte d'Ivoire.

Note

1. Cf. National Plan for Good Governance.

References

AIDS. (1981–1982). *Report, theme: The environment*. Stockholm: AIDS Information Division.

Evans, A. (2010). *Resource scarcity, climate change and the risk of violent conflict*. Washington, DC: World Bank.

Guesnet, L., Schure, J., & Müller, M. (2009). Natural ressources in Côte d'Ivoire: Fostering crisis and peace? The cocoa, oil, diamond and gold sector. *Brief* 40. Bonn: BICC.

Hendrix, C., & Saleyman, I. (2012). Climate change, rainfall, and social conflict in Africa. *Journal of Peace Research, 49*, 35–50.

Hyden, G., Olowu, D., & Okoth-Ogendo, H. W. O. (2000). *African perspectives of governance*. Trenton, NJ: Africa World Press.

Le Billon, P. (2007). The political ecology of war natural resources and armed conflict. In UNEP (Ed.), *From conflict to peacebuilding. The role of natural resources and the environment*. Geneva: UNEP.

Muhammed, S. (2003). *African political parties: Evolution, institutionalization and governance*. London: Pluto Press.

Olukoshi, A. (2008). *Governance trends in West Africa: A synthesis report*. Dakar: Codesria and OSIWA.

Sane, A. (2016). The environmental issue in Africa: A general statement. In E. Vubo (Ed.), *Environment, agriculture and cross-border migration* (pp. 5–18). Dakar: CODESRIA.

Witness, G. (2007). *Hot chocolate: How cocoa powers was the conflict in Ivory Coast*. Washington, DC: Report 2007.

Kabran Aristide Djane holds a PhD degree in Social Sciences (with an option in Environmental Sociology) from the University of Cocody, Côte d'Ivoire. He is presently a lecturer in qualitative and quantitative research at the Sociology Department, Peleforo Gon Coulibaly University of Korhogo, northern Côte d'Ivoire. His research focuses on the socio-construction of environment and environmental behaviour modelling.

5

Land Degradation and Sustainable Human Development in Nigeria

Tolulope Adetayo Fajobi

Introduction

Land degradation remains a recurring social challenge with negative implications for socio-economic and human development, particularly in the developing countries of the world. Today, the task of estimating the extent of the problem posed by the damaging utilisation of land resources is huge. There is no doubt that people around the world are suffering from the impact of poor land use. Rosenberg (2007) posited that a staggering 70% of all dry lands (or non-tropical regions) are already classified as degraded across the globe. This represents 14% of the entire earth's land surface. With 73% of its agricultural dry lands currently estimated as degraded, Africa remains the most affected continent globally. The number of people affected is vast, because more than about 70% of Africans depend directly on the land for their livelihoods (Rosenberg

T.A. Fajobi (✉)
Department of Sociology and Anthropology, Obafemi Awolowo University, Ile-Ife, Osun State, Nigeria

A. Adeniran, L. Ikuteyijo (eds.), *Africa Now!*,
https://doi.org/10.1007/978-3-319-62443-3_5

2007). The economic fortune of most developing countries, including Nigeria, however, revolves largely around human resources and the exploitation and use of land resources, especially in the primary industries such as agriculture, for development and the reduction of poverty (Titilola and Jeje 2008). Consequently, the loss of productive land affects farming and development. As a result, farm workers and others may be forced to move to towns and cities, only to face unemployment and poverty. Hence, land degradation distorts the process of urbanisation in Nigeria—unexplained settlements emerge with attendant implications for rising food costs, increasing levels of criminality, over-stretching of social infrastructures and so forth.

In Nigeria, there has been significant growth in recent years, with per capita income crossing the $1000 mark in 2006; there is still a prevalence of extreme poverty, with agriculture, the driver of growth accounting for about seven out of ten case of poverty (AfricaFocus Bulletin 2010). In spite of the economic growth recorded in recent years, progress in human development has been unimpressive in Nigeria on various indicators such as poverty incidence, inequality and access to basic social services. The manifestation of disparities includes inequality in asset distribution, such that 20 % of the population own 65 % of the national assets. Additionally, there is unequal access to basic needs and social infrastructure by sector (urban and rural) and within sector (urban poor versus urban non-poor, and rural poor versus non-poor) (United Nations Development Programme 2009).

The phenomenon of poverty has generally been assessed in terms of economic downturn. Degraded soil and dwindling forests, with resultant rural–urban migration, has led to a conspicuous utilisation of land resources in Nigeria. Every year, many hectares of land are mismanaged for other purposes, including uses such as urban expansion, which affects human development in Nigerian communities. Moreover, some individuals or groups are in the habit of excavating and mining the land surface in order to generate income; they leave the explored land surface bare, thereby exposing the land to damaging use which increases poverty in the society and affects the quality of life. Poor land utilisation has spurred poor people to migrate elsewhere because they lack the skills needed in such a society to survive situations which threaten their existence. At this rate, in a number of years, for example, there will no more than a few productive hectares of land per person available to produce food in Ogun State, Nigeria.

However, the threat of land degradation to sustainable humanity, among other current global development concerns, is worsened by its immediate distortive role vis-à-vis the world food security programme and the quality of the environment. It is worth noting that high population density is not essentially related to the problem of land degradation: 'it is what people do to the land that determines the extent of degradation.' (Eswaran et al. 2001) By and large, as individuals can be potential facilitators of the problem of land degradation, so also can they serve as potent 'resolvers' of associated challenges to the problems this could pose to human existence (Eswaran et al. 2001). Studies have shown that due to the absence of any functional regulation to discourage destructive use of land resources, Nigeria has remained one of the countries with the worse records of land mismanagement in Africa. This has posed a serious threat to food security, the quality of life and the quality of the environment in Ado-Odo/Ota Local Government Area (LGA) of Ogun State. Earlier studies have focused on how law can be used to curb land degradation with little attention paid to issues bordering on land degradation from a human development perspective. This chapter essentially seeks to draws attention to the connection between land degradation and the incidence of poverty, the involvement of land degradation in rural–urban migration and the function of institutional intervention towards mitigating the challenges of land degradation in Nigeria.

Theoretical and Methodological Frameworks

Marxism theory posited that all land in society belongs to the government, but the distribution of the land is carried out by various investors after collecting their taxes, kickbacks and commission from the investors, with no measures put in place to check their activities on the land. Thus, Marxist theory has emphasised the lack of ownership of land as the bane of holistic human development, while vicious cycle poverty theory postulates (Oscar 1959; Ascher and Healy 1990) that poor rural households have only land and unskilled labour as their major assets, and thus few human or financial capabilities. The poor feel compelled to do what they often recognise is harmful to their own survival interests yet they feel they have no alternative by virtue of their situation. Simply put, the vicious cycle of poverty has been central in affirming that, even in certain situations where individuals have had direct access to land, it could still be wrongly utilised due to subsisting limited human basic needs.

The study was carried out in the Ado-Odo/Ota LGA in Ogun State, south-western Nigeria with latitude 6° 40′ 59.99″ N and longitude 3° 40′ 59.99″ E. The study employed a mixed method research design. Primary sources of data collection were engaged. This data was collected among people (farmers, traders, building contractors, relevant government officials, timbers dealers, sand excavators and community leaders in-charge of land matters—Baale and Olori-Ebi) in the Ado-Odo/Ota LGA in Ogun State, Nigeria. Data was collected through the administration of questionnaires, in-depth interviews (IDI) and focus group discussions (FGD). Data collected was subjected to descriptive statistics and content analyses.

Human Development: The Case of Nigeria

In the Nigerian federation, the 36 states, the Federal Capital Territory as well as 774 local governments, share about 45 % of consolidated revenue, while the federal government controls the rest. However, the federal government has no statutory powers to control the sub-national governments for the purposes of macro-stabilisation. Apart from this, areas such as primary and secondary education, primary and secondary healthcare services, rural roads and infrastructure, water and sanitation and community services, with direct implications for human development, are assigned to these lower tiers of government in Nigeria. However, the federal government often provides services in most of the areas where the sub-national governments have responsibilities, thus resulting in conflict, waste and inefficiencies (AfricaFocus Bulletin 2010).

The overall goal of economic development is improvement in human well-being. Nigeria possesses a stark dichotomy of wealth and poverty. Although the country is rich in natural resources, its economy cannot yet meet the basic needs of the people. Such disparity between the growth of gross domestic product and the increasing poverty is indicative of a skewed distribution of Nigeria's wealth. A majority of Nigeria's 140 million (2006 census) citizens live below the poverty line and have limited or no access to basic amenities, such as potable water, good housing, a reliable transport system, affordable healthcare facilities, basic education, sound infrastructure, security, and sustainable sources of livelihood. Given the nation's history of wide income disparities, which have manifested themselves in large-scale poverty, unemployment and poor access to healthcare, the disconnect between the country's economic growth and human development has to be addressed to increase the well-being of its people (Okon 2012).

The United Nations Development Programme has stated that Nigeria is not one of the African countries recording remarkable improvements in its human development index. The agency in its 2013 Human Development Index Report listed Angola, Burundi, the Democratic Republic of the Congo, Ethiopia, Liberia, Mali, Mozambique, Niger, Rwanda, Sierra Leone and Tanzania as among the African countries that have made the greatest strides since 2000. According to the report, Nigeria was positioned among the countries with a low development index, ranking 153 out of 186 countries. Life expectancy in Nigeria is 52 years, while other health indicators reveal that only 1.9 % of the nation's budget is expended on health; 68 % of Nigerians are stated to be living on less than $1.25 a day while the adult illiteracy rate is 61.3 %. The report comes despite the reported growth in the Nigerian economy, with the country recording a gross domestic product growth rate of 6.99 % in the fourth quarter of 2012. The country's economy has been described as robust and resilient (United Nations Development Programme Report 2013).

Connection Between Land Degradation and the Incidence of Poverty

According to Titilola and Jeje (2008), land degradation has far-reaching effects on the earth's surface and therefore on human life. In Nigeria, the linkages between the threat of continued soil erosion and productivity, income generation, and poverty alleviation—and more importantly on future generation—resulting from the consequences of misuse or inadequate land management practices are alarming. With approximately 98 million hectares, Nigeria is liberally endowed. However, soil erosion constitutes a major threat to land resources in several parts of the country. This limits the utility of the land, adversely affecting food supply and food security at a national and household level. The severity of the effects of soil degradation is perhaps better appreciated when cognisance is taken of the fact that some communities in Nigeria have had over 10% of their land mass wasted by gully erosion and other types of erosion (Titilola et al. 1996).

Titilola and Jeje (2008) posited that poverty and resource misuse is linked because of the pattern of land distribution. The rich usually have access to land which is less prone to degradation or erosion, while the poor farmers continue to till a marginal resource base despite increases in their number. Projected population increases superimposed on existing land holding patterns will result in an increase in poverty-derived pressure on the environment with accelerated erosion, deforestation and desertification. In Nigeria, 80 to 90% of the poor live outside cities. Half of these rural poor live on resources with the potential to increase production and income in environmentally friendly ways. Increasingly, however, as development raises incomes in the more productive region of these countries, rural poverty tends to concentrate in environmentally fragile areas (Titilola and Jeje 2008).

Impact of Land Degradation on Rural–Urban Migration

Environmental or ecological degradation accompanied the birth of the industrial revolution, which also brought along with it mass exodus of people to urban areas. The industrial revolution with its complex technological and economic changes produced spectacular increases in world population and cities. According to Leitmann (2005), in developing countries, the rural poor migrated to cities and accelerated urban population growth, which usually led to a chaotic urban atmosphere. Sub-Saharan Africa has long been one of the least developed and least urbanised regions of the world with most sub-Saharan African economies still heavily dependent on subsistence agriculture. Nevertheless, the region has absorbed relatively high rates of urban growth over the past 50 years. In 1950, only 15% of the Africa population was living in towns or cities, while 39.9% lived in urban areas in 2000 (United Nations 2002; Satterthwaite et al. 2010). By 2030, about 53% of Africa's population is expected to be living in urban areas (Cohen 2004). However, cities have been identified as engines of development and the proportion of environmental damage will continue to increase relative to the rate of physical development, of which building and road construction plays the greatest part.

In another view, Oduwaye (1998) attributes part of the problem of rapid urbanisation to the rapid rate of migration from rural areas to urban areas, among other causes. All the growth and rapid urbanisation has not been without its problems. He claimed that the tremendous pressure brought on by unplanned urbanisation has put pressure on land, resulting in urban sprawl and largely unmet demand for basic urban services and infrastructure such as water, electricity, sanitation, access roads, public transport, effective drainage and waste disposal. He explains further that the inevitable result has been the rapid development of slums both on land and along the waterways, including the lagoons where the tidal waves are too weak to sweep away the temporary shanties that have been constructed by people who are desperate to live near the city centre but are unable to afford the cost of scarce accommodation. He concluded that rapid unplanned growth was characterised by over-population of some areas and neglect of others. Ajayi et al. (2014) accentuated that unplanned settlements in developing countries are the product of rapid urbanisation, which is bringing migrants from rural areas to industrialised centres and the main cities in increasing numbers.

Institutional Intervention on the Challenges of Land Degradation

As noted in the Nigerian Ministry of Environment (2012) report, as part of the implementation strategies of the United Nations Convention to Combat Desertification, a National Coordinating Committee to combat desertification control was established in 1993, with its Secretariat in the then Federal Environment Protection Agency. The Committee, which is comprised of representatives of line ministries, agencies, relevant research institutes, and non-governmental organisations, has responsibility for coordinating the implementation of the Convention to Combat Desertification. It must be noted that most states of the Federation have legislation relating to environmental protection, such as edicts on bush burning and deforestation.

The creation of the Department of Drought and Desertification Amelioration in the new Federal Ministry of Environment in 1999 strengthens the existing institutional arrangements for more effective coordination of activities by the government towards the implementation of the Convention to Combat Desertification in the country. The move was to ensure a sharper focus on the rehabilitation and restoration of desertified and desert-like conditions in the areas affected. The upgrading in 2003 of Nigerian Metrological Services from a department in the Ministry of Aviation to a semi-autonomous agency, the Nigerian Meteorological Agency, represents yet another major step in addressing drought and desertification in the country. There is also a National Council on the Environment, made up of representatives of governments at the federal and state levels. The Council meets at regular intervals to take stock of the state of the environment in Nigeria (Federal Republic of Nigeria, Ministry of Environment 2012).

Nevertheless, Akinola and Adewale (2012) posited that at the level of federal government, there is the question that centres on the status of the Climate Change Commission, under which ecological funds are to be disbursed to the affected areas within the country. The 17th Conference of Parties to the United Nations Framework Convention for Climate Change held in Durban, South Africa, from 28 November to 9 December 2011, sidelined Nigeria's participation as the country lacked a Climate Change Commission. Nigeria's participation remained at the ministry level with no legal and structural commitment for implementation.

In the light of recent developments, Simire and Abutu (2011) revealed that Nigeria's limited prominence in climate change negotiation erodes her position as a leading voice seeking justice for the African continent. For instance, recent indications show that out of over 200 negotiators appointed as Africa's representatives under the platform of the African Group, only one Nigerian was acknowledged, as against 29 for South Africa; 18 for the Democratic Republic of Congo; 8 for Senegal; 7 for Algeria; 7 for Tanzania; 6 for Egypt; 6 for Gabon; 6 for Ghana; 5 for Sudan; and 5 for Kenya. Similarly, the recently constituted transitional committee charged to design the Green Climate Fund, which had members from eight African countries excluded Nigeria, but included South Africa, Gabon, Egypt, Ethiopia, Morocco, the Democratic Republic of the Congo, Burkina Faso and Zambia.

Association Between Land Degradation and the Incidence of Poverty

As shown in Table 5.1 below, the income generated by the people in the study location was not self-sustaining while the majority of the respondents argued that the standard of living of people in relation to the management of land resources in their communities was not improving—higher percentages of the respondents revealed that the standard of living of people in their community was low and inadequate.

Nevertheless, poor land utilisation facilitates declining standard of living in the host community. This implies that the standard of living of

Table 5.1 Association between land degradation and the incidence of poverty

Variable	Frequency	%
Respondents average monthly income		
N10,000–N30,000	130	43.3
N31,000–N50,000	56	18.7
N51,000–70,000	85	28.3
N71,000 and above	29	9.7
Total	**300**	**100**
Improvement of standard of living of people in relation to the management of land resources in the community		
Yes	105	35.0
No	195	65.0
Total	**300**	**100**
Respondents' rating of standard of living of people in the community		
Poor	213	71
Average	83	27.7
Good	1	0.3
Satisfactory	3	1.0
Total	**300**	**100**

Source: Fieldwork

the host community has stagnated and is not improving the socio-economic conditions of the people in the host community vis-à-vis the problem of land degradation. Specifically, the vicious cycle of poverty theory is useful in this regard. It could be affirmed from the interviews that generating income to sustain people in the host community is not an easy task. It was revealed in the interviews that as a result of continued poor land utilisation, income expectation is hindered, creating challenges for them and the consumers of their product in the host community.

This was revealed from the participant excerpts:

> [It is] quite difficult to generate income to sustain oneself in this environment...standard of living of people in this community is poor because higher number of the people in this area depend on land resources for their survival. As land is being degraded daily in this area, there is an increase in price of food commodity here because the farm products are not that available as we want it....so the standard of living of people is poor. (FGD, male, sand tipper driver, Atan, Ado-Odo/Ota LGA)

> [The] degradation of land has severe effect on standard of living of people because the farm will no more be yielding as it is supposed to be yielding... the farmers will be losing and there will not be sufficient crops or foods for the consumers because where someone is expecting certain metric tons, is getting lower of it and eventually the cost of food will be high and income will be low. (IDI, male, Baale, landowner, Ado-Odo, Ado-Odo/Ota LGA)

In another view, it was reported by the participants that the standard of living in the host community in decline was also facilitated by the activities of herdsmen on land and the indiscriminate felling of trees, which constitutes a great challenge to food production in the host community. The participant makes clarification on this, that:

> It is not easy to generate income to sustain oneself in this environment at all because the land we depend on is not yielding as we expected...[the] living condition of people in this community is poor because when those herdsmen with their cows come around January and February which was the planting period for us, they always destroy our land and farm produce which affects our income. The little we save from land was what we use to survive. (FGD, female, farmer, Ado-Odo, Ado-Odo/Ota LGA)

Clarifying the argument further, the participant explained:

> [G]enerating income to survive here is quite difficult because the economy here is stagnant and it benefits the elite or the politicians...The way the land is being degraded here through the activities of industries and indiscriminately felling of trees threaten the standard of living here. Living conditions [are] very poor and people hardly eat three square meals. We have high rate of crime in this community because no meaningful employment for the youths. (FGD, male, sand tipper driver, Atan, Ado-Odo/Ota LGA)

Association of Land Degradation and Rural–Urban Migration

The causes of poor land utilisation in the study area include the health crisis, declining agricultural productivity and urban sprawl. It was revealed by the respondents that they could not continue in their locality with the way land resources are being utilised presently, which will make them migrate to other community or neighbouring state. Most of the respondents revealed that migrating to another area was a way to find a means of survival which was not accessible to them again due to the prevalent poor land utilisation in their locality.

Nevertheless, poor land utilisation encourages unbridled rural–urban migration in the host community. This ultimately affects the socio-economic stability severely, which is facilitated by the problem of land degradation and results in an exodus of human resources out of the host community. The finding equally validated aspects of the vicious cycle of poverty and Marxism theory in particular. It was also revealed by the interviewees and participants that survival is the major consideration for migrating in and out of the host community due to prevalent poor utilisation of land resources in the study area. This was revealed from the participant excerpts:

> [T]here will not be anyone that will want to live in a degraded environment and unproductive land…I will surely relocate because my survival as a human is very important to me. (IDI, male, Baale, landowner, 65 years, Ado-Odo/Ota LGA)

Supporting the views above, the participant explained further that:

> I have to provide for my family and if this degradation of land continues, I will have to relocate to Lagos in order to survive and provide for my family…you cannot expect me to stay in this situation and allow my family to starve because of how land is being mismanaged here…God forbid. (FGD, male, timber seller, Ado-Odo, Ado-Odo/Ota LGA)

Moreover, it was revealed by the people that business development is hampered by the damaging utilisation of land in the host community, which make majority relocate to neighbouring communities and states in order to secure their means of livelihood in the study location. This was revealed further from the excerpts thus:

> I have already relocated to Iyana Ipaja in Lagos state but some of my property is still here…during raining season, this area is always flooded and I

> cannot risk losing my family to floods but I work here as [a] sand tipper driver to make...income to survive. (FGD, male, sand tipper driver, Atan, Ado-Odo/Ota LGA)

However, a lack of employment opportunities and an increasing crime rate discouraged the majority of the people in the host community to continue staying in the study area, which was a result of persistent degradation in the area. One participant commented:

> I cannot continue to stay here again and will relocate to Lagos in the next three months because the crime rate is persistent here as result of degradation of land and no employment opportunities for the youths...I cannot die now because the situation was disturbing and I need my family to be safe. I will surely relocate soon. (FGD, male, sand tipper driver, Atan, Ado-Odo/Ota LGA)

The Functionality of Institutional Intervention Towards Mitigating the Challenges of Land Degradation

The people in the host community stated that governments, even at the local level, have consideration for generating revenue with no attention as to how to curb the poor use of land in the study area. The reality of the situations was reported from the excerpts:

> Nigeria has vast land and the moment [the] government realises land degradation is taking place in one area, they will shift their attention to another place to continue their business. They are not interested in the area and people affected by the degrading of the land, if they realised the damage is too much. To me, governmental effort toward land degradation especially in this area of Ado-Odo/Ota is not effective. (FGD, male, timber seller, Atan, Ado-Odo/Ota LGA)

> [The] governmental role is not functional at all here, they politicise fertilisers here and those that are close to them are the ones that get the fertilisers and sometimes they brought few fertilisers when we face all the challenges

> of planting in the farm alone and harvested the crop which is not useful again…To me, [the] governmental effort is not functional here. (IDI, male, 45 years, Olori-ebi, landowner, Ota, Ado-Odo/Ota LGA)

Despite the fact that there are governmental structures in the host community to combat the menace of poor land utilisation in the study area, there is still a misuse of land resources in the area. The people revealed that taxes and levies from industrial companies are what the government focuses on, whereas the affected areas with poor utilisation of land are neglected. It was also revealed that despite the challenges met by herdsmen in the process of the utilisation of land resources in the host community, the government's efforts to reduce the problem for them could be improved. Some of the participants explained further that:

> [The] government is not doing anything meaningful here especially with the concentration of industries that damage land and the way tree felling is rampant here…[The] government [is] only interested in collecting taxes from these industries and thereby neglecting us here. They only focus on the state capital in Abeokuta. Their role is not in any way functional at all. (FGD, male, sand tipper driver, Atan, Ado-Odo/Ota LGA)

Supporting the view, another interviewee stressed that:

> [T]he herdsmen scared us anytime they are here and when they degraded our land, we try to confront them…but with that they killed one of our members and we tried to report them to the police. The police act on it, to some extent but without that, [the] governmental role here is not functional at all. (FGD, female, farmer, Ado-Odo, Ado-Odo/Ota LGA)

Concluding Comments

From the entire study, it is clear that an effective response to mitigating land resources mismanagement calls for a proper understanding of the problems, especially at the local level. Specifically, the study showed that land degradation has contributed to the incidence of poverty within the study area, which has resulted in declining standards of living, lack of job opportunities and an increase in the crime rate among young people.

Essentially, the problem of land degradation has been a driving force encouraging rural–urban migration from the study area, engendering instability among the local economics with an unplanned exodus of the human resources which should have contributed to the sustainability of human development in the host community. Hence, the study affirmed that existing institutional interventions towards mitigating the challenges of land degradation were such that, in spite of the existence of governmental structures in the host community, the laxity in environmental law and ineffective development control has encouraged the degradation of land in the host community and the inherent challenges of poor land management have not been met with commensurate institutional interventions. The study concluded that the resultant implications of poor land utilisation within the host community had contributed to a decline in food security, the quality of the environment and the well-being of people within the study area. The vicious cycle of poverty and Marxism theories accord precisely with this standpoint. Theoretically, the framework as a model for understanding the impact of poor land utilisation on the sustainability of human development has been presented in this study.

However, in order to combat the challenges of land degradation in Nigeria at large, so as to achieve some of the developmental agendas being embraced around the world, such as the need to eradicate extreme poverty and hunger and ensure environmental sustainability, which are the indices of sustainable human development, there is need for the following courses of action.

Land resources protection should be implemented to alleviate poverty in the host community since the poor are the hardest hit by land degradation when it occurs. Essentially, when people can no longer survive on depleted land, they move into neighbouring towns and cities, though there are few chances of a better life. Hence, appropriate management for land resources can be used sustainably by increasing people's livelihood options through the development of the agricultural sector in generating income for individuals, thereby reducing influxes and the emigration of people away from the host community, and reducing loss of human resources needed in the area. Legislative and functional institutional reforms will therefore be required in Nigeria community in order to deal effectively with the menace of land degradation.

The study therefore proposes a massive law reform exercise paying specific attention to the land resources usage in areas such as crop rotation, shift farming, flooding and erosion; deforestation and sand excavating should be adequately addressed in the host community.

References

AfricaFocus Bulletin. (2010). Nigeria: New Human Development Report. Retrieved from http://www.africafocus.org/country/nigeria.php

Ajayi, O. O., Oviasogie, F. O., Azuh, D. E., & Duruji, M. M. (2014, June). Urban design and sustainable development: A case of Makoko area of Lagos state, Nigeria. *European Scientific Journal, 10*(2), 105–112.

Akinola, S. R., & Adewale, B. A. (2012). *Climate change, urban degreening and flooding in Nigerian cities: Reducing vulnerability through polycentric planning and urban greenery strategy (Ppugs)*. Paper presented at the International Research Conference on Sustainable Development at the Faculty of Technology Complex, University of Ibadan, Ibadan, 7–9 November 2012.

Ascher, W., & Healy, R. (1990). *Natural resource policymaking in developing countries: Environment, economic growth, and income distribution*. Durham: Duke University Press.

Cohen, B. (2004). Urban growth in developing countries: A review of current trends and a caution regarding existing forecasts. *World Development, 32*(1), 23–51. UK: Elsevier Ltd. Retrieved from www.elsevier.com/locate/worlddev

Eswaran, H., Lal, R., & Reich, P. F. (2001). Land degradation: An overview. In E. M. Bridges, I. D. Hannam, L. R. Oldeman, F. W. T. Pening de Vries, S. J. Scherr, & S. Sompatpanit (Eds.), Responses to land degradation. *Proceedings of Second International Conference on Land Degradation and Desertification*, Khon Kaen, Thailand. Oxford Press, New Delhi, India.

Federal Republic of Nigeria, Ministry of Environment. (2012). National strategic action plan. Great green wall for the Sahara and Sahel initiative.

Leitmann, J. (2005). *Sustaining cities—Environmental planning and management in urban design* (p. 78). McGraw-Hill.

Oduwaye, A. O. (1998). Urban landscape planning experience in Nigeria. *Landscape and Urban Planning, 43*, 133–142.

Okon, E. O. (2012). Five decades of development aid to Nigeria: The impact on human development. *Journal of Economics and Sustainable Development, 3*(1), 32–42. Retrieved from www.iiste.org.

Oscar, L. (1959). Culture of poverty. Fieldwork among the urban poor in Mexico and Puerto Rico. Retrieved from http://www.google.com.ng/m?q

Rosenberg, E. (2007). *Land degradation: Environpaedia rethinking reality*. Department of Water and Environmental Affairs, South Africa.

Satterthwaite, D., Gordon, M., & Tacoli, C. (2010). Urbanization and its implications for food and farming. International Institute for Environment and Development (IIED). *Philosophical Transactions of the Royal Society B, 365*, 2809–2820.

Simire, M., & Abutu, A. (2011). Nigeria's waning status in climate negotiation. Retreived July 26, 2011, from http://www.independentngonline.com/daily-independent/article.aspx?id=37731

The United Nations Development Programme (UNDP). (2013). The United Nations Development Programme (UNDP), Report.

Titilola, S. K., Phillips, A. O., & Ajakaiye D. O. (1996). Population, pressure and environmental degradation. *A pilot study of the economic effects of soil erosion in Efon Alaaye"* In *Population Environment Interactions in Nigeria* (Eds). The Nigerian Institute of Social and Economic Research (NISER) Ibadan, Nigeria.

Titilola, S. T., & Jeje, S. K. (2008). Environmental degradation and its implication for agricultural and rural development: The issue of land erosion. *Journal of Sustainable Development in African, 10*(2), 122–139.

United Nation (2002). United Nation Report 2002.

United Nations Convention to Combat Desertification (UNCCD). (2008). Land degradation module. Retrieved from http://www.cehi.org.lc/SLMLDC-SIDS/cehi_slm_main.htm

United Nations Development Programme (UNDP). (2009). UNDP Human Development Report, 2008–2009. Retrieved from http://www.ng.undp.org

Tolulope Adetayo Fajobi is a PhD candidate/researcher in Development Sociology, specifically land conflict, knowledge transfer, regional integration, development studies and migration, at the Department of Sociology and Anthropology, Obafemi Awolowo University, Nigeria, and Laureate at Council for the Development of Social Science Research in Africa, Senegal, and the Centre for African Studies, Basel. Recent publications include 'Societal Stratification', in T. Falola (ed.), *Culture and Customs of Yoruba*, Indiana: Indiana University.

6

Land, Natural Resources and the State in Kenya's Second Republic

Connor Joseph Cavanagh

Introduction

Across sub-Saharan Africa, various rural populations are currently embroiled in a wide range of struggles that foreground rights of access, control, and ownership of land and natural resources. These contestations are, however, much more than 'mere resource conflicts' (Coombes *et al.* 2012). At issue is not simply the costs and benefits of environmental management, or apprehensions about 'who gets what', but also broader concerns about suitable forms of property and authority, the nature of the contemporary African state, and perhaps also the very substance of citizenship itself (see also Boone 2007; Lund and Boone 2013; Wily 2016).

These broader dimensions of land and resource governance are certainly visible in the struggles and mobilisations that led to the promulgation of Kenya's new constitution in 2010, and the subsequent context of

---

C.J. Cavanagh, PhD (✉)
Department of International Environment and Development Studies (Noragric), Norwegian University of Life Sciences (NMBU), Aas, Norway

A. Adeniran, L. Ikuteyijo (eds.), *Africa Now!*,
https://doi.org/10.1007/978-3-319-62443-3_6

what some have called the country's 'Second Republic' (e.g. Zeleza 2014). Laudably, sections of both the constitution and the related National Land Policy of 2009 are explicitly oriented towards addressing 'present and historical land injustices', as well as the ways in which such injustices are bound up in the broader institutional legacies of colonial rule. Kenya's own 2009 National Land Policy puts it as follows:

> Historical land injustices are grievances which stretch back to colonial land administration practices and laws that resulted in mass disinheritance of communities of their land, and which grievances have not been sufficiently resolved to date. Sources of these grievances include land adjudication and registration laws and processes, and treaties and agreements between local communities and the British. The grievances remain unresolved because successive post-independence Governments have failed to address them in a holistic manner. (Government of Kenya [GoK] 2009: para. 178)

As one means of rectifying this legacy of injustice, both the National Land Policy and the constitution now enshrine a category of 'community land' tenure, which elevates collective forms of land ownership to a legal status equivalent to that of their public and private counterparts (see GoK 2010). This is significant given that, effectively, this amounts to the first such acquisition of legal equivalence—in principle, at least—for common property rights since the declaration of British sovereignty over Kenya in 1895 (Okoth-Ogendo 2003; Odote 2013). Under previous legal regimes, collective forms of land ownership had been consigned either to the quasi-legal domain of 'customary rights' or subordinated to local government ownership in the form of 'trust land' (Musembi and Kameri-Mbote 2013). Further, the constitution also embeds these land reforms within those targeting wider systems of governance, providing for an enlarged Bill of Rights (which includes socio-economic rights), essential checks on the power of the executive to dispose of public land, and a devolved system of governance. Understandably, therefore, these reforms were initially met with significant celebration both in Kenya and throughout networks of activists, human rights advocates and social justice organisations in East Africa and beyond. Indeed, comparisons between this document and South

Africa's post-apartheid constitution of 1996, no less, were initially quick to emerge (e.g. ITCJ 2011).

Examining the translation of these provisions into both law and practice, this essay adopts an explicitly 'decolonial' perspective (Mignolo 2011; Ndlovu-Gatsheni 2015a), or an approach that perceives decolonisation as a lengthy, ongoing process rather than a singular historical event of 'flag independence'. Accordingly, I suggest that the struggle to implement the spirit as well as the letter of the 2010 constitution is in many ways a *decolonial* initiative; that is, a struggle to exorcise or otherwise reform certain legal and institutional legacies of the colonial era, as well as the ways in which this foundation has been used and abused by Kenya's ruling classes since independence. This is a struggle, in other words, not simply over a constitution and its accompanying legislation, but for a transition to what some have called a 'Second Republic' in Kenya (e.g. Githongo 2008; Zeleza 2014), one in which formal political decolonisation might finally be accompanied by substantive forms of economic and social decolonisation as well.

The chapter proceeds as follows. First, I introduce the concepts of coloniality and decoloniality, and examine their interaction with conflicts over land and natural resources over the course of Kenya's recent history. In doing so, I investigate: the implications of colonial state formation for land and resource governance; the political economy of the transition to independence; the ways in which this political-economic foundation cast its shadow over both the post-election violence of 2008; and the subsequent promulgation of a new constitution. Second, in this context, I outline the provisions of Kenya's 2010 constitution: for the devolution of land and resource governance; for a new category of collective land tenure; and other institutional reforms related to the access and ownership of natural resources. Throughout, I juxtapose the substance of these provisions with ongoing challenges to their implementation, including possible instances of 'bureaucratic sabotage' (McAuslan 2003) or efforts to obstruct the advancement of these reforms with the substance of laws and policies meant to implement them. I conclude with a discussion of related efforts to undermine the spirit of the 2010 constitution alongside broader struggles for a transition to a 'Second Republic' in Kenya.

(De)coloniality, Natural Resources, and the Politics of Indigeneity in Kenya

In the first instance, the concepts of coloniality and decoloniality emerge primarily from the work of Walter Mignolo (2011) and associated members of the Modernity/Coloniality/Decoloniality network (Mignolo and Escobar 2013). Initially, this entailed a primary focus on Latin America and its histories of highly unequal power relations with various European and North American powers from the sixteenth century onwards. As Mignolo (2007: 39) succinctly puts it:

> The basic thesis is the following: 'modernity' is a European narrative that hides its darker side, 'coloniality'. Coloniality, in other words, is constitutive of modernity—there is no modernity without coloniality. Hence, today the common expression 'global modernities' imply 'global colonialities' in the precise sense that the colonial matrix of power (coloniality, for short) is being disputed by many contenders…Consequently, de-colonial thinking and doing emerged, from the sixteenth century on, as responses to the oppressive and imperial bent of modern European ideals projected to, and enacted in, the non-European world.

In the most general sense, then, coloniality and decoloniality constitute opposite sides of the same metaphorical coin. In other words, whereas coloniality refers to the historical process of the imposition—*inter alia*—of imperial norms, institutions, laws and forms of governance upon various colonised populations and territories, decoloniality refers to widespread attempts to resist, abolish, reconfigure or otherwise reform these. Usefully, this enables us to understand coloniality as a process or relative *property* more so than an event or historical period—one whose effects potentially linger and evolve long after formal political independence. Indeed, for this reason, Mingolo (2011: 53) is frequently at pains to distinguish *decolonisation* (a formal political transition or acquisition of 'independence') from *decoloniality* or ongoing intellectual and political struggles to reconfigure the epistemologies, institutions and forms of governance imposed via processes of colonisation.

More recently, however, these concepts have notably been reworked and extended in relation to the African context by Sabelo Ndlovu-Gatsheni (e.g. 2013, 2015a, b) and associated members of the Africa Decolonial Research Network. In short, such critical attentiveness to the historical and geographical particularities of coloniality and colonial history in Africa is in many ways quite necessary. Indeed, although Africa of course shares with the Americas a common history of colonisation by European powers—including by the Spanish and Portuguese empires, to varying degrees—the trajectories and temporalities of these processes are perhaps more distinct than is commonly understood. Whereas, for instance, Bolivia had won its formal independence from Spain as early as 1825, Kenya—to take but one example—was not *declared*, much less territorialised on the ground, as a British protectorate until 1895 (Lonsdale 1992a). Though undoubtedly moulded by their histories of conflict and accommodation with European powers, in other words, the contemporary states of eastern and southern Africa in particular were forged decidedly within comparatively idiosyncratic modes of both capitalism and late colonial governance (e.g. Berman and Lonsdale 1992). The objective of any careful analysis of (de)coloniality specifically in Africa, in other words, is necessarily to account for the effects of such idiosyncrasies while nonetheless tracing their imbrication with much wider and historically recurring patterns of domination in the broader world system.

As Ndlovu-Gatsheni (2015b: 16) notes—following the seminal writings of the Congolese scholar Mudimbe (1988)—such patterns must be continuously interrogated, given that they have repeatedly yielded a 'paradigm of difference' that tends to undermine Africa's 'stature in world affairs and questions its contribution to human civilisation, progress and development', albeit via a shifting range of discourses and idioms. Such a paradigm of difference, I would add, is not merely confined to the realm of culture or identity. Indeed, such a downgrading of status is also evident in the historical subjugation of customary institutions that regulate the collective ownership of land and natural resources, as I examine with reference to the history of state formation in Kenya below.

Political Ecologies of Rule in Kenya Colony

In the first instance, one could argue that the substance of the concept of decoloniality is perhaps already organically well-represented within Kenyan public discourse. Since independence, a recurring theme in both national political debates and popular criticism has concerned the enduring legacies of the inequalities and other forms of exploitation instituted under British colonial rule in the relatively brief period between 1895 and 1963.[1] At issue, in other words, is a widespread sense that formal political *decolonisation* has not been coupled with practices of *decoloniality* within the state itself, or that the departing colonial administration was simply replaced with a new class of elites from majoritarian communities, ones that have generally upheld the prevailing status quo for their own benefit (see, for example, Boone 2012: 91).

Here, perhaps the most fraught dimension of these debates concerns the problem of managing the country's land and natural resources—their ownership, inequalities in their distribution, and the suitability of the forms of tenure that have governed their acquisition since the colonial era, among other sources of grievance (see, *inter alia*, Okoth-Ogendo 1991, 2003; Anderson 2002; Steinhart 2006; Kanyinga 2009; Boone 2012). The territory that is now Kenya is in some ways predisposed to intense competition over fertile agricultural land in particular, given that approximately 80 % of the country's surface area is classified as arid or semi-arid, with the remaining and relatively fertile 20 % concentrated in the Central and Western Highlands (World Bank 2005). Not coincidentally, the Highlands were the focus of most land alienations for white settler agriculture, amounting to some 31,000 km^2 at their most extensive, and comprising an area that the Njonjo Commission (2002) would later estimate to encompass approximately 75 % of the highest potential agricultural lands in the country.

The legal basis for these alienations arose from the Crown Lands Ordinances of 1902 and 1915, which declared all land in Kenya to be Crown Land, and its occupants simply 'tenants-at-will of the Crown' (Okoth-Ogendo 1991). Although these ordinances prohibited the alienation of land 'in the actual occupation of natives' to settlers or the colonial state, this did not affect the *underlying* dispossession of property rights. Indeed, following the passing of these ordinances, Kenya's African population effectively became simple 'occupants' (e.g. in possession of use rights) rather than 'owners' in full possession of property rights to land (Okoth-Ogendo 2003; Ojienda and Okoth 2011). Such legal sleight-of-hand would have long-lasting effects for the legal status of what would become known, first, as 'customary' property rights, and later as 'trust lands' under the ownership of local government—both throughout the colonial period and afterwards.

Moreover, in addition to lands alienated for white settler agriculture—which became explicitly racialised as the 'White Highlands' from the early twentieth century onwards (see Morgan 1963)—a further 22,000 km^2 was alienated as forest reserves and other protected areas (East African Statistical Department 1964). This had the effect of enclosing a number of important water catchments that were crucial for commercial agriculture in the Highlands, as well as important grazing areas and water sources in pastoralist-dominated regions of the country (Collett 1987). In particular, these enclosures of fertile agricultural lands, pasture and forest resources brought about a decidedly tumultuous disruption of the region's pre-colonial social and political-ecological milieu. Firstly, this amounted to the de jure dispossession of Kenya's upland hunting and gathering communities—such as the Ogiek and Sengwer—who had historically inhabited both forests and Highland glades, and who were later often slated for eviction and forcible assimilation into larger groups of agriculturalists or pastoralists (GoK 2009: para. 199).

Likewise, given that forest reserves were often territorialised horizontally by altitude, forbidding human residence above the newly established boundaries, such protected areas also wreaked havoc with the land tenure and livelihood systems of many Highland agro-pastoralists, whose customary territories were typically vertically organised to encompass segments of Highland glades, forests and lower-lying plains (e.g. Lonsdale 1992a: 20). Indeed, as Anderson (2002: 21) puts it, '[i]n taking possession of these lands the European settler community had seized seasonal grazing areas and watering points that were crucial to the survival of African herders' in lower-lying areas. Similarly, in the plains of the Rift Valley, the creation of wildlife reserves further limited the territories of pastoralists such as the Maasai, who had already lost more than half of their customary lands to European settlers and the colonial state, and likewise found their seasonal grazing patterns and adaption strategies increasingly further restricted (Collett 1987; Peluso 1993: 203). Just as the process of colonisation entailed a reworking of both economic production and systems of governance, then, so too did it entail a reworking of prevailing relationships between rural populations and the non-human environment.

Perhaps even more contentious than the initial alienation of these lands and ecosystems under colonial rule, however, is the manner in which they were to be redistributed after independence. First, the rise of settler agriculture—as well as the forms of capitalism that came with it—had brought about significant changes to the demographic composition of the Highlands, and especially so via the demand for African labour on settler farms (Berman and Lonsdale 1980). At least initially, many settlers were unable to put the full extent of their newly acquired properties to productive use, and especially so in the absence of the necessary machinery and other inputs for commercial-scale production. The solution, variously, was either the importation of sufficient African labour as such, or—more infamously in Kenyan historiography—the use of African families as 'tenant' or 'squatter' farmers on land owned by Europeans (e.g. Furedi 1974; Youe 1988; Anderson 2000).

Such internal migrations created the basis for a situation in which—following the redistribution of formerly European-owned lands after independence—claims to the legitimate receipt of these properties were highly complex. On one hand, communities of Kalenjin and other (agro-) pastoralists in the Western Highlands, in particular, demanded that the 'stolen' lands be returned to them, on the grounds that they were the 'indigenous' or 'autochtonous' inhabitants of the Rift Valley province, and thus the rightful beneficiaries of land redistribution schemes in these areas. On the other hand, groups of Kikuyu and other farmers from Central Province—who had been brought to the Rift Valley as labourers or tenant farmers—felt that they had acquired a claim to these areas by virtue of their labour, residence and, in some cases, their willingness to organise armed opposition to the British in the form of the Mau Mau rebel movement (see, especially, Furedi 1974, 1976).

Moreover, underpinning these latter claims was a complicated intersection of the Kikuyu moral economy and the colonial political economy.

Especially prior to the early twentieth century, impoverished Kikuyu would routinely seek to acquire new land or livestock through a system of patron–client relations with more powerful or wealthy clans, gradually improving both their status and socio-economic standing through the investment of both skill and hard labour in newly acquired farms. In the early twentieth century, it at least initially would have appeared that Europeans were offering a similar opportunity for a mutually beneficial set of such relations, and perhaps a solution to the rapidly closing agricultural frontier in the increasingly territorialised native reserves (see Lonsdale 1992b, c). In this sense, the investment of labour and several decades of residence in the Rift Valley would have struck many Kikuyu migrants as sufficient grounds for an enduring claim to land in these areas.

Moreover, given both growing populations and the formalisation of individual tenure in the native reserves, there may have been few options for emigrants to return to an agrarian livelihood on their 'ancestral lands'. Though the colonial state's intervention in the African agricultural sector had grown steadily since the Depression of the 1930s (Anderson 1984), it greatly accelerated this process of individuation, most directly throughout the 1950s in the form of the Swynnerton Plan (Swynnerton 1954). The latter was underwritten by two interrelated assumptions: that intensive commercial agriculture was the most productive use of African land, especially relative to either pastoralism or hunting and foraging; and that agricultural production could be optimised only via the formalisation of private tenure rights, which would in turn improve access to credit and other services (Hornsby 2012: 48–50).

In practice, however, this process of individuation and the formalisation of tenure in the reserves connected with a process of socio-economic differentiation that had been ongoing since the early colonial era (Furedi 1974; Berman and Lonsdale 1980; Lonsdale 1992b, c). Particularly in the agrarian regions of Central Province, some farmers had done quite well by supplying the rapidly growing capital of Nairobi with primary commodities, as well as by provisioning the colonial state more broadly. Likewise, 'loyalist' populations of Kikuyu and other farmers had been rewarded with special privileges, such as the right to grow coffee and other cash crops normally reserved for white settlers, as well as favourable terms of trade (Furedi 1974: 498–499). Coupled with the formalisation of tenure, these growing differences of wealth and status made it significantly easier for wealthier farmers to expand their operations, or to consolidate existing ones. Conversely, however, for those not so blessed by the colonial state's beneficence, the terms of access to land, as well as the price of acquiring it, became increasingly insurmountable.

Majimboism and the Political Economy of Independence

As a result of such dynamics, the stakes of claims to contested lands throughout the country were decidedly high. Moreover, such claims were increasingly phrased within an idiom of ethnicity or even autochthony in the lead-up to independence—and especially so as membership in Kenya's nascent political parties appeared to ossify along ethnic lines. Whereas the Kenya African National Union (KANU), led by Jomo Kenyatta, drew its membership largely from the agriculturalist and majoritarian Kikuyu and Luo communities, the opposition Kenya African Democratic Union (KADU) was primarily comprised of 'indigenous' pastoralist and agro-pastoralist communities in the Rift Valley, and was explicitly organised around protecting their interests. In pursuit of the latter, KADU advocated for a system of devolved federal regionalism, or *majimboism*, as a model of governance in the post-independence era; this would entail a greater administrative role for regional—but not necessarily national—majoritarian communities in the governance of their own 'ancestral lands' or former native reserves (see, for example, Anderson 2005). By contrast, Jomo Kenyatta and KANU propounded a strong unitary state, ostensibly in the interests of modernisation and the avoidance of divisive forms of 'political tribalism', which were rhetorically framed as a relic of the divide-and-rule strategies of colonial governance.

Following the victory of KANU in Kenya's first elections in 1963, however, it quickly became apparent that ethnicity would continue to play a role in the political economy of land in the ostensibly 'unitary' post-independence state, though perhaps in more informal ways than those envisaged by the advocates of *majimboism*. In 1964, KADU was dissolved into

KANU, effectively making Kenya a de facto one-party state. Yet, despite this apparent display of 'solidarity', nowhere were longstanding ethnic grievances more evident than in struggles over the redistribution of formerly European-owned lands. Following independence, Kenyatta negotiated several of these initiatives with the British government—most notably the 'Million Acre Scheme'—which entailed the Kenyan state's purchase of 1.5 million acres from former European settlers, a transaction facilitated by loans from the British government. These lands were slated for redistribution via a range of schemes intended to foster both 'low-density' commercial agriculture and the settlement of otherwise landless families on 'high-density' and relatively subsistence-oriented farms (Hornsby 2012: 74–75). Though available in principle to anyone meeting the requisite economic criteria—in other words, on a 'willing buyer, willing seller' basis—not all Kenyans possessed the same means, nor benefited from the same degree of state support, to buy land. Indeed, it gradually became clear that the KANU state was allocating land as a means of solidifying its support among its core Kikuyu and Luo constituencies, as well as informally enriching members of the political elite (Klopp 2000). To this day, for example, the Kenyatta family itself allegedly remains one of the largest landowners in Kenya (Schutzer 2012: 349).

With the election of Daniel arap Moi in 1978, the use of land and natural resources as patronage certainly continued, albeit in ways that now seemed to favour the Kalenjin-related communities and other pastoralists in the Rift Valley. Crucially, however, Moi was faced with the problem of de facto land scarcity, given that much of the former White Highlands had already been redistributed by Kenyatta or seized by members of the elite. As Suzanne Mueller (2008: 188) notes, '[u]nlike Kenyatta, who could give without taking away, Moi had to take away before he could give'. In the first instance, this was accomplished by informally state-sanctioned displacements in the Rift Valley, in which 'settler' farmers were violently evicted—before elections in 1992 and 1997 in particular—and their lands subsequently allocated to Kalenjin families in many cases (see Kanyinga 2009; Boone 2012). Secondly, however, Moi's regime also opened up a new agricultural frontier within the country's protected forested area, distributing newly deforested land to various political clients in the Rift Valley (see Klopp and Sang 2011;

Klopp 2012). Finally, such 'taking away' was accomplished by the plunder and onward redistribution of state-owned properties—such as Agricultural Development Corporation farms—both to favoured constituencies and to elites (e.g. Klopp 2000; Southall 2005).

When Things (Almost) Fell Apart: Post-election Violence and the New Constitution

Given this precedent, the widespread violence that followed Kenya's contested 2007 elections struck many analysts as predictable, if not preventable indeed (Anderson and Lochery 2008). Despite exoticised representations in the international media of 'tribal clashes' and 'primordial hatreds', this violence was once again patterned along the same long-standing grievances about the rightful occupation of land in the Rift Valley and elsewhere. What distinguished these events from similar clashes in 1992 and 1997, however, was the degree to which violence was anticipated by both the public and the political establishment, resulting in significant preparations for conflict, and the willingness of both political elites and incumbent officials to advocate for violence either as a means of obtaining contested lands or—to a greater degree than in the 1990s—defending occupancy of them. Consequently, the mobilisation of Kalenjin 'warriors' in the Rift Valley was countered by the mobilisation of criminal gangs—such as the infamous Mungiki—to protect Kikuyu and allied communities' interests elsewhere in the country. When the dust had settled, more than 1100 had reportedly been killed, and more than 600,000 displaced (Human Rights Watch 2011).

In many ways, however, the crisis of the 2007–2008 post-election violence was also the crucible in which the 2010 constitution was definitively forged. In the words of the Kenyan historian Paul Zeleza (2014: 17), one might conclude, ultimately, that 'the crisis begat the constitution.' This is not to say that initiatives for constitutional reform were not afoot in Kenya well before 2007—indeed, a similar draft constitution was debated and ultimately rejected in a 2005 referendum, following a considerable degree of elite attempts to derail the process (Murunga *et al.* 2014). Yet after the post-election violence of 2007–8, an uncommon consensus emerged across a broad swathe of the Kenyan population—fostered and cultivated by the tireless efforts of Kenyan civil society organisations—that a durable resolution of the crisis would not be possible without a substantive attempt to address broad-based grievances concerning the nature of governance in Kenya from the colonial era to present. In other words, this was an organically 'decolonial' moment in the country's history: at issue was not only the perceived injustices of the colonial era, but also the ways in which the institutions imposed under colonisation had evolved in ways that continued to produce decidedly malign effects in the present.

In such a context, practices of decoloniality inevitably entail not only attempts to reformulate national political or economic institutions, but also the prevailing means of governing land and natural resources. Indeed, as Escobar (2008: 154–155) writes, 'a decolonial view of nature and the environment calls for seeing the interrelatedness of ecological, economic, and cultural processes…Sustainability may thus become a decolonial project: thinking from existing forms of alterity towards worlds and knowledges otherwise.' Forms of alterity, in this instance, related to prevailing notions of property rights and the types of authority suitable for their governance. Hence, I turn to efforts to rework Kenya's land and resource governance institutions both via the widespread mobilisations that gave rise to the country's new 2010 constitution, as well as contemporary efforts to translate the constitution's provisions into workable forms of legislation.

Land and Resource Governance in Kenya's 'Second Republic'

In what follows, I discuss, in turn, three of the simultaneously most promising and most fraught aspects of the 2010 constitution related to land and resource governance, namely: (1) the devolution process; (2) the implementation of a new tenure category of 'community land'; and (3) related institutional reforms influencing ownership of and access to other natural resources.

Devolution: *Majimboism* Reborn?

Among the most striking of the governance reforms implemented in accordance with the 2010 constitution is the substantive reorganisation of Kenya's administrative geography. In particular, this concerns the dissolution of the country's eight provinces—whose prefectural administration was in itself a lingering symbol of the colonial era—and their replacement with a system of 47 counties. Each of these counties retains its own leadership in the form of a governor and executive committee, as well as a county assembly that consists of elected members. In some ways, this might at first seem like a return to the model of *majimboism* envisaged in Kenya's independence constitution, but which was quickly abandoned by Jomo Kenyatta's KANU government after 1963 (Branch and Cheeseman 2006). Interestingly, however, whereas some of these counties are populated overwhelmingly by—and perhaps also named after—ethnic groups associated with former native reserves, others are much more ethnically 'cosmopolitan', and still others are highly diverse yet characterised nonetheless by a clear ethnic majority. In this sense, the politics of ethnicity under this system of devolved governance is perhaps now significantly more complex than may have been imagined by previous generations of *majimboism* proponents. In its simplest versions, a *majimbo* system would have entailed communities of the former 'native reserves' governing themselves within a broader federal polity (see, variously, Anderson 2005, 2010; Maxon 2016). In practice, however, such fantasies of 'purified' ethnic territoriality are of course often complicated by the realities of necessary interaction, accommodation and collaboration—and particularly so in the context of economic growth and capitalist development across the contemporary system of 47 counties.

Further, while this model of devolution was generally mooted as a means of rectifying the excesses of centralised one-party rule in Kenya during the regime of President Daniel arap Moi in particular (1987–2002), devolution appears to present challenges in this regard as well. For example, though one might assume that such reforms would impede prevailing forms of patronage and rent-seeking among central government officials (Mwangi 2012: 58–59), devolution in Kenya may simply be rescaling or otherwise reconfiguring these practices. As D'Arcy and Cornell

(2016) argue, devolution is perhaps resulting in a prevailing sense of 'everyone's turn to eat' as new streams of rent are made available for redistribution via the administrations of 47 counties rather than a central state and its eight provinces, and on a scale equivalent to more than 30 % of the national budget. Moreover, it would appear that certain officials responsible for implementing Kenya's devolution process agree. As the former head of Kenya's Transition Authority—a taskforce created to implement the devolution process—recently wrote in Kenya's *The Star* newspaper, devolution to date has apparently been 'a case of unmitigated impunity', which foregrounds 'the massive corruption...we see in Kenya on a daily basis' (Wamwangi 2016).

Given these emerging concerns about the misuse of state resources at the county level, similar questions arise in relation to the exercise of county government authority over land and natural resources. For instance, the 2010 constitution's provisions of the governance of state or 'public land' hold that any areas 'in respect of which neither individual nor communal ownership can be legally established' will become the property of county governments to hold 'in trust' for their populations (GoK 2010: §61(1), (2)). This also includes any forests that are neither individually nor community owned, such as those previously managed by local governments, as opposed to state agencies such as the Kenya Forest Service (KFS). Given that legislation on 'community land' (discussed in more detail below) is yet to be passed, in practice this means that county governments are now assuming control over substantial areas of land and forest that are likely simultaneously claimed by local communities. If past experiences with the disposal of land held 'in trust' by local governments in Kenya are any indication (Klopp 2000; Ndung'u Commission 2004; Southall 2005), this is a rather ominous dimension of the devolution process.

Finally, we can infer from these emerging contours of county governance that both new 'winners' as well as 'losers' are emerging from the devolution process. On one hand, a larger number of communities now have access to state resources, which is particularly significant for national minorities that have little hope of winning control of the centralised state itself via national elections. On the other hand, this may simultaneously entail the double marginalisation of what D'Arcy and Cornell (2016:

258) call 'trapped minorities', or minority communities *within* the newly created counties, who may have few possibilities for obtaining rents redistributed from either the central state or their local county government. This is especially the case for minority communities—such as the Ogiek and Sengwer—that also claim a uniquely 'indigenous' status, but whose small population sizes challenge their ability to benefit from prevailing flows of patronage in exchange for political support (Tiampati 2015). In principle, however, the interests of such minority communities could at least potentially be mitigated by the implementation of community land reforms, the opportunities and challenges for which are discussed below.

Community Land: Formalising the Customary?

While the 2010 constitution certainly presents a new institutional framework for governing access to state rents, so too does it promise to shift the terms of access and ownership of land—another crucial aspect of Kenya's political economy since at least the nineteenth century. Here, at least in de jure terms, the land reforms enshrined in both the 2010 constitution and the accompanying National Land Policy of 2009 seem inherently quite promising. As the land policy itself frames the issue:

> The genesis of the land question can be traced to the colonial times when the objective was to entrench a dominant settler economy while subjugating the African economy through administrative and legal mechanisms. For example, the process of colonisation introduced an alien concept of property relations in Kenya, where the State or the protectorate as a political entity came to own land and grant to property users subsidiary rights…Politically the land question is related to the administrative controls of the economy that use land as leverage for political support. (GoK 2009: para. 21)

In response to both this legacy of historical injustice and its manipulation after independence, the land policy recommends that Kenya's existing two forms of 'statutory' land tenure—namely, public and private tenure—should be complemented by a third category: 'community land', or land that is held collectively by members of a community rather than either a private entity or the state. In turn, this recommendation is paralleled by a corresponding section of Kenya's 2010 constitution, which provides for a category of 'community land', or land that 'shall vest in and be held by communities identified on the basis of ethnicity, culture or similar community of interest' (GoK 2010: §63). Rather than apparently overlapping forms of 'statutory' and 'customary' law, therefore—as theorists of so-called 'legal pluralism' examine in detail (see, for example, Peluso and Vandergeest 2001: 765)—Kenyan law now provides for alternative forms of statutory tenure that allow for either private, state or community ownership.

Although previous incarnations of 'trust land' have occasionally enabled the collective use of local government-owned land (Ojienda and Okoth 2011: 165–166), the 2010 constitution elevates community land tenure to the legal equivalent of both public and private tenure for the first time—and effectively so since the colonisation of what is now Kenya by Britain in 1895 (Odote 2013, see also Okoth-Ogendo 2003). The genealogy of such trust land *as such* in Kenya dates to the colonial Native Trust Land Ordinances of 1930 and 1938, which distinguished 'trust land', land within native reserves, from 'Crown land', land otherwise claimed by the colonial state on behalf of the metropole (Ojienda and Okoth 2011: 162). In practice, however, trust lands were legally subordinated to the categories of public and private tenure. Under the prevailing 'evolutionary' theory of property rights, collective forms of ownership were seen merely a transitional stage, one that would eventually give way to the universal registration of lands under either public or private tenure (Musembi and Kameri-Mbote 2013: 7).

After independence, remaining trust lands were subsequently transferred to local governments to manage on behalf of citizens within their jurisdictions (Ogolla and Mugabe 1996). While local governments sometimes allowed trust lands to be utilised for subsistence purposes, communities themselves had no title to these as collective property, and government officials were in practice easily able to 'set apart' land for allocation to individuals or private firms, or for use by government organisations (Ndung'u Commission 2004; Southall 2005). The now-repealed constitution itself could hardly have better facilitated such transfers if it had been explicitly designed to do so. As it states, trust land may be transferred to:

> any person or persons for a purpose which in the opinion of that county council is likely to benefit the persons ordinarily resident in that area or any other area of Trust land vested in that county council, either by reason of the use to which the area so set apart is to be put or by reason of the revenue to be derived from rent in respect thereof. (GoK 2008 [1998]: Section 117(1)(c))

In practice, therefore, the degree to which trust lands actually benefited rural communities depended entirely on the benevolence of specific trustees, as well as the position of the former in relation to prevailing flows of patronage (Klopp 2000, 2012). As a result, this situation was particularly disadvantageous for the above-mentioned 'trapped minorities', whose small populations would have often entailed de facto marginalisation from such ostensible beneficence, and perhaps also the loss of their customary lands to more 'politically favoured' groups.

Similarly, the new category of 'community land' tenure is also potentially distinct from the forms of collective land use organised as 'group ranches' in arid and semi-arid regions of Kenya. Following the 1968 Land (Group Representatives) Act, lands typically managed by pastoralists in accordance with customary laws have been registered in the names of between three to ten representatives. But here again, elected group ranch representatives became the 'trustees' of the land in question. In practice, this has opened up the internal dynamics of group ranch management to highly asymmetrical power relations, and has frequently resulted in the trustees' unauthorised use of group ranch land and resources for private benefit (e.g. Odote 2013). In a number of cases, this has also precipitated the fragmentation and deliberate sub-division of group ranches to check the abuses of representatives (Galaty 1994; Kimani and Pickard 1998). These experiences with dysfunctional group ranches provide us with perhaps the clearest indication of the pitfalls to avoid in the implementation of new legislation around the tenure category of community land. It is imperative, for example, that legislation on community land tenure also includes clear and effective guidelines for community land *governance* that neither resorts to local government ownership nor replicates the experiences with misappropriation and misuse prevalent in the management of group ranches.

Conversely, however, it is here that we perhaps encounter the source of long delays with the enactment of legislation on community land. Although the 2010 constitution stipulates deadlines for the drafting of laws to support its provisions—and specifically allocates five years for the enactment of legislation on community land (GoK 2010: Schedule Five)—such legislation was not signed into law until August of 2016. Moreover, the constitution's provisions on community land are decidedly vague concerning how it envisions these to be operationalised. It provides no definition of a 'community' as such, simply alluding to 'ethnicity, culture or similar community of interest'. Such ambiguity creates substantial difficulties for equitably determining precisely who is and is not a member of a certain community, as well as adjudicating disputes in this regard when they (inevitably) arise. Moreover, it is likewise unclear specifically how certain communities will ultimately come to own specific lands under new forms of collective ownership. This is particularly ominous given that there are innumerable cases throughout the country where land ownership is simultaneously claimed by several different communities: by communities and the state, as well as by communities and private individuals or firms.

For its part, the constitution simply notes that, following the adoption of the relevant legislation, lands of the following types will be converted to community land:

> (i) land lawfully held, managed, or used by specific communities as community forests, grazing areas or shrines; (ii) ancestral lands and lands traditionally occupied by hunter-gatherer communities; or (iii) lawfully held as trust land by the county governments. (GoK 2010: §63(d))

Neither here nor elsewhere in constitution, however, do we find clear guidance on the adjudication of overlapping land claims, except for the deferral of such issues to the National Land Commission. And again, the specific provisions themselves are rather vague, simply mandating the Commission 'to initiate investigations, on its own initiative or on a complaint, into present or historical land injustices, and recommend appropriate redress', as well as 'to encourage the application of traditional dispute resolution mechanisms in land conflicts' (GoK 2010: §67 (e), (f)).

These ambiguities leave considerable space for what the legal scholar Patrick McAuslan (2003: 342) once termed 'bureaucratic sabotage', or efforts from within the executive branch of government to dilute or undercut the substance of constitutional or policy reforms. For instance, the 2012 Land Act and 2012 Land Registration Act, which were meant to establish a framework for the administration of each of Kenya's three categories of land tenure, are largely silent on issues of community land, deferring questions of its administration to the recent Community Land Act of 2016. Moreover, as Manji (2014: 126) notes, the process of drafting and debating these bills appears to have been inordinately rushed, as well as marred by insignificant public or civil society consultation, denoting that there is 'a disjuncture between a decade-long struggle to achieve equitable land policies and the resulting land laws'. Lamentably, these challenges seem to be clear and present in relation to the implementation of institutional reforms of other types of resource management as well.

Community Forest and Watershed Management: From Access to Ownership?

Though much has been written about the relatively substantive nature of the reforms associated with Kenya's 2010 constitution, less has been said about comparative silences or omissions within these reforms. This is particularly the case in relation to biodiversity conservation and the governance of Kenya's protected area estate, which appears slated to remain within the control of centralised state agencies such as the Kenya Forest Service and the Kenya Wildlife Service.

For example, in relation to forest reserves, the prevailing piece of legislation until 2016 was the 2005 Forests Act, which defines protected forested areas as state-owned lands, and stipulates that they will be governed primarily by the KFS. That said, the Act provides grounds for the KFS to enter into collaborative management partnerships with private firms, non-governmental organisations, and communities in cases where the latter form and register interest groups known as community forest associations. Importantly, however, one should note that these collaborative arrangements did nothing to alter the underlying *ownership* rights to forested protected areas—they simply allowed access or co-management privileges under certain prescribed conditions.

Hence, it could be argued that the 2005 Forest Act now did not pass the test of constitutionality. As noted above, the constitution's definition of community land includes 'land lawfully held, managed, or used by specific communities as community forests, grazing areas or shrines' (GoK 2010: §63(d)(i)). This clause is admittedly somewhat unwieldly and difficult to operationalise, given that virtually all of Kenya's forest reserves hold great value for nearby communities for both economic and cultural reasons (e.g. Mwangi 1998). Moreover, the Act also seems poised to conflict with the constitution's related provision that community land encompasses 'ancestral lands and lands traditionally occupied by hunter-gatherer communities' (GoK 2010: §63(d)(ii)). This provision appears to arise from the recommendations of the 2009 National Land Policy, which makes specific reference to the ways in which 'hunter-gatherer communities' were marginalised when 'the colonial Government alienated their lands through forest preservation policies, which effectively rendered them landless as they were denied the right to live in the forests' (GoK 2009: para. 199). Yet, like the concept of community itself, the constitution does not define what a 'hunter-gatherer' community is, nor does it refer to those whom we might regard as Kenya's hunter-gatherer communities by known ethnonyms, such as those specified in the most recent census of 2009.

Nonetheless, the constitution effectively establishes a legal category of persons known as hunter-gatherers, both via the provisions on community land, and via its targeted protections for historically 'marginalised communities.' Indeed, the legal definition of the latter is now framed to encompass any 'indigenous community that has retained and maintained a traditional lifestyle and livelihood based on a hunter or gatherer economy' (GoK 2010: §260). Likewise, the newly expanded Kenyan Bill of Rights includes explicit provisions for these marginalised communities, including the right to 'develop their cultural values, languages and practices' (GoK 2010: §56(d)). It would thus seem, at first glance, that this new institutional context might entail significant promise for both forest-dependent groups and communities of current or former hunter-gatherers who are currently struggling to maintain access to customary lands now within forested protected areas. Such groups include the Sengwer of the Cherangani Hills, the Ogiek of Mount Elgon and the Mau forest complex, as well as the Aweer of Boni forest in Lamu County, all of whom have been subject to recurring bouts of eviction over recent decades (Kitelo 2014).

Unfortunately, recently passed legislation suggests a degree of 'bureaucratic sabotage' here as well. For instance, the 2012 Land Act forbids the conversion of public land to community land when the area in question is deemed to constitute a watershed, or is otherwise demonstrated to be a protected area of significant ecological importance (GoK 2012: §12(2)). This directly contradicts the previous provisions on the rights of hunter-gather communities, given that they have historically inhabited what are Kenya's most significant water catchment areas, conventionally known as the nation's 'water towers' (Kenya Water Towers Agency 2015).

Likewise, the draft Forest Conservation and Management Act of 2016, which has recently replaced the 2005 Forests Act, appears to be riddled with contradictions that enable it to simultaneously acknowledge the constitution's provisions for community forest rights, while rendering it virtually impossible for those rights to be upheld in practice. For instance, the Act creates a category of 'Community Forest' that parallels the constitution's definition of 'Community Land', encompassing:

> (a) forests on land lawfully registered in the name of group representatives; (b) forests on land lawfully transferred to a specific community; (c) forests on any other land declared to be community land by an Act of Parliament; (d) forests on land that is lawfully held, managed or used by specific communities as community forests; (e) forests on ancestral lands and lands traditionally occupied by hunter-gatherer communities; and (f) forests lawfully held as trustland by the county governments. (GoK 2016: §30(3))

On one hand, this is a commendable development, given that the Act at least hypothetically allows for collective *ownership* of forests, rather than mere community *access* to forests on terms defined by the

KFS. Likewise, the Act's definition also acknowledges the rights of 'hunter-gatherer' communities to their customary territories (GoK 2016: §30[3][e]), and would thereby solidify these provisions of the constitution in law. Further examination of the Act, however, evinces a certain dovetailing or slippage between the concept of *community forest* (e.g. forests owned in common by communities) and *community forest associations,* which are essentially interest groups with limited privileges for drafting co-management plans and accessing forest resources (Wily 2015). Tellingly, moreover, the Act enables the KFS to terminate such arrangements when, among other circumstances, 'the association breaches the terms and conditions thereof' or it is otherwise 'necessary for the purpose of protecting or conserving biodiversity' (GoK 2016: §51(1)).

From the perspective of Kenya's historically forest-dwelling communities, such broad discretionary powers leave them in a very precarious situation, and quite clearly undermine the spirit of the constitution's provisions for upholding their rights. The real problem, however, is this: the predicament of Kenya's forest dwellers is in many ways the predicament of the Kenyan citizenry itself. Though the 2010 constitution contains many laudable provisions for redressing the institutional legacies of both colonialism and authoritarian rule in the land and resource management sectors, these provisions appear to remain limited by bureaucratic resistance. The result, as I discuss below, is that the promulgation of the new constitution may be simply the beginning rather than the end of broader decolonial struggles for a transition to Kenya's 'Second Republic'.

Conclusion

'The vision for change and what the new Kenya should be is there,' Chief Justice Willy Mutunga recently told a reporter for Kenya's *Daily Nation*, '[b]ut there are forces that are making it difficult to implement the Constitution...They will always be there, and they are not going to disappear' (Lang'at 2016). Investigating such struggles for and against reform, this chapter has examined the provisions of Kenya's 2010 constitution that formally seek to elevate collective forms of land and resource tenure to a legal status equivalent to that enjoyed by their public and private counterparts, effectively so for the first time since the declaration of Kenya as a British protectorate in 1895. In doing so, I juxtaposed the substance of these protections with ongoing challenges to their implementation, surveying resultant implications for the devolution process, the fate of a new tenure category of 'community land', and related changes to institutions for access to and ownership of natural resources.

At the time of their adoption, the 2010 constitution and the related 2009 National Land Policy were met with considerable salutation both in Kenya and throughout networks of activists, human rights advocates, and social justice organisations in East Africa and beyond. Initially, comparisons were quick to emerge between Kenya's constitutional moment and South Africa's post-apartheid constitution of 1996. Indeed, both documents laudably contain substantial protections for citizens' rights (including socio-economic rights), essential checks on the power of government, and measures to ameliorate the fraught legacies of land-related historical injustices in both countries.

Yet a closer look at emerging attempts to translate the spirit of the 2010 constitution's provisions related to devolved governance, community land tenure and other forms of collective resource management evince both enduring challenges and perhaps also significant resistance from within the executive. In itself, however, this is neither surprising nor constitutive of grounds for despair. The histories of both the colonial and the republican state in Kenya are replete with evidence of state repression and 'bureaucratic authoritarianism' (Branch and Cheeseman 2006). The fate of more recent reforms will thus depend upon the ability of activists, citizens, civil society organisations, and a committed segment of the political class to realise the spirit as well as the letter of its provisions in the face of substantial opposition. The promulgation of the new constitution is a substantial battle won, yet the war for Kenya's Second Republic goes on.

Note

1. For a classic example, see Oginga Odinga (1967), *Not yet Uhuru*.

References

Anderson, D. (2002). *Eroding the commons: The politics of ecology in Baringo, Kenya, 1890–1963*. Oxford: James Currey.

Anderson, D. (2010). Majimboism: The troubled history of an idea. In D. Branch, N. Cheeseman, & L. Gardner (Eds.), *Our turn to eat: Politics in Kenya since 1950* (pp. 23–52). Berlin: Lit Verlag.

Anderson, D., & Lochery, E. (2008). Violence and exodus in Kenya's Rift Valley, 2008: Predictable and preventable? *Journal of Eastern African Studies, 2*(2), 328–343.

Anderson, D. M. (1984). Depression, dust bowl, demography, and drought: The colonial state and soil conservation in East Africa during the 1930s. *African Affairs, 83*(332), 321–343.

Anderson, D. M. (2000). Master and servant in colonial Kenya, 1895–1939. *The Journal of African History, 41*(03), 459–485.

Anderson, D. M. (2005). 'Yours in struggle for majimbo': Nationalism and the party politics of decolonization in Kenya, 1955–64. *Journal of Contemporary History, 40*(3), 547–564.

Berman, B. J., & Lonsdale, J. M. (1980). Crises of accumulation, coercion and the colonial state: The development of the labor control system in Kenya, 1919–1929. *Canadian Journal of African Studies, 14*(1), 55–81.

Boone, C. (2007). Property and constitutional order: Land tenure reform and the future of the African state. *African Affairs, 106*(425), 557–586.

Boone, C. (2012). Land conflict and distributive politics in Kenya. *African Studies Review, 55*(1), 75–103.

Branch, D., & Cheeseman, N. (2006). The politics of control in Kenya: Understanding the bureaucratic-executive state, 1952–78. *Review of African Political Economy, 33*(107), 11–31.

Collett, D. (1987). Pastoralists and wildlife: Image and reality in Kenya Maasailand. In D. Anderson & R. H. Grove (Eds.), *Conservation in Africa: Peoples, policies, and practice*. Cambridge: Cambridge University Press.

Coombes, B., Johnson, J. T., & Howitt, R. (2012). Indigenous geographies I: Mere resource conflicts? The complexities in Indigenous land and environmental claims. *Progress in Human Geography, 36*(6), 810–821.

East African Statistical Department. (1964). *Economic and statistical review*. No. 16. Nairobi: East African Statistical Department.

Escobar, A. (2008). *Territories of difference: Place, movements, life, redes*. Durham: Duke University Press.

Furedi, F. (1974). The social composition of the Mau Mau movement in the white highlands. *The Journal of Peasant Studies, 1*(4), 486–505.
Furedi, F. (1976). The Kikuyu squatters in the Rift Valley: 1918–1929. In B. A. Ogot (Ed.), *Hadith 5: Economic and social history of East Africa* (pp. 177–194). Nairobi: East Africa Literature Bureau.
Galaty, J. G. (1994). Ha(l)ving land in common: The subdivision of Maasai group ranches in Kenya. *Nomadic Peoples, 34–35*, 109–122.
Githongo, J. (2008). Kenya–Riding the tiger. *Journal of Eastern African Studies, 2*(2), 359–367.
Government of Kenya. (2008 [1998]). *Constitution of Kenya* (Repealed). Nairobi: Government Printer.
Government of Kenya (GoK). (2009). *Sessional paper no. 3 of 2009 on national land policy*. Nairobi: Ministry of Lands.
Government of Kenya (GoK). (2010). *Constitution of Kenya*. Nairobi: National Council for Law Reporting.
Government of Kenya. (GoK). (2012). *Land Act, No. 6 of 2012*. Nairobi: National Council for Law Reporting.
Hornsby, C. (2012). *Kenya: A history since independence*. London: I.B. Tauris.
Human Rights Watch. (2011). *"Turning pebbles": Evading accountability for post-election violence in Kenya*. Washington, DC: Human Rights Watch.
International Centre for Transitional Justice (ICTJ). (2011). *Breathing life into a new constitution: A new constitutional approach to law and police in Kenya—Lessons from South Africa*. New York: International Centre for Transitional Justice.
Kanyinga, K. (2009). The legacy of the white highlands: Land rights, ethnicity and the post-2007 election violence in Kenya. *Journal of Contemporary African Studies, 27*(3), 325–344.
Kenya Water Towers Agency. (2015). *Kenya water towers status report 2015*. Nairobi: Kenya Water Towers Agency.
Kimani, K., & Pickard, J. (1998). Recent trends and implications of group ranch sub-division and fragmentation in Kajiado District, Kenya. *Geographical Journal, 164*, 202–213.
Kitelo, P. (2014). The constitution of Kenya and the land question of the forest indigenous communities. Retrieved August 26, 2016, from http://www.forestpeoples.org/region/kenya/news/2014/08/constitution-kenya-and-land-question-forest-indigenous-communities
Klopp, J. M. (2000). Pilfering the public: The problem of land grabbing in contemporary Kenya. *Africa Today, 47*(1), 7–26.

Klopp, J. M. (2012). Deforestation and democratization: Patronage, politics and forests in Kenya. *Journal of Eastern African Studies, 6*(2), 351–370.

Klopp, J. M., & Sang, J. K. (2011). Maps, power, and the destruction of the Mau Forest in Kenya. *Georgetown Journal of International Affairs, 12*(1), 125–134.

Lonsdale, J. (1992a). The conquest state of Kenya, 1895–1905. In B. Berman & J. Lonsdale (Eds.), *Unhappy valley: Conflict in Kenya and Africa, part one—State and class* (pp. 13–45). Oxford: James Currey.

Lonsdale, J. (1992b). The moral economy of Mau Mau: The problem. In B. Berman & J. Lonsdale (Eds.), *Unhappy valley: Conflict in Kenya and Africa, part two—Violence and ethnicity* (pp. 265–314). Oxford: James Currey.

Lonsdale, J. (1992c). The moral economy of Mau Mau: Wealth, poverty, and civic virtue in Kikuyu political thought. In B. Berman & J. Lonsdale (Eds.), *Unhappy valley: Conflict in Kenya and Africa, part two—Violence and ethnicity* (pp. 265–314). Oxford: James Currey.

Lund, C., & Boone, C. (2013). Introduction: Land politics in Africa–constituting authority over territory, property and persons. *Africa, 83*(1), 1–13.

Manji, A. (2014). The politics of land reform in Kenya 2012. *African Studies Review, 57*(1), 115–130.

Maxon, R. M. (2016). The demise and rise of *majimbo* in independent Kenya. In M. M. Kithinji, M. M. Koster, & J. P. Rotich (Eds.), *Kenya after 50: Reconfiguring historical, political, and policy milestones*. New York: Palgrave Macmillan.

McAuslan, P. (2003). *Bringing the law back in: Essays in land, law, and development*. Aldershot: Ashgate.

Mignolo, W. (2007). Coloniality: The darker side of modernity. *Cultural Studies, 21*(2–3), 155–167.

Mignolo, W. (2011). *The darker side of western modernity: Global futures, decolonial options*. Durham: Duke University Press.

Mignolo, W. D., & Escobar, A. (Eds.). (2013). *Globalization and the decolonial option*. London: Routledge.

Morgan, W. T. W. (1963). The 'white highlands' of Kenya. *The Geographical Journal, 129*(2), 140–155.

Mudimbe, V. Y. (1988). *The invention of Africa*. Bloomington: Indiana University Press.

Mueller, S. D. (2008). The political economy of Kenya's crisis. *Journal of Eastern African Studies, 2*(2), 185–210.

Murunga, G. R., Okello, D., & Sjögren, A. (2014). Towards a new constitutional order in Kenya: An introduction. In G. R. Murunga, D. Okello, & A. Sjögren (Eds.), *Kenya: The struggle for a new constitutional order* (pp. 1–16). London: Zed Books.

Musembi, C. N., & Kameri-Mbote, P. (2013). Mobility, marginality, and tenure transformation in Kenya: Explorations of community property rights in law and practice. *Nomadic Peoples, 17*(1), 5–32.

Mwangi, D. I. (2012). The state of constitutionalism in Kenya, 2010. In M. Ohiambo (Ed.), *The annual state of constitutionalism in East Africa 2010* (pp. 29–67). Kampala: Fountain Press.

Mwangi, E. (1998). *Colonialism, self-governance, and forestry in Kenya: Policy, practice, and outcomes.* Retrieved August 26, 2016, from http://dlc.dlib.indiana.edu/dlc/bitstream/handle/10535/5706/Colonialism%20self%20governance%20and%20forestry%20in%20Kenya.pdf?sequence=1

Ndlovu-Gatsheni, S. J. (2013). *Coloniality of power in postcolonial Africa: Myths of decolonization.* Dakar, Senegal: CODESRIA Books.

Ndlovu-Gatsheni, S. J. (2015a). Decoloniality as the Future of Africa. *History Compass, 13*(10), 485–496.

Ndlovu-Gatsheni, S. J. (2015b). Genealogies of coloniality and implications for Africa's development. *Africa Development, 40*(3), 13–40.

Ndung'u Commission. (2004). *Report of the commission of inquiry into the illegal/irregular allocation of public land.* Nairobi: Government Printer.

Njonjo Commission. (2002). *Report of the commission of inquiry into the land law system of Kenya.* Nairobi: Government Printer.

Oginga, O. (1967). *Not yet Uhuru.* London: Heinemann.

Ogolla, D. B., & Mugabe, J. (1996). Land tenure systems and natural resources management. In C. Juma & J. B. Ojwang (Eds.), *In land we trust: Environment, private property, and constitutional change.* London: Zed Books.

Ojienda, T., & Okoth, M. (2011). Land and the environment. In P. Lumumba, M. K. Mbondenyi, & S. Odero (Eds.), *The constitution of Kenya: Contemporary readings* (pp. 153–180). Nairobi: LawAfrica Publishing.

Okoth-Ogendo, H. W. O. (1991). *Tenants at will of the Crown: Evolution of agrarian law and institutions in Kenya.* Nairobi: ACTS.

Okoth-Ogendo, H. W. O. (2003). The tragic African commons: A century of expropriation, suppression and subversion. *University of Nairobi Law Journal, 1*(October), 107–117.

Peluso, N. L. (1993). Coercing conservation? The politics of state resource control. *Global Environmental Change, 3*(2), 199–217.

Peluso, N. L., & Vandergeest, P. (2001). Genealogies of the political forest and customary rights in Indonesia, Malaysia, and Thailand. *The Journal of Asian Studies, 60*(3), 761–812.

Southall, R. (2005). The Ndungu report: Land & graft in Kenya. *Review of African Political Economy, 32*(103), 142–151.

Steinhart, E. (2006). *Black poachers, white hunters: A social history of hunting in colonial Kenya*. Oxford: James Currey.

Swynnerton, R. J. M. (1954). *The Swynnerton report: A plan to intensify the development of African agriculture in Kenya*. Nairobi: Government Printer.

Tiampati, M. (2015). Kenya. In *The indigenous world 2015*. Copenhagen: International Working Group on Indigenous Affairs (IWIGA).

Wamwangi, K. (2016, August 18). A case of unmitigated impunity. *The Star*. Retrieved August 26, 2016, from http://www.the-star.co.ke/news/2016/08/18/a-case-of-unmitigated-impunity_c1405463

Wily, L. A. (2015). The National constitution and forest dweller land rights in Kenya. Retrieved August 26, 2016, from http://www.forestpeoples.org/topics/rights-land-natural-resources/publication/2016/national-constitution-and-forest-dweller-land-

Wily, L. A. (2016). Customary tenure: Remaking property for the 21st century. In M. Graziadei & L. Smith (Eds.), *Comparative property law: Global perspectives*. Cheltenham: Edward Elgar.

World Bank. (2005). *Agricultural productivity and sustainable land management (KAPSLM) project*. Report No. 41908 v.2. Washington, DC: World Bank.

Youe, C. P. (1988). Settler capital and the assault on the squatter peasantry in Kenya's Uasin Gishu District, 1942–63. *African Affairs, 87*(348), 393–418.

Zeleza, P. T. (2014). The protracted transition to the Second Republic in Kenya. In G. R. Murunga, D. Okello, & A. Sjögren (Eds.), *Kenya: The struggle for a new constitutional order* (pp. 17–43). London: Zed Books.

Connor J. Cavanagh is a Research Fellow in the Department of International Environment and Development Studies (Noragric) at the Norwegian University of Life Sciences. Recent articles have appeared in *Environment and Planning D*, *Antipode*, the *Journal of Peasant Studies*, and *Geoforum*.

Part III

Governance, Economy and Sustainable Development

7

Urban Agriculture, Food Security and Poverty Alleviation in Post-Apartheid Metropolitan Durban, South Africa

Sultan Khan

Introduction

Urban agriculture, according to Mougeot (2003: 11), is an agricultural activity that takes place within or on the fringe of a town, city or metropolis that grows, processes and distributes a diversity of food and non-food products to that urban area. It is an activity that increases during periods of economic or political crisis as it is considered by the poor and destitute as the only means of survival (Tambwe 2010). Urban agriculture in the developing world is viewed as a panacea for food insecurity given the high rates of urbanisation and poverty. Since that many poor live in urban environments, it makes logical sense to look at food security within the urban context.

In developed countries, the agricultural sector is well advanced and often urban agriculture takes the form of pleasure and recreational purposes. Given high population density, characterised by high rise

---

S. Khan (✉)
School of Social Sciences, University of KwaZulu-Natal, Durban, South Africa

A. Adeniran, L. Ikuteyijo (eds.), *Africa Now!*,
https://doi.org/10.1007/978-3-319-62443-3_7

buildings, urban agriculture involves diverse methods of growing food (planting in pots, roof garden tops and the backyard if space permits). In Europe and North America, urbanisation took centuries to unfold. Rapid industrialisation and steady increases in per capita income helped to provide food security and a better quality of life. In so far as the developing world is concerned, such growth is expected to take place over two or three generations in the future. Urban growth in developing countries is being driven not by economic catalysts but by increasing birth rates. There is a tendency for a mass influx of rural people to urban centres to meet their food security needs and escape the scourge of poverty. It is estimated that by 2020, the proportion of the urban population living in poverty could reach 45 %, or 1.4 billion people. By then, 85 % of poor people in Latin America, and almost half of those in Africa and Asia, will be concentrated in towns and cities (Food and Agricultural Organisation of the United Nations 2010: 1–2).

Considering the high population densities that characterise urban centres in developing countries, conformity to land usage and plans are always a challenge given urban sprawl characterised by informal housing slums are constructed on valuable arable land. There are certainly those who will always garden—for food, pleasure, the feeling of productivity, health and exercise, nutrition or love of the outdoors. There are thousands of reasons that people find a relationship with plants important. To survive, millions of slum dwellers resort to growing their own food on every piece of available land—in backyards, along rivers, roads and railways, and under power lines.

In South Africa, since the dawn of democracy in 1994, the country's cities have undergone massive restructuring to accommodate the national programme for reconstruction and development. This was an attempt to break away from apartheid-style planning and development, which was fragmented, wasteful and designed to serve the needs of the minority white capitalist class at the expense of the majority of the black underclass, which was confined to the periphery of urban areas. One of the major attempts by the new national government was to consolidate the boundaries of the peripheral areas to form a megapolis representing an inclusive city. In Durban, the boundaries of the former apartheid city were extended by 68 % amounting to 2297 kilometres, as compared to the original boundary of 1366 kilometres during the apartheid regime. As a consequence, former under-developed black townships and native reserves under the tutelage of tribal authorities characterised by persistent levels of poverty were included within the post-apartheid city (Khan and Lootvoet 2003: 9).

Notwithstanding the persistently negative effects of poverty inherited from apartheid, many of these areas are known to be endowed with natural resources conducive to agriculture, which holds out of promise of providing food security opportunities to the urban poor. It is in this context that the Durban metropolitan government, in partnership with different stakeholders, has pioneered urban agricultural projects throughout the city. This chapter seeks to evaluate the success or otherwise of urban agricultural projects in providing food security to the urban poor in Durban.

History of Agriculture in South Africa

The history of agricultural activities in South Africa has been and is largely dominated by the white population. Its history can be divided into distinct phases—the pre-colonial, colonial, apartheid and post-apartheid periods. For the pre-colonial period, very little has been recorded about the history of agriculture among the indigenous people of South Africa. What is known is that they were largely pastoralist and moved from one area to another with their herds of cattle. Those that remained on the land cultivated sorghum and millets, herded cattle, sheep and goats, and manufactured iron tools and copper ornaments. These homesteads were sited near water and good soil that could be cultivated with iron hoes.[1] In the pre-colonial era, various tribes, such as the Khoi-Khoi and the Nguni, occupied land in the Western Cape and Eastern Cape, KwaZulu-Natal and Mpumalanga provinces respectively.

Since colonisation in 1652, land use by the Dutch, British and the Afrikaaner (Boer) varied. It dates back to the early colonial period, when slaves were introduced to work on the commercial farms in and around Cape Town. Some of the land use outcomes were different under Dutch, British and Boer rule due to the different objectives pursued by the governing authorities during each phase of colonisation. The Dutch were primarily interested in providing fresh supplies for passing ships on the lucrative spice trading routes. They opted for conservative trade and agricultural expansion away from the Cape. However, in their 150-year rule of the colony between 1652 and 1803 they began penetrating the interior of the country in search of agricultural land, displacing the San tribes (Jooste 2013: 7–8).

The rapid expansion into the interior was aided by the fact that the colonisation process took place at the end of the Mfecane, a period of time between the 1780s and the late 1820s when large-scale upheavals were taking place in the interior of South Africa, displacing settled tribes in different parts of the country. In this period, there was widespread chaos among the different African ethnic groups; they began migrating to safer areas and adjusting their sense of community. As a scattered community, they became weaker in their ability to withstand the Afrikaaner and British penetration of their valuable land. The arrival of Dutch-speaking farmers in what is referred to the Great Trek was a response to the abolishment of slavery by the British government in 1833. The abolishment of slavery meant a lack of labour to work on the farms that emerged in the Cape.

The conquest of new land by the British and the Afrikaaners led to the displacement of the indigenous African population. The British and Afrikaaners occupied the most arable land suitable for agriculture, around which small towns began to emerge. Despite being forcibly removed from arable land and confined to locations which came to be known as native reserves, the indigenous Africans managed to sustain agricultural activities on a small scale. The scale of agricultural activities was restricted, resulting in the emergence of surplus labour. The discovery of diamonds in 1867 and gold in 1886 further impacted on the shape of the economy in the country. Many indigenous Africans were made dependent on wage labour as migrant workers in these mines while the white agricultural activities began to emerge as commercial enterprises through the introduction of machinery spurred by rapid industrialisation (Jooste 2013).

In the Province of KwaZulu-Natal, which is home to the city of Durban, sugar cultivation was the mainstay of this part of the country. In Natal, the indigenous population refused to work on the farms, resulting in a large contingent of indentured Indians being engaged on white-owned sugar plantations from 1860 onwards. In this period, right until the birth of apartheid in 1948, South Africa experienced appreciable levels of industrialisation and growth in the manufacturing sector, which needed unskilled and semi-skilled labour in the different urban manufacturing sectors of the country. This trend may be attributed to the booming production of raw materials by white commercial farmers in the country, which invigorated production within the manufacturing sector.

The presence of people considered black in the urban centres was rigidly controlled. Their presence was tolerated on a pass system that ensured that they would be temporarily permitted in the urban centres to work and not to live. The Natives' Land Act of 1913 effectively reserved 87 % of the country for whites, while the vast majority were confined to the homelands under a self-governing system. In the urban areas, black people were confined to racially defined townships and native reserves. In essence, the Act sought to extract cheap black labour for the manufacturing sectors located in urban centres and deprived the majority of South Africans of the right to own land (Mahlangeni 2013: 2–3).

Agricultural Activities in the Urban Context

Within the urban context agriculture emerged as an important economic activity among the poor who lived on the periphery of the city. For example, in Durban, the Indians post-indenture cultivated vast portions of land, either renting these from white farmers or simply occupying vacant plots adjacent to their homes. An important example of urban agriculture among Indians in Durban was Cato Manor, where they grew vegetables for both domestic consumption and sale to the public.

However, with the coming of apartheid in 1948, settled communities were forcibly removed and relocated in dormitory-type townships designed for the respective race groups (Indians, Coloureds, Africans). These townships were densely populated and provided very little room for urban agriculture. The apartheid regime, in an attempt to segregate the different race groups, left vacant huge tracks of environmentally sensitive land around riverines, preserving land for the future needs of the minority white population. These vacant plots of land came under threat around 1987 when the enforcement of the notorious Group Areas Act of 1950 became no longer enforceable. Rapid migration from the rural areas and population growth in the townships resulted in the emergence of shack dwellers on the land that served as buffer zones to prevent racial mixing. Given the high poverty rates in these shacklands, the poor used every piece of vacant land within and outside to cultivate food for personal use. It should be noted that while the poor indigenous population struggled to make a living from the limited ground space available to them, white commercial farmers were heavily subsidised by the apartheid government.

Land usage management planning and the misappropriation of land both during colonialism and apartheid had a profound effect on how the urban poor could make a living by cultivating the land. In the post-apartheid era, new policies have been formulated to restore land to those who have been victims of colonialism and apartheid. Urban land claims were awarded at a greater rate than rural land claims. However, in total, by the end of 2012, 76,229 out of 79,696 claims received since 1994 had been settled. An analysis of the actual expenditure on land restitution from 1994 to 2013 reveals that over the period approximately R25.72 billion has been spent on land claims settlement, both in the urban and rural areas of the country (Ellse 2013: 3–4). Many of these claims have been settled in the rural areas of the country where land is under the jurisdiction of tribal administration and/or occupied by white commercial farmers, whose land is under dispute in terms of the land claim settlement.

Although the large claims for land settlement relate to the rural fringes of South African cities and provinces, the urban areas, which comprise only 8 % of the total surface area, are where 60 % of the population resides (National Planning Commission 2012). This settlement pattern suggests that the South African urban areas are densely populated and that any move towards urban agricultural activity within the urban context will be restricted by this. The urban agricultural needs of the poor have to be mitigated by the competitive need for housing and other important basic services, such as schools and healthcare infrastructure. The post-apartheid spatial planning at a macro level is largely dominated by a compact city model, given the segregated, fragmented and inefficient city structure that was inherited from the apartheid era (Gordon 2013). Notwithstanding the spatial prerogative of retaining a compact city structure, there is much research that asserts the failure of spatial planning in post-apartheid South Africa due to the persistence of urban sprawl (Gordon 2013).

Urban Sprawl, Poverty and Food Security Issues in South African Cities

Since the demise of the Group Areas Act in 1990, the urban and peripheral areas have witnessed large-scale mushrooming of informal settlements. Informal settlements are essentially shacks erected using a variety of materials for shelter. These are largely occupied by poor migrants who have moved into the city or the periphery of the city in search of a livelihood.

Statistics analysed by the Housing Development Agency (2013) provide greater insight into the extent of homelessness among the very poorest in South Africa. The data indicates that there is a total of 1,249,777 households, containing 3,306,697 individuals, living in shacks and not in backyards (Housing Development Agency 2013: 16–18). Additionally, the Census 2011 indicates a total of 712,956 households living in shacks in the backyards of formal housing, as compared to 459,526 in 2001. This trend suggests that the informal housing market is now grafted on formal housing in the form of backyard shacks. Grafting backyard shacks onto the formal housing stock provides easy access to municipal services such as electricity, water and sanitation.

Post-apartheid, the country's Reconstruction and Development Plan sought ways in which to eradicate high levels of poverty. A variety of programmes were put in place—the most effective was the social wage, which was paid to those who found themselves below the poverty line. South Africa's social assistance system has expanded tremendously since 2000, growing from around 3 million grants to 15 million by 2011. Despite these interventions, the most up-to-date headcount of persons living in extreme poverty in South Africa indicates that there are 10.2 million such persons (Statistics South Africa 2014: 12).

Rural–urban migration has to a large extent increased poverty levels within the urban context. It has been estimated that some 3.7 million farm workers have been displaced, and that 1.6 million of those evicted between 1984 and 2004 moved off farms altogether. Of those evicted some 67 % settled in and around urban areas, primarily in backyard shacks or poorer parts of the townships (38 %), or in informal settlements (29 %). Bigger urban centres in Gauteng and KwaZulu-Natal had large numbers relocating due to the need to find suitable jobs (Todes *et al.* 2008).

The reasons for relocating from farms are often work-related. The financial strain on commercial agriculture has led to casual or seasonal employment within the agricultural sector. It is estimated that regular employment in the agricultural sector declined by 170,000 in 1993 to 480,000 in 2002, while casual work increased by 14,000 to 460,000 over the same period. Between 1993 and 2002, there was a large consolidation of agricultural activities, resulting in the decline of the number of farms from 57,980 to 45,818. The expectation that land reform would provide an alternative to urban migration, however, was not realised. The scale of land reform has not matched the extent of evictions from farms, nor have farm workers benefited through the programme. The provision of land or improved tenure rights by July 2005 to some 164,185 households resulted in only 7543 farm worker households benefiting (Todes *et al.* 2008).

Amid the skills that rural migrants bring to the urban centres are farming techniques and other indigenous knowledge systems that can be better utilised within urban food gardening projects. In South Africa, indigenous and older females are known to engage in agricultural practices in the rural areas. Hence the transfer of their skills to the urban context can improve the quality of their life by providing food security and at the same time provide them with some degree of economic independence.

Urban Agriculture in Metropolitan Durban

Historically, Durban's climate and soil was found suitable for a wide variety of agricultural activities. Located on the east coast line of the Province of KwaZulu-Natal, the earlier British settlers cultivated the first sugar cane crop in 1848, which led to the rise of the sugar industry. The sugar industry was beset by serious labour challenges as the indigenous people were uninterested in engaging in this form of agricultural activity. Given the success of the sugar industry in Mauritius through the use of indentured Indians from India, the Natal Colony followed suit. In 1860, the first batch of indentured Indians arrived in Port Natal and since then the sugar industry was on the rise with rapid industrialisation taking place through the establishment of various mills around the city.

On completion of indenture after five years, the labourer had the option of renewing the contract for a further five-year period, which would have provided him/her the privilege of settling permanently in the colony, with a portion of land bequeathed to him/her. Many chose to remain in the colony and began migrating from the sugar plantations to the fringes of the city. Some leased lands while others squatted on vacant land. In order to make a living these new migrants to the city began market gardening activities on the fringe of the city and sold their produce in the city, either through 'squatting' on the street or hawking from door to door. Market gardening became a lucrative part of the city's economy, resulting in the all-white city council establishing a formal market in Victoria Street, Durban by 1910. Sixty-three years later, this famous market was tragically burnt down; a new one developed in Warwick Avenue (Mohamed 1999). This market is still in existence despite recent attempts by the post-apartheid metropolitan government to demolish it to make way for a new mall. This initiative has seen wide resistance from the community and the market is now considered an historical artefact.

The Durban Indian market gardeners was dealt a severe blow when the Group Areas Act (1950) was enforced in the city. This Act was used by the city council to curb the competition that Indian traders provided to their white trading counterparts. With the Afrikaaner nationalist government coming into power during this time, territorial apartheid was enforced more aggressively, resulting in the forced eviction, relocation and resettlement of all race groups other than white to monolithic, racially defined townships in the city. Hence market gardeners that were settled in places such as Cato Manor, Clairwood, Magazine Barracks and Riverside were forcibly removed and relocated to the townships. The effects of forced removal and resettlement had a severe impact on the number and extent of market gardeners in the city. According to Mohammed (1999), the number of market gardeners in the city declined from 2887 in 1936 to around 1000 in 1963.

Resettlement into townships meant that blacks, comprising Indians, Africans and the Coloureds, were restricted in terms of the land they occupied. These townships were dormitory-like and mono-functional—the average black sold his/her labour in the city and returned in the evening to a dormitory township just to eat and sleep. The townships over the years became densely populated and dependent on the wage labour. In the same period, rural poverty began to emerge due to underdevelopment, resulting in a mass influx of poor on vacant land in the city. This was the first sign that apartheid was no longer sustainable. Squatter settlements began to mushroom in the city on land left for the development of future infrastructure for the white population, on environmentally sensitive land and on land that was not suitable for development used as buffer zones to separate the different racially defined townships. In 1990 the Group Areas Act was repealed and more rural migrants began to move to the city, resulting in the greying of the all-white city.

Rural poverty came to bear heavily in the city due to unplanned urbanisation. In order to make a living, those living in the informal settlements began to engage with the informal economy to sustain livelihoods, apart from engaging in menial work as, for example, gardeners, domestic workers, cleaners and the like. In the informal economy, many set up street stands with fruit and vegetables for sale to the public. Most of these are sourced from wholesale fruit and vegetable outlets. Some sell indigenous food to the general public, including medicinal plants.

In informal settlements in the city, one would note that small pockets of vacant land are used for urban agriculture, which may just be sufficient to meet the basic needs of the family. Such miniscule activity suggests that land accessibility and ownership is essential if the poor are to have food security. To this extent, municipalities are required to be responsive to poverty and integrate food security measures into their plans. Food security and poverty alleviation programmes have been highlighted in the municipality five-year Integrated Development Plan for 2012/2013 to 2016/17 by promoting community gardens and nurseries in the city (Integrated Development Plan 2012: 179).

The nature and extent of agricultural projects in the city was derived from an analysis of the 2010 data base of the Urban Agriculture Department in the city of Durban, which contained 660 projects involving 5950 persons at its inception. In fact, the number of persons that are actively involved in these projects amounts to 1170. In terms of distribution of projects, 93 of these were located in the central areas of the city, 156 to the east, 117 to the north, 142 to the south and 139 to the west. A total of 593 (89.8 %) of the projects in the aforementioned localities were active while the remaining were redundant (0.03 %), inactive (4.1 %) or had not commenced (2.1 %). Only 3.6 % of the projects had been started and later abandoned.

With regard to water sources for the project, 46.8 % depended on natural sources, 39.1 % on potable water, 10.8 % on a combination of natural resources, potable and rainfall. Only 4.2 % of the projects had no access to water altogether. Most of the land (74.9 %) on which these agricultural projects are located has been categorised as unsuitable for arable agriculture. As consequence, 51.8 % of the urban agricultural projects cultivate organic plots of ground. Only 25.1 % of the land was found to be suitable for agriculture. The produce was largely (74.4 %) sold within the local community, while the remaining was sold at clinics, schools and at retail shops and local fresh produce markets. Animal husbandry did not feature prominently in the projects analysed. Only 3.1 % of the project engaged in poultry while 0.8 % dealt with livestock. The remaining projects primarily focused on fruit and vegetables. Many of these vegetables, including indigenous food products, form the common diet of locals.

A quarter of the projects undertook certain social responsibility activities by supporting creches, the disabled, orphanages, poor households and sick individuals in the community. In terms of rating the urban agricultural projects, only 17 % were rated poor—the rest ranged from good to excellent. Many of the projects (81.1 %) have been classified as semi-commercial, suggesting that the agricultural gardens are used for both personal consumption and to grow food for the open market, bringing in some form of income to the household. A significant 4.1 % of the agricultural gardens have already reached commercial status, suggesting that they are competing on the open agricultural market.

Conclusion

This paper highlights the role of urban agriculture in providing food security and poverty alleviation in post-apartheid metropolitan Durban, South Africa. In a context where the general contribution of agriculture to gross domestic product has declined from 10 % in the 1960s to less than 3 % in 2009 (Department of Finance 2010: 2), it becomes imperative to look at other alternatives, such as urban agriculture, to provide food security for the urban poor. Agriculture has historically come to be associated with rural productivity for the economy. However, given that there is large-scale migration of the rural poor to the urban centres, contributing to urban growth, poverty is automatically transferred to the urban context. It is in this context that poverty has to be addressed. The metropolitan government in Durban has been responsive to the value that urban agriculture can add to food security among the urban poor and hence ensures that it becomes part of its Integrated Development Plan, although agriculture does not fall within the purview of local governments.

It appears from this study that the issue of land is a contested one both for the metropolitan government and the country as a whole as it competes against land being allocated to the more pressing need of alleviating homelessness within the urban sphere. While this may be an issue of concern, use of vacant land that was allocated as buffer zones among the different race groups in the country during apartheid could possibly provide an opportunity to integrate urban agriculture as part of a programme to provide sustainable livelihoods for the poor. In its current form, where land is used for compact, dense housing settlements in the city, very minimal space is left for any such agricultural activities. In this instance, where land is available along riverine and waterways, this should be reserved for urban agriculture to be used by co-operatives.

From the analysis of projects that are currently on record, it appears that urban agriculture is well on its way in Durban and shows promise for the future. Although a significant (74.4 %) number of agricultural gardens are located on unsuitable land, it appears that using organic farming methods has helped to make the ground more productive. This is alluded to on the grounds that 51.8 % of the agricultural projects are organic in nature. Organic farming methods involve the use of environmentally friendly materials for agriculture. Given the fact that the urban context is dense in population, the use of non-organic materials, fertilisers and pesticides poses a major risk to the safety of the residents in the city. If non-organic materials are to be used in urban agricultural projects, it would require greater municipal regulation and policing.

Insofar as land availability is concerned, much land in the city is under the protection of traditional leaders and lies on the fringe of the city. The participation of traditional leaders in urban agricultural projects holds the potential to use land for different animal husbandry projects—these did not feature prominently on the database analysed for urban agriculturalists. Tribal land constitutes a large section of the urban landscape in Durban and is to be found on its fringes, where there is a lower population density. Animal husbandry such as the raising of pigs, cattle and poultry could offer opportunities for the urban agriculturalist. Given that 46.8 % of the urban agriculturalists have access to natural sources of water, diversifying into areas such as aquaculture could be a possibility. Adding value through products and by processing these products can have possibilities for stimulating the manufacturing sector, which in turn could create employment opportunities and eradicate poverty among the urban poor. The fact that 81.1 % of the projects on the database are classified as semi-commercialised enterprises suggests the potential of these projects advancing to fully commercialised status sometime in the future. At this point, it would be strategic to form co-operatives that can invest in the agro-processing sector, supporting manufacturing and adding value to the products harvested. By forming co-operatives, the urban agriculturalists can compete in the market with other commercialised farmers and cut down on competition among themselves. As co-operatives, they can gain new ground in the agriculture chain and provide support to the emerging urban agriculturalists.

Note

1. http://www.sahistory.org.za/early-history/pre-colonial-history-sa

References

Department of Finance. (2010). Economic analysis directorate Gauteng review and outlook of agriculture industry in South Africa and Gauteng.

Ellse, B. (2013). Costing land restitution 1994–2013 research unit, Parliament of the Republic of South Africa.

Food and Agricultural Organisation of the United Organisation. (2010).

Gordon, T. (2013). Good intentions: The production of urban public space in eThekwini, South Africa in cities to be tamed? *Standards and Alternatives in the Transformation of the Urban South Conference Proceedings*, Milan.

Growing Greener Cities. Retrieved February 14, 2016, from www.fao.org/ag/agp/greenercities/

Integrated Development Plan. (2012). eThekwini municipality.

Jooste, J. (2013). Changing land use patterns in South Africa with special reference to the Natives Land Act (1913) and its implications. Research unit, Parliament of the Republic of South Africa.

Khan, S., & Lootvoet, B. (2003). Tribal authority and service delivery in the Durban Uni-City in Antheaume, B.; Giraut, F. and Maharaj, B. Rencontres scientifiques franco- Sud-Africaines de l'innovation territoriale, Jan 2002, Grenoble–Avignon, France.

Mahlangeni, B. M. (2013). Reflections on the impact of the Natives Land Act, 1913, on local government in South Africa. Research Unit, Parliament of the Republic of South Africa.

Mougeot, J. A. (2003). Urban agriculture: Concepts and definitions. *Urban Agriculture Magazine no. 1* [Online]. Retrieved January 24, 2013, from www.ruaf.org/node/138

National Planning Commission. (2012). National Development Plan 2030.

South African History Online Precolonial History of South Africa. Retrieved January 10, 2015, from http://www.sahistory.org.za/early-history/pre-colonial-history-sa

Statistics South Africa. (2014). Poverty trends in South Africa: An examination of absolute poverty between 2006 and 2011, Pretoria.

Tambwe, N. (2010). Urban agriculture and food security in the city of Lumbumbashi (DRC) [Online]. Retrieved January 30, 2016, from wiredspace.wits.ac.za/handle/10539/8176?show=full

The Mercury Friday. (1999). The 'market' is a story of survival against the odds Sayed-Iqbal Mohamed, August 20, 1999.

Todes, A., Kok, P., Wentzel, M., & van Zyl, J. (2008). *Contemporary South African urbanisation dynamics*. Paper for UNU-WIDER Conference: Beyond Tipping Point. African development in an urban world, Cape Town, June 1998.

Sultan Khan is an Associate Professor in Sociology at the University of KwaZulu-Natal, South Africa. His research interests lie in the fields of urban sociology and social issues. He is the co-editor of the books *UnDressing Durban*, *Contemporary Social Issues in Africa* and *(De)Monopolising Paradise*.

8

Internal Democracy and Leadership Crisis in the Nigerian Legislature

Taofeek Oluwayomi Gidado

Introduction

This chapter investigates the nexus between internal democracy and the leadership crisis in the Nigerian legislature. Since the return to civil rule in 1999, the crisis of leadership commonly heralding the birth of a legislative assembly is usually spawned by the spoils of office. We contend that: (1) the impunity that is largely associated with the primaries of political parties reflects the character of politics in a rentier state; (2) public office is exploited for personal enrichment; and (3) the structure of the state promotes impunity in governance. This chapter concludes that the pursuit of leadership is unconnected with the desire to serve, but is largely driven by the opportunities that such exalted office provides in the distribution of patronage.

T.O. Gidado (✉)
Department of Sociology and Anthropology, Obafemi Awolowo University, Ile-Ife, Osun State, Nigeria

A. Adeniran, L. Ikuteyijo (eds.), *Africa Now!*,
https://doi.org/10.1007/978-3-319-62443-3_8

Contextual Review

After 16 years of military adventurism in the political governance of Nigeria (1983–1999), Nigeria returned to civil rule and constitutionalism in 1999 with the commencement of the fourth republic. Given the characteristics of Nigerian society, multi-party democracy appears to be the appropriate means to guarantee the inclusion of the different cultures that are meshed together in the formation of the Nigerian society. It is, however, instructive to emphasise that the adoption of multi-party pluralism does not on its own promote the cause of democracy without the entrenchment of institutional parameters to advance and engender due process in theory and practice (Bratton and Van de Walle 1997, cited in Omotola 2007: 612).

The pathetic reality in the last 18 years (in spite of multi-party democracy) reveals that democracy has been an abysmal failure in terms of the index of human development. This is evident in the plummeting economy, growing insecurity, ethnic suspicion, impunity in governance, absence of internal democracy among the political parties and the growing spectre of state failure, with the ascendancy of centrifugal forces such as the Oduduwa Peoples' Congress in the south-west, the Arewa People's Congress in the north, the Indigenous Peoples Republic of Biafra in the south-east and the Niger Delta Avengers in the south.

The failure of democracy to engender development cannot be extricated from the malfunction of the political parties in the fourth republic. In spite of the entrenched multi-party system, Nigeria runs a one-party system that means that the ruling party controls state resources at the expense of the other parties. At the inception of the fourth republic, Nigeria had three political parties that were registered by the military. As a result of litigation initiated by the human rights and constitutional lawyer, the late Chief Gani Fawenhinmi, the Supreme Court gave a judgment that allowed for the process of registering political parties to be liberalised. As a result of this judgment, three strata of political parties emerged in Nigeria, namely: (1) the party that controls the state machinery at the centre; (2) those that have some representation in the National Assembly and control some states; and (3) those that are amorphous and provide a haven for politicians whose ambitions have been truncated by either the party in government or the main opposition party. Instructively, political parties with an executive presence either in the state or at the centre have either the Governor or the President as the leader of the party. This is the signpost for party subordination to the government, and provides the context for the destruction of party discipline and cohesion as party organisational structure are completely annihilated.

However, it needs be stressed that the judgment of the Supreme Court in liberalising the registration process for political parties could neither translate into qualitative growth for the political parties nor enhance the form of the political parties, moving them away from an authoritarian and paternalistic nature and giving them a democratic character. This is evident in the way political parties are still controlled by godfathers and barons who wield leverage as a result of wealth and connection. Political party members are still unimportant in the power matrix of the political parties. They are only important when they are used as pawns to settle scores through violence. As a result, the character of relationship promotes a perverse network of patrons and clients in which the clients are organised to promote self-interested sentiments to gain pecuniary advantage.

Instructively, political parties constitute the nucleus of any democracy as they provide the forum for those aspiring to elective office in order to gain power. Also, it aggregates the interest of the people and constitutes a viable opposition (when out of government) as they proffer alternative programmes to government policies. In spite of these responsibilities, political parties can only be effective if they have democratic structures, discipline and organisational cohesion, and if they are distinctly autonomous and imbibe internal democracy (Egwu 2005).

In general terms, political parties provide the fulcrum for the expression of democratic values as they aggregate the interest of the public, articulating them in policy options and providing the structures for political participation. One of the critical responsibilities of political parties is to educate and mobilise voters in a general set of interests. In many countries, parties are instrumental to the democratic process as they formulate policy, choose candidates, initiate election campaigns and guide the conduct of their elected representatives, both in the legislative and executive arms of the government. Political parties connect citizens to the government as they provide avenues for the expression of voices.

In the light of this, political parties not only provide a screening platform for the selection of candidates but also a training ground for leaders who contest on their platforms and hone their ability in the art of governance. Once a political party wins an election, it provides the skeleton of a government, and when in a minority, serves as the conscience of a nation. As the United Nations Development Program correctly observed in its *Handbook on Working with Political Parties*, parties 'provide a structure for political participation; serve as a training ground for future political leaders and seek to win elections in order to enter government… [they] exist to transform aggregated social interests into public policy. In the legislature, political parties play an important role in shaping the relationship between the executive and the legislature and in prioritising the legislative agenda' (Deme 2013).

Against this background, the strength and quality of political parties determine the strength and quality of the democratic growth in any country. The tragedy of our democracy is the failure of the political parties to appreciate the nuances of good governance. Indeed, democracy is not conceivable without political parties as they nurture democracy and provide the anchor on which democracy hinges. It is indisputable that political parties have the onerous responsibility of delivering governance with a view to promoting democracy and the well-being of citizens. Hence, political parties can neither be extricated from the political process nor from the democratic process. Political parties distinguish a liberal and representative democracy from a military regime. Also, they provide a distilling characterisation from an authoritarian regime as it constitutes the core of liberal democracy. It is incumbent on political parties in any democracy to hold the vital task of aggregating the interests of citizens and providing the fulcrum for competing for political office.

Methodology

This article derives its form from experiences and personal observations while I was on staff of the National Assembly. My experience was largely influenced by interactions and conversations with members and Senators in the seventh legislative Assembly. Incidentally, my views are equally framed by the outcomes of the quarterly workshops that I participated in while I was on staff of the Assembly.

This chapter also draws its strength from my observations on the deliberations of the numerous sessions of the Senate and the House of Representatives. My narrative was influenced by my interactions with some members and Senators in the Seventh National Assembly. This was a period in which the National Assembly was largely fragmented as a result of the emergence of the new Peoples Democratic Party (PDP) that left the party to join the opposition and the covert intense ethno-religious division that also played out between members and Senators of the south-west, north-west and north-east and some states in the north-central area, on the one hand, and the south-south, south-east and a few of the states in the north-central area on the other.

Political Parties and Democratisation

Political parties constitute a critical segment of relevant institutions in the democratisation process. There has been the problem of the absence of internal democracy among the political parties in post-colonial Nigeria's fourth republic. This problem constitutes a threat to the consolidation of democracy and, consequently, obstructs the Parliament in carrying out its statutory responsibility. This is because members of the legislature contest on the platforms of their respective political parties. It is somewhat problematic to have a robust Parliament when the political parties are ineffectual. In Nigeria, the constitution does not allow for independent candidacy. Therefore, membership of a political party is required for access to the Parliament once the party agrees to select, nominate and present a member for an elective post to the Parliament. Expectedly, the character of the parties and their politics has significant implications for participatory democracy.

All the political parties in the fourth republic have been hijacked by the wealthy political elites, commonly referred to as 'godfathers' and 'barons', who use their wealth to influence the party machinery at the expense of the members. This is reflected in the imposition of candidates who have never won in the primaries of the political parties. Incidentally, the imposition of candidates has created a deepening crisis in the political parties, and has disconnected the political parties from the electorate. Regrettably, this has meant that the political parties usually connect with the people around election period when they are campaigning for votes but disconnect with the people once the elections are over. This is why Nigerian democracy has existed without the real participation of the people.

In Nigeria, the political class is largely corrupt and self-interested. Their dominance over party formation has meant that democracy has lost its meaning, and has made good governance a mirage. Mostly, political parties promote the interest of the business class in Nigeria, against the interests of the country as a whole. There is hardly a fundamental distinction between the interests of the political class and those of the business class. This development has ruptured, disarticulated and distorted democracy so that it has become a plutocracy. This is reflected in the way the institutions of the government have been made to promote the interest of the tiny business class at the expense of the people as a whole.

In spite of the fact that a democracy enjoys institutions that complement one another through the principles of the separation of powers and checks and balances, the political class and the politicians have weakened the statutory institutions of checks and balances which inhere in the separation of powers among the three arms of the government. To maximise electoral gains, politics has become expensive—to the point that aspiring candidates for the Parliament need the support of the wealthy business class and those that wield power over the political parties. This is sometimes reflected in the way political leaders collect large donations for the party and resort to the extraction of 'rent' by way of making decisions that unduly favour the business class.

In Nigeria, politics is not about service. It is a means for the primitive accumulation of wealth. It is only access to government that guarantees wealth, as Nigeria does not have a productive economy. Nigeria is essentially a rentier state. The government relies on rent to meet its responsibilities to the people, and individuals rely on the state to become wealthy. Indeed, the state, through the institutions of government, determines what happens to both the public and the private realms as the economy is largely structured by politics that usually allows for the distribution of the spoils through contracts by those who wield power. The character of the state has definitive consequences for politics. Those who seek power are largely motivated by the desire to use power for primitive accumulation, and as a means of building a network of clients and loyalty to ensure their continued relevance in the matrix of power relations. As noted by (Joseph 1987: 1), 'In Nigeria…the state has increasingly become a magnet for all facets of political and economic life, consuming the attention of traders, contractors, builders, farmers, traditional rulers, teachers, as much as that of politicians or politically motivated individuals in the usual sense of these terms.' This development has serious deleterious consequences for governance.

The Nigerian state remains essentially the same in terms of its sovereignty. But the political parties differ nominally in terms of their structures, size and access to the state. In very substantial terms, the political parties in the fourth republic are essentially the same. They have the same ideological complexion and are largely driven by the same character of people—who are obviously unable to craft policies that would launch the country on the path of development and nation-building. The attitude of the politicians, regardless of their political parties, is essentially the same in terms of their objective of maintaining their hold on power (Musa 2005) for primitive accumulation. To worsen this crisis, ideology is not an explanatory tool in differentiating the political parties in Nigeria. It is best an assemblage of people with a similar base interest. This has greatly impacted on the quality of governance. Even from the intra-party selection or election process, it is usually the interests of those who control the party machinery that mediate either a selection or an election process, embodied by the delegate system, which is overwhelmingly influenced by power and money.

It is therefore not surprising that political parties in Nigeria have botched the effort to deepen democracy, strengthen the institutions of government, expand the democratic space and engender good governance. In broad terms, political parties have failed to achieve the purpose for which they were created: good governance. Perhaps this realisation may have made Schmitter (1992) contend that parties have gone through a worldwide decline in recent years (a contention some would dispute). However, there appears to be some justification for

Schmitter's postulation in the light of the inability of the political parties to engender good governance in Nigeria's fourth republic.

Putting the decline in the quality of political parties in Nigeria into perspective, there is no substitute for political parties in the democratic trajectory in Nigeria. Political parties continue to dominate the contest for electoral positions and at least influence policy formulation. The 1999 constitution does not give legal force to independent candidacy. However, a challenge for political parties in Nigeria would be to institutionalise policy processes that would deepen internal democracy in the party machinery.

Political Parties and Representation in the Parliament

Political parties provide the anchorage for the Members of Parliament. All members of the legislature are members of the different political parties. In fact, the Parliament is composed of those who are the direct representatives of the people. In its elemental form, it constitutes the elected core of the people, unlike the executive branch, where the President is the direct symbol of representative government. If political parties largely lack internal democracy, it is difficult to think that the Parliament will be proactive and carry out its functions without hindrance or lethargy. The inaction of the Parliament is a consequence of the impunity that political parties reflect in their different domains.

Therefore, democracy requires political parties to internalise democratic procedures in how decisions are reached and to allow the voices of members to direct the affairs of the party, with a view to deepening democracy. The essence of democracy is effective representation. Once the people are voiceless within the framework of the machinery of the political parties, the Parliament can neither be proactive nor effective. Thus, the criteria used to resolve whether or not a political system is reliably democratic are to be defined by those factors that establish whether or not the structure and tradition of that system allow the people of the country to be effectively represented. This is why the Parliament is the expression of people's mandate, and defines the hallmark of representation to governability. It is really through the purview of political parties that Seward's (1998) definition of democracy as 'a political system in which citizens themselves have an equally effective input into the making of binding collective decisions' becomes saliently instructive.

As of March 2017, Nigeria had 29 registered political parties. With this number of political parties, one would expect the Parliament to be robust in terms of ideas that would deepen governance. Regrettably, the lacklustre performances of the political parties which represent an aggregation of narrow interests have adverse implications for the growth of a robust Parliament. Hence, Nigerians are gradually losing confidence in the ability of the political parties to deepen democratic governance. Indeed, a survey of Nigerian public opinion by the International Republican Institute (2010) showed only about 37% of respondents having some confidence in political parties, while 26% have no confidence at all, and 35% do not have very much confidence in them (Simbine 2002).

In Nigeria's democracy, the critical question that is raised by the practice of multi-party democracy is whether competitive party politics, the inclusion of citizens in inter- and intra-party affairs, and whether elected official are reflections of the voice of party members are sufficient to engender democracy, or whether party competitive politics, inclusive participation and representation within parties are required as well to engender democracy. At both levels, intra- and inter-party democratisation are important to the process of engendering effective representation and functional democracy. However, it is difficult to establish whether the existence of multi-party system is sufficient to engender democracy.

The reality in Nigeria does not portray multi-party system is sufficient to engender democracy. However, the merits of multi-party system allow people to have choices as the parties encourage moderation among people. Consequently, they also ensure that individuals and social groups steadily move away from extreme views on issues and policies that could lead to anarchy. However, the strength of this political contrivance is not reflected in the capacity of the Parliament to be robust and proactive.

Political parties in the fourth republic are no more than an association of people who hold similar views about what should be a community or society's social and economic priorities, and come together to establish these priorities by gaining control of the machinery of government (Derbyshire and Derbyshire 1991: 139). There is nothing that has suggested that ideology either has a place in the formation of political parties in Nigeria or in characterising political parties from one another. Perhaps the apt description of Chief Bola Ige[1] of Nigeria's political parties as the fingers of a leprous hand has some true in it. Sadly, Chief Bola Ige was murdered in December 2001 by unidentified assassins.

In Nigeria, the reality is that the party-political domain is replete with a large number of parties representing the tapestry of the Nigerian society. All the political parties are welded together by the crude ambition to gain power for primitive accumulation and not on the basis of proposing alternatives to public policy that could initiate participatory democracy and engender development. It is, therefore, not surprising that some of the parties have become concrete expressions around social and linguistic identities, thus, reinforcing ethno-religious politics in Nigeria. In spite of the gains in the 2015 presidential elections (at least in terms of the transparency of the election process), political parties exploited the fault lines as they whipped up controversial issues underscoring ethno-religious sentiments in an effort to mobilise the electorate. Instructively, these parties across the political spectrum (as a result of the ideological void) have tended to unite on macroeconomic policy, but continue to disaggregate on social policies and the larger issues that confront Nigeria's path to nationhood.

Political Parties and Good Governance

Political parties are crucial to the processes of engendering good governance. This is because political parties anchor democracy. A political party that adheres to constitutionalism and ensures that its motives are transparent and its policies are driven by the public good will engender good governance. The World Bank (1994), quoted in the *Handbook for Parliamentarians No. 5* (2003: 23) entitled 'Parliamentary Oversight of the Security Sector: Principles, Mechanisms and Practices' defined good governance as follows:

> Good governance is epitomized by predictable, open and enlightened policy making, a bureaucracy imbued with a professional ethos acting in furtherance of the public good, the rule of law, transparent processes, and a strong civil society participating in public affairs. Poor governance (on the other hand) is characterized by arbitrary policy making, unaccountable bureaucracies, unenforced or unjust legal systems, the abuse of executive power, a civil society unengaged in public life, and widespread corruption.

An assessment of political parties in post-colonial Nigeria reveals that it has circumscribed democratic growth. This is evident in the way that the Parliament has lacked independence, sovereignty and the political will in holding the executive accountable and ensuring that it is transparent; it is unable to be proactive in creating good laws for the governance of the country. At the heart of this crisis is the absence of internal democracy among the political parties in post-colonial Nigeria, culminating in impunity in governance.

Political parties in post-colonial Nigeria hardly allow room for decisions to be reached in a democratic way. This is manifested in the imposition of candidates in party primaries, with grave consequences for party cohesion and discipline. The most recent was the imposition of candidates in the primaries of the ruling All Progressive Congress (APC) party in Ondo State. This led to a serious division in which some party stalwarts left the party for the Labour Party, while others were alleged to have worked against the interests of the party. However, the party still won the governorship election in the state. But the victory of the party has become a subject of litigation in the court.

If democracy is a political system that allows people to have a sense of participation and inclusion in governance, political parties in Nigeria hardly demonstrate this. In its elementary form, democracy is a government that is representative of the people as a result of their participation in forming the government. Thus, the core virtue of democracy is to advance the interests of all the members of a politically sovereign state. Unfortunately, the reality is that the practice of democracy has unleashed intense disillusionment among Nigerians. In a recent interview on Sunrise Daily (an on-air TV programme in Nigeria), a constitutional lawyer and Senior Advocate of Nigeria, Robert Clark, bluntly called for a military takeover of government for two years if the APC ruling government could not resolve the governance crisis. This is because the political class has destroyed democracy. The heart-rending reality in Nigeria is that power hardly belongs to the people, and that democracy has been turned into plutocracy as governance has more meaning for the wealthy people and less for the poor people.

Contextualising the travails of democracy, the sanctity of the ballot box is usually defiled by politicians who seek elective offices with the connivance of their godfathers, who provide the logistics to subvert the will of the people in both intra- and inter-party elections. The truncation of the will of the people at the party level provides the context for the selection of legislators who violate their oaths, and pay homage to their godfathers and those who wield power to decide who represents political parties. Indeed, the selection and election of candidates to represent the party violates the policy thrust and guidelines of the party in a manner that does not strengthen the ideals of democracy, highlighting the trajectory of the absence of internal democracy among the political parties (Simbine 2002: 8). It is largely the crafty and undemocratic method of selecting legislators that account for the lacklustre performance of the National Assembly.

Discussion of the Findings

Political Parties and the Eighth National Assembly

The eighth National Assembly of the Federal Republic of Nigeria operates a bicameral legislature, as the country has had since the return to civil rule in 1999. The inauguration of the eighth legislative assembly was done on 9 June 2015, following the reading of the proclamation letter of President Muhammadu Buhari, the President of Nigeria, directing the inauguration of the National Assembly. This letter was read by the Clerk of the National Assembly, Alhaji Salisu Maikasuwa. The life of the eighth Assembly will come to an end on 9 June 2019.

The National Assembly has 469 members elected across the six geopolitical zones of Nigeria and the Federal Capital Territory (FCT). In specific terms, the House of Representatives is composed of 360 members and the Senate is composed of 109 members. The election to the leadership of both arms of the National Assembly produced a leadership that was not approved by both the leadership of the ruling APC and the executive arm of the government controlled by the party. The election into the eighth Senate of the National Assembly witnessed the participation of 57 Senators. This is more than one third of the constitutional requirement to form a quorum. Section 54, subsection 1 of the 1999 constitution (as amended) stipulates that only one third of the total Senators would be needed to form a quorum.

Thus, the election that produced Senator Bukola Saraki, who was nominated as candidate for the office of the Senate President by Senator Ahmad Sani Yerima, representing Zamfara Central Saraki's nomination, was seconded by Senator Dino Melaiye, representing Kogi West. Saraki's election conformed to the constitution regardless of the disposition of the leadership of the party. In line with a strategic arrangement that was perfected to achieve its agenda, the pro-Saraki group under the banner of Senators of 'like mind' had ostensibly entered into alliance with the PDP Senators. The tactical mistake was made by the Senate Unity Forum—a pro-Senator Ahmed Ibrahim Lawan, representing Yobe North, who was the party official nominee for the office of the Senate President, was to be busy attending a meeting outside the National Assembly building when the proclamation letter was written by President Buhari, directing the inauguration of the eighth Assembly.

Consequently, Senator Saraki took the oaths of office and allegiance led by the Clerk of the National Assembly, Alhaj Maikasuwa. In this respect, Senator Saraki assumed office as the Senate President, while the Assembly proceedings continued, leading to the emergence of Senator Ike Ekweremadu, the Deputy Senate President of the sixth and seventh National Assembly to be re-elected following his nomination by Senator George Thompson Sekibo. Senator Ike Ekweremadu had polled a total vote of 54 to defeat Senator Mohammed Ali Ndume, who polled only 20 votes. However, Senator Ndume was made the Senate Leader after he lost to Seantor Ike Ekweremadu. In less than two years into the life of the

Senate, Senator Ndume was relieved of his position as the Senate Leader, and Senator Ahmed Ibrahim Lawan was made the Senate Leader, obviously to appease the echelons of the APC and calm frayed nerves.

The election into the leadership of the House of Representatives saw a similar pattern of revolt against the directive of the party leadership to make Senator Femi Gbaja Biamila the Speaker. After the voting process, the Honourable Yakubu Dogara emerged as the Speaker of the Nigerian House of Representatives by 8 votes, having polled 182 votes with 1 vote above the simple majority stipulated by the 1999 constitution, against the Honourable Femi Gbaja Biamila, who, in spite of being the party official nominee, polled 174 votes and lost to the Honourable Yakubu Dogara. Instructively, two ballots were voided in the election. The Honourable Dogara was subsequently sworn in by the Clerk of the National Assembly, Alhaji Maikasuwa, and he assumed office as Speaker of the Nigerian House of Representatives, while the Honourable Femi Gbaja Biamila was compensated with the position of the Leader of the House of Representatives. But the outcome of the election in the House was grudgingly accepted by the leadership of the party with the excuse that no member of the opposition party had been made the Deputy Speaker of the House.

The election into the Nigerian Senate in 2015 was to draw more ire from the party hierarchy, obviously because of the emergence of Ike Ekeremadu, a member of the People's Democratic Party, as the Deputy Senate President. Instructively, for the 16 years that the People's Democratic Party leveraged on the government at the centre, no opposition member became a Deputy Senate President. But the PDP members leveraged on the intense division within the ruling APC members to clinch a key position in the Senate. Incidentally, the constitution was explicit that any member of either the Senate or the House could aspire to become the Senate President, Deputy Senate President, Speaker or the Deputy Speaker. But the party machinery and forced cohesion among members were employed by the PDP to get their members to occupy these offices for 16 years. Section 50 subsection 1(a) and (b) of the 1999 constitution had unambiguously emphasised that the Senate President, Deputy Senate President, Speaker and Deputy Speaker 'shall be elected by the members of the House from among themselves'. There is nothing that suggests that members of the party with the highest numbers of Senators and members have the exclusive right to occupy these offices. In the eyes of the law, the emergence of Ike Ekeremadu was not an affront to the constitution. It would have been if Senator Ike Ekeremadu had been made the Senate Leader or the Majority Leader of the Senate. Also, the procedure for the election conformed to Section 54 (1) of the 1999 constitution as amended since 57 Senators were in attendance for the election. Section 54(1) had explicitly emphasised that t

> he 'quorum of the Senate or of the House of Representatives shall be one-third of all the members of the legislative house concerned'.

It was in appreciation of the limitation imposed on the All Progressive Party by the constitution that made the party seek out a political solution to the crisis generated by the outcome of the election into the Senate. This reasoning seems to provide a basis for why Senator Ali Ndume was sacrificed to placate the party hierarchy; Senator Ahmed Ibrahim Lawan was compensated with the position of the Senate Leader after he had been outwitted by the incumbent Senate Leader, Bukola Saraki, who was never the party's anointed candidate for the position.

The 1999 constitution does not give room for independent candidates. Therefore, political parties are the only legitimate association through which one can seek election into the legislature or into the executive arm as either President or Governor. It is within this context that democracy remains a sine qua non for the existence of the political parties. Thus, political parties become platforms for the representation of people in the Parliament. The reason for this is that all members of the legislature have their membership of the different political parties. Hence, the incapacitation of the Parliament to engender good governance is a reflection of the lethargy that has characterised the political parties. One needs to emphasise that the cohesion of the party has always allowed the executive to use the machinery of the party (once it forms a majority in the legislature) to push policies that are sometimes detrimental to the national interest.

Though political parties differ in name and sometimes a thin line separates one manifesto from the other, they can also be differentiated by several criteria—by the internal organisation of the parties, and on the basis of the strength of the members and structure (Duverger 1969: 79). But the reality is that they are not fundamentally different from each other once they form the government. For example, it was hard to conceive that the APC-led government of President Muhammad Buhari would increase the pump price of fuel considering its opposition to such an increase that was initiated by former President Goodluck Jonathan's administration. It was also inconceivable that almost two years into the administration of President Buhari, the economy has further plummeted with inflation hovering around 18 % and the National Bureau of Statistics revealed that in the third quarter of 2016 unemployment rose to 13.9 % and underemployment to 19.7 %. Yet the APC launched a blistering attack on the PDP-led Jonathan administration for its inability to engender good governance—it was as a result of the change mantra of the APC that Nigerians voted for the party. Two years into the tenure of the APC government, it is difficult to see the effect of good governance and the promise the party offered to Nigerians, though the government seems to be doing well in the fight against corruption and the fight against the Boko Haram insurgency in the north-east. But the government has scored below average in the way it has handled the economy. The failure of the APC government to be ideologically different from the PDP is a clear demonstration of what the Italian political theorist Norberto Bobbio has termed the 'broken promises'—the 'contrast between what was promised and what has actually come about' (David, B. 2006, 1).

The Parliament has with full responsibility to make appropriate laws and carry out oversight functions on the executive. This is contained in Sections 58 and 59 of the 1999 constitution (as amended). Thus, the legislature is a critical institution in the democratisation process in Nigeria. Disappointingly, the lethargy associated with the party system in Nigeria has made the Parliament less open, transparent and accountable. Indeed, the Nigerian Parliament has lost five crucial qualities that were identified by (David 2006) as ones that every Parliament should have: representativeness, transparency, accessibility, accountability and effectiveness. This could not have been otherwise considering the comatose state of the party system in Nigeria. If the executive arm of the government wields considerable influence on the machinery of the political parties, the Parliament will become lacklustre as the party uses its caucus in the Parliament to guide deliberations in. The caucus ensures individual members' compliance to party directives—this is coupled with the fact that political parties are weakly institutionalised. This is apparent in the frequent splits and mergers as well as the formation of new parties at regular intervals. This development is further exacerbated by voters' tendency to shift from one party to the other, as reflected in the high rates of electoral volatility, which is sometimes reflected in the way politicians cross-carpet from one party to the other.

The tragedy of the Nigerian situation is that the country condones the excesses of the party system and demonises the Parliament for its prostrate state. Some would even argue that the haven of corruption is the Nigerian Parliament, which takes less than 5% of the national budget. Recently, former President Obasanjo lampooned the Parliament in this regard. But the reality is that the executive arm of government is not insulated from corruption. Regrettably, the coercive institutions of the government are controlled by the executive, and there is no arm of government that can successfully open up the haven of corruption in the executive better than the legislature. The underlying reality is that the constant friction between the Parliament and the executive is fuelled by the morbid desire of the executive to always use the party machinery to impose a pliant leadership on the Parliament with a view to ensuring that government is made less accountable and transparent. Once there is a compliant and subordinate leadership, the Parliament will be constrained to enact laws that would make the executive accountable and transparent.

However, it is not in doubt that Parliament has a vital role to play in engendering good governance. This is because it is the core institution of democracy and embodies the will of the people in government. The Parliament is the direct representative of the people and it carries the burden of conveying to the government the expectations of the people. Yet the Parliament has little or no confidence of the people in spite of its huge responsibility to carry out oversight on the executive arm of the government and to ensure it is accountable to the people. However, members of the Parliament are members of political parties that are weakly institutionalised with their caucuses in the Parliament. In addition, the executive often uses the party machinery to ensure that members of the Parliament that failed to toe the party line (usually dictated by the executive) does not get re-elected on the platform of the party. This is a strategic way of ensuring that the Parliament reflects the thinking of the executive. This has made the responsibilities of the institution more cumbersome.

Political parties cannot exist outside the framework of a democracy. Indeed, there are two critical institutions that are central to the deepening of democracy and the expansion of the public space, namely, the political parties and the Parliament. Essentially, political parties provide a framework and an anchor for the expression of competition in the form of elections. The Parliament, on the other hand, is the aggregation of the elected representatives of the people; it has the statutory responsibility of making laws for the good governance of the country. It is in this context that adherence by the political parties to the democratic ethos and the spirit of the constitution becomes important to the democratisation process, with a view to allowing the legislature to carry out its statutory responsibility of making laws for the good governance of the country.

The diversity of the country is reflected in the legislature. This is to ensure a sense of belongingness and to forge a sense of unity among the heterogeneous cultures that have formed the Nigerian state. This is reflected in the way the Senate is structured to have equal numbers of representatives of states, while the House of Representatives is a fair representation of the population strength of each of the 36 states and the Federal Capital Territory. In spite of this constitutional arrangement, the crisis of democratisation and the consequent low quality of governance in Nigeria is inextricably tied up with the absence of internal democracy among the political parties. The weak nature of political parties is reflected in the shrinking political space, one devoid of ideological complexity, culminating in a frail National Assembly that is usually involved in a leadership crisis and serial allegations of corruption.

Concluding Comments

Political parties provide the bedrock for any democracy. It is within the context of a political party that democracy flourishes and is sustained. Therefore, democracy can only be meaningful for the people when the political parties adhere to a democratic ethos by ensuring that the people becomes the centre of how decisions are reached within the party machinery. Hence, the trajectory of democracy in Nigeria lies in the arbitrariness with which political parties pummel the democratic foundations. The perversion of democracy is revealed by the wanton disregard for the powerless voices that dominate the membership of the political parties. This has prevented the voiceless members taking ownership of how leaders emerge within the party, and how decisions are reached. The excessive influence of money in who determines what happens within the political parties has given rise to the replacement of democracy by plutocracy. This is gradually what government is turning into in Nigeria.

Political parties constitute a critical institution in the search for participatory democracy. For democracy to flourish, political parties must embrace inclusivity, justice, fairness, transparency, accountability, respect for the sanctity of the electoral process and unquestioning obedience to the constitution, as well as the rules that govern the political parties. Once political parties show a commitment to democratic values, the Parliament will become more proactive and robust in carrying out its constitutional responsibilities to make good laws and make the executive more accountable and transparent and to strengthen the institutions of the state. Democracy will continue to be hollow if political parties merely constitute a vessel by which the political class access power.

The intense schism in the National Assembly and the irreconcilable differences specifically between the leadership of the House of Representative on the one hand and the former President Jonathan-led executive and the mainstream PDP leadership on the other was a consequence of how the leadership emerged without the support of the party and the PDP-led government. The main opposition party (the APC) played a critical role in the way the leadership emerged. The inability to resolve these differences heightened the tension and provided a context in which the executive mobilised the security forces (especially the Department of State Security Services (DSS) and the police) to invade the National Assembly and assaulted its independence in a botched effort to remove the Speaker of the House of Representatives, the Right Honourable Aminu Waziri Tambuwwal.

In conclusion, the methodology was also largely shaped by the literature on the political party system, the legislature and democratisation.

This influenced the analytical narrative here and the theoretical issues that explain why the political parties are incapable of deepening democracy and why the National Assembly is ineffectual in carrying out its responsibilities.

Note

1. A prominent Yoruba nationalist and one of Nigeria's finest politicians, who became Nigeria's Attorney General and Minister of Justice and later Minister for Power on the return to civil rule in 1999.

References

Bratton, M., & Van de Walle, N. (1997). *Democratic experiment in Africa regime transitions in comparative perspective*. Cambridge: Cambridge University Press.

David, B. (2006). *Parliament and democracy in the twenty-first century*. Geneva: Inter-Parliamentary Union.

Derbyshire, J. D., & Derbyshire, I. (1991). *World political systems*. Edinburg: Chambers.

Duverger, M. (1969). The concept of membership. In J. Blondell (Ed.), *Comparative government-A reader*. London: Macmillan.

Egwu, S. G. (2005). Preface. In Citizens Forum for Constitutional Review (CFRCR), Freedom of Association/Political Parties: A reform Agenda. A Memorandum, by CFRCR. Abuja: CFRCR, pp. 1–3.

Inter-Parliamentary Union. (2003). *Parliamentary Oversight of the Security Sector: Principles, mechanisms and practices*. Handbook for Parliamentarians No. 5. Geneva: IPU.

Joseph, R. (1987). *Democracy and prebendal politics in Nigeria: The rise and fall of the Second Republic* (p. 1). Ibadan: Spectrum Books.

Musa, Y. (2005). Party politics, electoral crisis and democratic consolidation in Nigeria. *International Journal of Academic Research and Reflection, 3*(1), 2015. ISSN 2309-0405.

Schimitter, P. (1992, March/June). The consolidation of democracy and representation of social groups. *American Behavioral Scientists, 35*, 422–449.

Seward, S. I. (1998). *The terms of democracy*. USA: Polity Press.
Simbine, A. T. (2002). *Political parties and internal democracy in Nigeria*. Centre for Democracy and Development Working Paper, CDD, Abuja.

Taofeek Oluwayomi Gidado teaches Sociology at the Obafemi Awolowo University, Ile-Ife. His areas of interest lies in political sociology, the sociology of development, criminology, sociological theories and research methods.

9

Exploring Social Interactions Between Chinese Immigrants and Corruption in Nigeria

Allen Hai Xiao

Introduction

Sometime in 2012, in the course of the fieldwork that resulted in this publication, the car of one Mr. Wang (a Chinese man living in Nigeria) was stopped by a policeman as he was driving back from a marketplace in Lagos. The policeman asked for the vehicle's documents one after another. Mr. Wang got angry and shouted, 'You just ask for money. How much do you want?' The policeman replied, 'Calm down, my friend.' Mr. Wang interrupted, 'I am not your friend. Money is your friend.' I witnessed this situation while I was sitting in the back seat of Mr. Wang's car, and I often heard similar stories from Chinese entrepreneurial migrants in Nigeria. They complained about suffering from the corrupt environment in Nigeria and the unexpected extra expense of negotiating with the Nigerian police. These Chinese representations of Nigeria's corruption, which imply complex social interactions between Chinese immigrants and

A.H. Xiao (✉)
Madison, WI, USA

A. Adeniran, L. Ikuteyijo (eds.), *Africa Now!*,
https://doi.org/10.1007/978-3-319-62443-3_9

Nigerian institutions, cannot be disregarded in the discourse about African development and China–Africa relations.

'Corruption' is always interpreted as a negative word. In popular discourse, Nigeria is regarded as a corrupt country by international institutions and societies. According to the 2015 *Corruption Perceptions Index* released by Transparency International, a non-governmental organisation that monitors corporate and political corruption in international development, Nigeria scores 27 out of 100 (a lower score indicates a higher rate of corruption); the country ranks 136th out of 168 countries. A number of scholars have noted various corrupt practices throughout Nigeria (Umez 1999; Nwankwo 1999; Akani 2002; Smith 2007; Okoosi-Simbine 2011, Pierce 2016). Instead of focusing on Nigerian actors, I will interpret the corrupt practices by both Chinese and Nigerians as social relationship practices and shed light on the socio-cultural implications of corruption in Nigeria.

In the Western media coverage of the Chinese presence in Africa, Chinese migrants are accused of bringing corruption to Africa: 'When these [Chinese] managers go abroad, they carry on bribing and undermine good governance in host [African] countries' (*Economist* 2011). This criticism has been rebutted in the Chinese state media that African corruption is pre-existing and China should not be blamed for it (Zhong 2011). The quarrels do not uncover the role of non-state actors at the micro-level and the complexity of everyday cultural and social interactions. Thus, this case study of Chinese immigrants in Nigeria will illustrate the nuanced Chinese–African relationships on the ground in the context of 'corruption' discourses and practices.

This chapter is based on fieldwork conducted in Lagos (July–August 2012 and July 2014) and Guangzhou (May–June 2012 and March–May 2014). I employed participant observation and open-ended interviews when I accompanied and interacted with Chinese businesspeople and their Nigerian business partners and clients. At the same time, I joined various internet communities where many Chinese businesspeople, who are involved in China–Nigeria businesses, and I conversed with them online extensively. I secured the informed consent of all key informants to disclose the information provided through either interviews or participant observation. All the names in this chapter are fictitious in order to guarantee the anonymity of the informants.

Corruption in Context

Corruption has been widely studied with regard to specific cases in many countries. When it comes to a universal definition of corruption, social scientists discuss it in different ways (Granovetter 2004; Smith 2007: 16–19; Uslaner 2008: 6–11). The main difference concerns whether corruption appears in different forms in different cultures and historical trajectories (Blundo and Olivier de Sardan 2006: 30–32; Larmour 2008; Ochonu 2011) or whether it can be measured against a standard across countries, as Transparency International does. Although I incline towards the former view, I treat the concept of 'culture' carefully, since it might suggest that corruption is rooted in some kind of traditional, and often romanticised, African past. Along with Smith (2007), I attempt to discuss corruption in the *contemporary* context of Chinese–Nigerian encounters.

Anthropologists specialise in analysing corruption from below (Haller and Shore 2005). Gupta's (1995) classic work examines the discourse of corruption and discovers how the state is conceptually represented at the lower levels of local bureaucracy. Due to the merits of ethnographic fieldwork, anthropologists not only collect narratives of corruption but also observe how corruption is practised (Blundo 2007: 27–52). Therefore, one of this chapter's objectives is to explore the implications of the Chinese representation of corruption in Nigeria and the strategic actions that Chinese migrants employ to cope with it.

Corruption in Nigeria is an everyday phenomenon. Some scholars emphasise the importance of vertically organised solidarity networks in Nigerians' relationship to the state, and argue that corruption hinges partly on the inseparability of individual and collective interests (Joseph 1987), even though anti-corruption campaigns have been initiated by political elites (Adebanwi and Obadare 2011). The fundamental social process in Nigerian authorities is characterised as 'prebendal' (Joseph 1987) or 'neopatrimonial' (Taylor and Williams 2008). On the other hand, Smith (2007) focuses on how ordinary Nigerians can be, paradoxically, active participants in the social reproduction of corruption, even as they are also its primary victims and its principal critics. This chapter will cover both officials, primarily at the lower levers, and ordinary Nigerians, and will moreover seek to shift the perspective from domestic Nigerians to migrant Chinese. In other words, it investigates whether Nigerian officials are able to grab personal profits not only from their national bureaucracy but also from incoming migrants, such as Chinese. Do ordinary Nigerians have another way of reproducing corruption when mediating Chinese migrants' interactions with Nigerian authorities? This exploration may help us to understand how Chinese migration to Nigeria worsens, betters or merely slightly alters the corruption in Nigerian society.

China has experienced corruption for decades (Lu 2000; Sun 2004; Meng 2013). Ordinary Chinese people are quite used to knowing that officials engage in corruption in many ways (Sun 2004: 26–37). One of major characteristics of Chinese corruption is the entanglement between corruption and *guanxi* (Luo 2011; Li 2011). *Guanxi* has a sense of 'social connection', a dyadic relationship that is 'based implicitly (rather than explicitly) on mutual interest and benefit' (Yang 1994: 1). The practice of *guanxi*, involving the exchange of gifts, favours and banquets, is ubiquitous in China. It is also referred to as the 'art of *guanxi*', 'because the style of exchange and the appropriateness of the performance are critical to its effectiveness' (Smart 1993: 399). Although social networking strategies are also widespread in other societies, *guanxi* is a 'specifically Chinese idiom of social networks, integrally linked to other building blocks of Chinese sociality' (Gold *et al.* 2002: 4).

There are negotiated boundaries between *guanxi* and corruption. Dunfee and Warren (2001) identify the close association of particular types of *guanxi* with corruption and bribery. Smart and Hsu (2007: 167–190) also note that tacts can serve to effectively disguise corruption and label the practice as 'reasonable' *guanxi* interaction. Following this, a relevant question is raised from this case study: how is Chinese *guanxi* in relation to domestic corruption affected by corruption in Nigeria when those Chinese businesspeople emigrate to Africa? International migration brings a new angle to looking at corruption in Nigeria. Much research has been done on the mutual influence between migrants and host countries (Portes and DeWind 2004). Nonini (2005: 167–200) indicates that Chinese petty capitalists encounter the varieties of everyday predation by the direct and indirect agents of the state of Malaysia, but he focuses on the state logics that restrict Chinese businesses to remaining small and hence in fierce competition among themselves. In this chapter, I do not show and have not clearly found such an intention on the part of the state. What I concentrate on instead are the strategies the Chinese migrants have developed to cope with everyday corruption in Nigeria.

Chinese migration to Nigeria began in the 1960s after Nigeria gained independence from the United Kingdom. Those early migrants from Hong Kong and Taiwan engaged in the manufacture of enamel, as well as textile and other light industrial production (Li 2000). Mainland Chinese migration began after Chinese reform and opening up in late 1970s, and increased greatly after 2000 (Xiao 2015). It has been reported that there are approximately 65,000 Chinese migrants in Nigeria (*Renmin Daily* 2014). In the past two decades, China has been investing extensively in many areas of the Nigerian economy, such as oil and gas, iron and steel, machine manufacturing, cement, electronics, motorcycle assembly, fishing, pharmaceuticals and telecommunications (Obiorah *et al.* 2009: 272–295).

It also should be noted that migration is a 'transitional process' with a 'recurring feature' (Waldinger 2008: 8). Migrants coming and going between two countries convey information and personal narratives to the people around them, who might become future migrants to the destination country. Moreover, the internet facilitates migratory processes. Apart from interviews with Chinese migrants, my investigation also collects testimony from blogs that share migrants' experiences in Nigeria and observes chat rooms where Chinese migrants exchange information with each other between China and Nigeria. All these data helps in analysing Chinese presentation and perception of Nigeria's corruption.

In summary, this chapter will discuss the implications of Chinese–Nigerian encounters with regard to corruption in Nigeria. Specifically, I will explore the questions below in following sections.

1. How do Chinese migrants represent Nigerian corruption?
2. How are Nigerians' corrupt practices affected by Chinese involvement?
3. How are Chinese practices of *guanxi* affected by Nigeria's corruption?

Chinese Representations of Nigeria's Corruption

Chinese migrants often get to know corruption in Nigeria before they go there, because many returned migrants relate their experiences to potential migrants around them and some migrants like to post their stories on the internet. Studying these narratives of corruption not only helps us understand representations and the constitution of the state (Gupta 1995, 2005), but also facilitates our understanding of corrupt practices and the strategies employed by 'victims' to cope with them (Perera-Mubarak 2012). In this section, I will identify three major ways in which Chinese migrants represent Nigeria's corruption; I will then discuss the implications of such representations.

The first way is to present personal diaries on the internet. I have found two main resources for reading the diaries that narrate how the authors experienced corruption in Nigeria and suggest how to cope with corrupt Nigerian officials: online fora, such as the Forum for Chinese in Nigeria (www.nigeriabbs.com/bbs/); and personal blogs maintained by experienced Chinese businessmen in Nigeria.

The Forum is the most active online community for Chinese migrants in Nigeria, and it covers business news relating to Nigeria and personal narratives posted by community members. I have collected 22 personal diaries in African Lives section and found that 11 diaries discuss corruption issues, narrating details of encountering immigration officials, customs and traffic police.

Mr Hu writes a popular personal blog (http://archerhuchina.blog.sohu.com/) among Chinese communities in Nigeria. He has been doing business in Nigeria since 2005. Instead of sharing how he suffered from Nigeria's corruption, he has offered suggestions for dealing with it. For instance, he wrote about how he got rid of a Nigerian policeman and avoided paying him a gratuity. At the end of his dairy, he summarised, 'we (Chinese) like to get out of trouble by paying money. The more easily we pay, *ogas* (bosses, referring to corrupt Nigerian officials) are more likely to stop us. If they find it troublesome to ask you for money, they will leave you.'

The second way Chinese migrants represent corruption is through instant messaging in the transnational process. QQ is extremely popular among Chinese businesspeople in both China and Nigeria. A QQ group is a chat room that people may enter only with the permission of the group's founder and administrators. When the word 'Nigeria' was searched as a keyword in the QQ system, almost 100 QQ groups came up with memberships in excess of 100 people. The largest group includes nearly 2000 members. According to preliminary text analysis of QQ chat records, I found that one of the more popular topics in these QQ groups is Nigeria's corruption. Some newly arrived Chinese businesspeople related their experience of being extorted by the Nigerian authorities and reminded other QQ group members not to make similar mistakes. Some more senior Chinese businesspeople shared their strategies for coping with corrupt officials.

WeChat is a mobile text and voice messaging communication tool popularly used by the Chinese. Compared with QQ, WeChat is more personal and is used among close friends. Through WeChat, one of my Chinese informants from a state company forwarded me this message when he knew I would go to Nigeria soon: 'Nigerian officials at the airport and customs are very greedy. They will come to ask for a gratuity if they suppose you are Chinese. I suggest you bring some Chinese herb balm or cigarettes. Try to give them this stuff before bribing. Usually, you should pay no more than 5000 naira[1] if you are lucky.'

The third way of sharing information about corruption in Nigeria, which is the most common, is to repeatedly mention Nigeria's corruption in daily face-to-face conversations. Compared to the methods described above, these narratives are more likely to shape people's perceptions. As I have indicated, Chinese migrants are quite mobile between Nigeria and China, so their stories about Nigeria's corruption are easily distributed among their colleagues, business partners and relatives who may go to Nigeria soon. Feige is a senior trader and likes to share his stories at gatherings every time he goes back to Guangzhou. When I attended a dinner party in June 2012, one young trader, soon to set off for Nigeria, asked for his advice. 'Keep a small sum of money at hand, as you get to the airport. If someone wants money, give him one dollar; if someone wants your life, give him five dollars', Feige joked. This exaggerative narrative reflects that corruption in Nigeria is being normalised by Chinese migrants in a way that is similar to what ordinary Nigerians do with the police at checkpoints (Smith 2007: 11–12).

These three ways of representing corruption in Nigeria by Chinese migrants have three implications. Gupta (2005) points out the constitution of the Indian state in representations of corruption. Likewise, I have detected the shadow of the states in the Chinese migrants' representations: on one hand, the Nigerian state has been stigmatised by those narratives and automatically labelled as a corrupt authority; on the other hand, the Chinese state's agents, the consulate and embassy in this case, are criticised by Chinese petty entrepreneurs due to their indifference to care about ordinary Chinese migrants' encounters with Nigeria's corruption (Xiao 2015).

Jeffrey (2002) indicates that low-level economic corruption in rural India causes material inequalities and the perpetuation of class divisions within rural society. The perspective of class indicates that we should examine different representations by Chinese petty traders and employees of big companies in Nigeria. In the field, I collected many narratives from petty traders about the privileges of big company employees, such as an escort service offered by Nigerian officials to exempt some individuals from checking bags and making inquiries, company reimbursement of gratuities and even extortion by Nigerian authorities, and much less harassment by Nigerian immigration staff due to agreements between the companies and the immigration authorities beforehand. From big company employees, I heard fewer narratives about suffering due to corruption, though they admitted that corruption is ubiquitous in Nigeria. That is to say, company employees depend on organisational negotiation with Nigeria's corruption, while indepedent entrepreneurs more likely develop individualised strategies, on which this chapter focuses.

Strategic practice has become a major theme in Chinese representations of Nigeria's corruption. Examining collected personal diaries, I found over time fewer accounts of suffering instances of it and more strategic suggestions — the main discourse gradually shifts from 'have to pay' to 'pay less', which manifests itself in the examples above. In other words, corruption in Nigeria has embedded itself in migratory lives and become an everyday phenomenon.

Strategic Interactions with Nigeria's Corruption

This section will illustrate how Chinese migrants strategically interact with corruption in Nigeria, and discuss how Nigeria's corrupt practices are affected by Chinese migration and how the Chinese practice of *guanxi* is affected in the Nigerian context.

On-the-Spot Coping Strategies

In Nigeria, the most 'notorious' and popularly recognised symbol of everyday corruption is the roadside police checkpoint, as in the story of Mr. Wang, which I related at the beginning of the chapter. On expressways or in some smaller cities, 'Nigerian police and soldiers routinely set up roadblocks, where vehicles are stopped, drivers are besieged, and passengers are hassled' (Smith 2007: 61). In metropolitan Lagos, traffic police often stand in the centre of road junctions and watch out for vehicles passing by. Many Chinese informants believe that they intentionally stop Chinese drivers and ask for a bribe. Thus, some of the Chinese would rather hire a Nigerian driver and even hide in the back seat. Nevertheless, cases of extortion by Nigerian police are still frequently narrated.

To cope with such corruption, many Chinese choose to send the police away as soon as possible with a sum of money. Qiu, a young petty entrepreneur, who had only been in Lagos for six months, was stopped by a policeman when he was driving home. Without clarifying the reason for being stopped, he gave the policeman 1000 naira and continued going home.

However, this strategy is sometimes criticised by senior migrants because they think it unnecessary. Mr Wang offered a suggestion to Qiu, 'If you meet the police, you just say "no English". Don't focus on the problem they mentioned.' He related a story to us: 'Once, a policeman stopped my car and claimed that I had violated traffic rules. He insisted on entering the car and asked me to drive to the police station. I gesticulated and indicated I couldn't understand a single word, and drove along with a circuit route. He found there was nothing he could do about me so I left without paying.' This story is similar to the strategic suggestion Mr. Hu offered in his diary. The central point here is to be patient when negotiating with Nigerian police, which makes them 'troubled' and tired of the effort of extortion.

If Chinese immigrants encounter Nigerian immigration officials, a swifter coping strategy is always used. No Chinese are willing to become entangled with immigration officials and patiently negotiate with them. An experience Chinese businessman told me, 'If you *dash* ('bribe' in the Nigerian context) on the spot, you can be set free immediately; if you are taken into cars, the price might be doubled; if you have been sent to immigration offices, you will definitely spend much more.'

The reason that Chinese migrants have different strategies to cope with Nigerian officials on the spot is due to their immigrant status. Most Chinese migrants enter Nigeria with a business visa, which restricts them to business visits and negotiations, and they are not allowed to work. In fact, they do work as traders and earn money by importing goods from China. Once this is discovered by Nigerian immigration officials, they have to be deported or pay *dash* to Nigerian officials. Thus, Chinese immigrants try their best to avoid being inspected by the immigration authorities. As for the traffic police, they have no right to check immigrant documents, so the Chinese are not afraid of negotiating with them. Therefore, the time spent on negotiation depends on whom they meet.

Then, how much should Chinese migrants pay for extortion? My informants like to exchange this information with one another when they meet up. They described the situation, and most importantly, shared how much they paid. People would discuss whether this case deserved this price and recalled the situations they encountered. Once, one of my informants was blamed for paying much more that he should have done, because some worried that this action would increase the normal price of *dash* and facilitate Nigerian authorities' greediness. The meeting helps Chinese immigrants avoid unnecessary expenses in specific cases and standardise encounters with Nigeria's corruption.

Chinese strategic interactions with corrupt Nigerian officials usually unfold according to a common script, with which many Chinese migrants are familiar with, even newly arrived migrants, not because of a similarity to corruption in China, but because they can learn from representations of corruption before they set off. Chinese corruption is manipulated secretly through symbolic gifts, private banquets and implicit *guanxi* networking (Hsing 1998; Sun 2004). Instead, Nigeria's corruption takes place in public with patrons overtly seeking rents from clients. Although expressed 'in the euphemism of a polite request' (Smith 2007: 62), it is a method widely recognised and regularly used by both patrons and clients in Nigeria.

Chinese immigrants are localising coping strategies. In order to reduce the expenses, they develop different strategies towards different Nigerian authorities. Instead of paying whenever they are asked, they prepare small sums for short encounters on the spot, as Feige suggested. Sometimes, learning local pidgin English, such as '*oga*' (boss) and '*wayo*' (cheat), helps them successfully negotiate with Nigerian officials. This negotiation on the spot, which is never seen in contemporary China, is actually an adaptation to the Nigerian context.

Nigeria's corruption would not be changed much if this localisation continues, because these Chinese immigrants act as ordinary Nigerians do with extortion on the spot. Moreover, standardisation of their strategies through representations of corruption helps to regularise corruption in the country. Patron-clientism is also normalised in the everyday interaction between Chinese immigrants and Nigerian officials. Despite complaints, the Chinese still participate in and repeatedly represent the corruption they encounter. All these interactions have developed into a seemly stable cycle.

Long-term Partnership

Not all corruption in Nigeria is temporarily encountered on the spot, and not all the Chinese immigrants passively cope with corruption. Some of them strategically establish long-term partnerships with corrupt Nigerian officials and benefit from them. Below show two cases of such Chinese-Nigerian relationships.

Zhao, a senior trading company manager, encountered a team of Nigerian immigration officials during a visa inspection. Zhao overtly proposed to establish a partnership in a visa business with Ade, one of the junior officials, after he was inspected. Now Zhao regularly collects passports from Chinese clients through his social networks in Chinese communities in Lagos and sends them to Ade for visa extensions. In July 2012, I visited him while he was answering a phone. After hanging up, he told me that he usually sent money, around 20,000–30,000 naira, to five senior immigration officials every month through Ade. He mentioned the phone call to me, adding 'this call is from one of the officials. He was expressing his appreciation. The money is not much, but I do it frequently.' Due to this regular bribery, the immigration officials never go to inspect him and his staff.

Min, a Chinese lady, is married to a Nigerian man and has lived in Lagos for three years. When she accompanied me to the airport in August 2012, at the customs counter, she actively chatted with a senior female Nigerian official about her Nigerian husband. The Nigerian felt so interested in Min's marriage experience and life in Nigeria that she did not inspect my luggage much and let me go quickly. Later, I learned that Min had exchanged phone numbers with the Nigerian lady after I left. In the following months, Min frequently interacted with the lady and would bring gifts to her from China every time she transported goods. In return, the lady offered a free escort service to Min at customs and allowed Min to bring in prohibited goods, such as certain food ingredients. Last year, Min even took the lady to Guangzhou and gave her free accommodation. Min's husband told me that the lady is now actually Min's friend and he felt comfortable with Min spending time with her.

Both cases show how a Chinese immigrant can conduct social networking in the Nigerian context and both of these interactions reflect the explicit nature of Nigeria's corruption. In China, social networking strategies (*guanxixue*) appear different: one must approach the person in charge through gatekeepers (Yang 1994: 99–101). However, in these cases, both Zhao and Min interacted with the officials directly and exchange benefits overtly without any symbolic or sentimental expressions.

The difference between these two cases is related to dimensions of *guanxi* practice, instrumentality and affectivity (Smart 1993; Yang 1994; Yan 1996). Zhao's strategy is instrumental, while Min's includes affective components, because her marriage to a Nigerian makes her incline towards a sentimental commitment to a Nigerian citizen. This relationship between Min and the Nigerian lady is rarely found in other cases of Chinese–Nigerian interactions. Like Zhao, most Chinese immigrants restrict the *guanxi* practice to instrumental actions.

However, being purely utilitarian and excluding an affective involvement sometimes does not make sense for Chinese immigrants in Nigeria. For instance, Mr. Yang employed a Nigerian policeman as his part-time driver, and paid him 55,000 naira per month. When the policeman was off work, Mr. Yang would require accompanying. Nevertheless, the policeman's performance did not satisfy Mr. Yang, because the policeman often asked Mr. Yang for *dash* as well. The policeman thought it was he who helped Yang to avoid more payments on the roads. Yang complained, 'If I want a driver, I can employ anyone on the street. I employ him to reduce expenses on the roads. I know he did me a favour, but this should not be paid. I actually treat him well in terms of salary and other businesses.'

The difference of opinion between Mr. Yang and the policeman reflects the different principles of reciprocity. In Nigeria, 'the patron is helping with the implicit expectation that their act contributes to retaining and strengthening the loyalty of their client…and the client will nonetheless frequently offer a "dash" to say "thank you"' (Smith 2007: 56). However, Chinese individuals would rather see such act as *renqing*, which means benefits outside a formal obligation that should be paid back as *renqing* on another occasion (Yang 1994: 122–123). The reciprocities, which imply different interactional logics, affect the mutual understanding of each "culture" of corruption.

In another situation, the Chinese practice of *guanxi* does not help towards a long-term partnership. Mr. Tang owns a trading company and employs four Chinese staff members in Nigeria. Instead of staying in Nigeria, as he had done when just starting the business, recently he has begun spending most of his time in China and has let his staff take charge of the Nigerian markets. He knew that keeping up a good relationship with immigration officials would be beneficial for his business in Nigeria, so he brought valuable gifts and even sent money to the officials every time he went to Nigeria (usually once a year). However, this did not seem helpful. Last year, his staff was frequently inspected by immigration officials and one of them was even arrested due to invalid documents.

One of his employees explained to me, 'Nigerian officials expect to hear from Mr. Tang regularly, even via phone calls, but Mr. Tang rarely gets touch with them as long as he is in China. The Nigerians felt offended so they take us as whipping boys.' I also collected similar narratives indicating that Nigerian patrons like to greet Chinese clients although there is no serious business to be done.

The failure of Mr. Tang's strategy indicates a misunderstanding of Nigerian clientism. The complete instrumentalist ignores the implicit affectivity in the Nigerian context. The patron–client relationship 'goes beyond a merely "commercial" one, becoming, for example, a joking relationship' (Blundo and Olivier de Sardan 2006: 62). Chinese migrants neglect such discursive affectivity but attach Chinese affective components of *guanxi* to gifts that symbolise and materialise interpersonal relationships (Yan 1996). Moreover, Yan (1996: 45) points out the temporal dimension of gift exchange in rural China: 'expressive gift exchanges… reflect a long-term relationship between a giver and a recipient; instrumental gifts are means to some utilitarian end and ordinarily indicate a short-term relationship.' Mr. Tang communicates long-term reciprocity via expressive gifts while Nigerian officials expect more frequent expressive communication, even daily interactions (Blundo and Olivier de Sardan 2006: 62–63).

Ordinary Nigerians as Agents of Corruption

Corruption in Nigeria is an everyday phenomenon in which ordinary Nigerians are involved. When it comes to Chinese immigrants in the context, some ordinary Nigerians interact with them as agents of corruption. The agency has two features: having local connections with specific persons or resources that Chinese migrants seek for; and having life experiences in China so that they have got social skills to interact with Chinese clients, which has been widely discussed (Xiao 2015).

In terms of the first feature, Nigerian agents can bridge corrupt officials and Chinese immigrants. Feige's neighbour Jeff is one such example. Feige became an agent of an Chinese electronic generator company and was assigned to sell a batch of stock. The Nigerian Jeff used to work in the local government in Lagos so he led Feige to a director in the educational department. Jeff helped Feige persuade the chief to purchase a generator for each member of staff in the department. Jeff and Feige suggested the director demand that his staff buy the generators and get commissions from each purchase. They made the price which the staff members had to pay a little bit lower than the price on the market so that the staff were happy to follow the director's proposal. Feige and Jeff also profited from this agreement. In this way, the generators were all quickly sold. Feige was delighted with this business: 'Jeff is so smart. It's easier to deal with corrupt officials in Nigeria than those in China.' For Chinese petty traders, it is difficult to get access to governmental procurement, thus Nigerian intermediaries must work in the process.

If ordinary Nigerian agents do not have connections with corrupt officials, they still have strengths in the act of brokering. The first is linguistic strength. A Chinese businessperson needs to submit an application to the Corporate Affairs Commission when they want to register a company. Nigerian agents can help draft these documents and submit them, because they know how to elaborate on ideas in accordance with the law and play tricks on the margin of the law. In July 2012, a Nigerian agent Udo helped Feige register a technical company, but included on the company registration documents such 'universal' phrasing as 'all the business of wholesale and retail sales of general goods'—the business scope obviously went far beyond what would normally fall under the remit of a technical company. However, this application was approved by the Corporate Affairs Commission, and Feige actually did all kinds of trading besides his IT business.

The second strength is identity. In some administrative processes, such as company registration and customs clearance, the involvement of Nigerians in Chinese business partnerships will facilitate the approval of the application. I know that one Nigerian agent played a role in Mr. Wang's company registration procedures. His name first appeared among those of the partners when Mr. Wang submitted an application. After it was approved, the Nigerian resigned his position and received a commission from Mr. Wang due to his involvement in the process. All the actions had been premeditated before the submission of the documents.

The third method of agency is based on power. The first and second strategies can be manipulated by almost all Nigerians, but this trick is only played by the Nigerians in power. Akin is a boss of a Nigerian hardware trading company. Although Ning, a Chinese trader, was not working in his company, Akin pretended to assign a position for Ning, because Ning wanted to get a one-year 'STR' visa (a working visa) and not be harassed by immigration officials due to problems working under a business visa. In return, Ning paid for this expatriate quota. According to Nigerian law, the approval of the expatriate quota in a Nigerian company is an essential condition for the application for STR visas.

Not only are Chinese migrants faced with Nigeria's corruption—ordinary Nigerian citizens also act as intermediaries in patron-clientism. Both Chinese and Nigerian individuals are ambivalent with regard to corruption, but they have different concerns. Nigerians' ambivalence about corruption is largely because 'they are pragmatic: the stakes for individuals in Nigeria are tied ideologically and materially to the social groups to which they belong' (Smith 2007: 225). Chinese immigrants are also pragmatic, especially petty entrepreneurs. They utilise the instrumentality of *guanxi* practice to cope with the Nigerians' pragmatism. That is why Chinese and Nigerian individuals can peacefully cooperate in the context of corruption. However, Chinese immigrants have a different social situation. The immigrant status, 'ideologically and materially', restricts an understanding of Nigerians' reciprocity and expressive interaction, which makes Chinese migrants pay much more attention to instrumentality.

The interaction between Chinese immigrants and ordinary Nigerians is permeating into many sectors in Nigeria, from trade to administrative paperwork, and results in a series of changes in the patron–client mechanism of Nigerian official sectors and Chinese immigrant communities. First, it facilitates social "equality" within Chinese migrant groups, which means that not only can big company employees approach Nigerian officials but petty entrepreneurs can do businesses with them as well. Second, it aggravates Nigerian officials' dependence on bribery and gratuities and probably leads to a tiered system of brokerage. Third, it allows ordinary Nigerians, especially those with lower levels of authority who have little connection to Nigerian officials, to profit from Chinese economic migration to Nigeria. As Smith (2007: 226) put it, 'individuals must make [the pragmatic choices] in the context of their obligations to deliver to their people whatever share of the national cake they capture', but I would add to this that now they have one more "cake".

Essentially, Chinese migrants develop interactional strategies to cope with corruption in Nigeria. They become accustomed to explicit on-the-spot extortion by Nigerian officials, which is different from implicit corruption in China, and localise their strategies that standardise how quickly to negotiate with Nigerian officials and how much to pay for *dash*. All these interactions normalise and regularise Nigeria's corruption for Chinese migrants as much as for ordinary Nigerians. Moreover, some of Chinese migrants try to establish long-term partnerships with corrupt Nigerian officials. They utilise the instrumentality of *guanxi* practice without any affective involvement, but they can misunderstand Nigerians' principles of reciprocity and discursive affectivity of interpersonal relationships. The participation of ordinary Nigerians in the interaction between Chinese immigrants and Nigerian officials impacts the patron–client mechanism, which is expected to continue in the future.

Concluding Comments

Returning to the popular debate on Chinese influence on corruption in Africa, it is meaningless to make the arbitrary judgement that Chinese migrants alleviate or aggravate the corrupt situation in Nigeria, but they do make the situation more complicated and involve entanglements with socio-cultural components.

Through the internet and face-to-face conversations, representations of Nigeria's corruption by Chinese migrants involve not only narratives of personal experiences but also knowledge transfer of specific coping strategies. Transnational mobility intensifies the normalisation of Nigeria's corruption in Chinese migrants' mind.

When this transnational migration happens, in other words, and when the implicit tact of *guanxi* practice encounters explicit corruption in everyday Nigerian lives, Chinese businesspeople adapt themselves to the Nigerian context by employing the instrumentality of *guanxi* practice, but the incompatible affectivities of interpersonal relationships between Chinese and Nigerians bring about misunderstandings. The strategic interaction situates Chinese migration well in the context of Nigerian patron-clientism, in which ordinary Nigerians also play a role as agents. This is how conviviality on the part of everyday African–Chinese encounters comes about (Lampert and Mohan 2014). Differences in interactional strategies between Chinese and Nigerians, on one hand, gives rise to the failure of long-term partnership in some cases, but, on the other hand, prevents further complicity between patron and clients.

To look forward to the future of Chinese–Nigerian interactions, can Chinese immigrants learn expressive reciprocity and can Nigerians learn the implicit tact of *guanxi*? Two facts I recently learned from my informants offer some clues. The first is that Ning is learning the Igbo language and has been given an Igbo name by his neighbour. He is also able to keep in touch with his partners on the phone in pidgin English. The second is that, besides Zhao, Ade has many more Chinese clients than before. He can distinguish to which interest group those Chinese businesspeople belonged and never mentions his connection with one group when he negotiates with their competitors. Thus, I expect that the mutual understanding of interactional strategies between ordinary Chinese and Nigerians will be deepened and strengthened in the future.

Does this mean Nigeria will become more corrupt as Chinese immigrants interact more easily and smoothly with Nigerians? I think this is a question for structural institutions in both Nigeria and China. As Smith (2007: 224) indicates, 'everyday life in Africa's giant is dominated by people's responses to corruption even more than by corruption itself.' We should not just blame ordinary people, whether they are Chinese migrants or Nigerian citizens, for corruption. It is only the change of societal institutions that might solve the problem. In the same vein, I do not agree that only the Nigerian state is responsible for it. An observer stated, 'The Chinese in Nigeria are incredibly corrupt, because the Nigerians in Nigeria are incredibly corrupt' (Sun 2013). This statement is partial because Chinese society has been struggling with corruption and has not fully handled it (Li 2012). If Chinese people were not familiar with the entanglement of *guanxi* and corruption in China, how would they practise it abroad? In this sense, both the Chinese and

Nigerian states, which shape structural institutions, are responsible for the issues of corruption.

Note

1. When I was doing fieldwork in Nigeria on this topic in 2012, the exchange rate between US dollar and the naira was 1: 170.

References

Adebanwi, W., & Obadare, E. (2011). When corruption fights back: Democracy and elite interest in Nigeria's anti-corruption war. *The Journal of Modern African Studies, 49*(2), 185–213.

Akani, C. (2002). *Corruption in Nigeria: The Niger Delta experience*. Enugu, Nigeria: Fourth Dimension Publishing Co., Ltd.

Blundo, G. (2007). Hidden acts, open talks. How anthropology can observe and describe corruption. In Nuijten & Anders (Eds.), *Corruption and the secret of law: A legal anthropological perspective* (pp. 27–52). Hampshire: Ashgate.

Blundo, G., & Olivier de Sardan, J.-P. (2006). *Everyday corruption and the state: Citizens and public officials in Africa*. London: Zed Books.

Dunfee, T. W., & Warren, D. E. (2001). Is guanxi ethical? A normative analysis of doing business in China. *Journal of Business Ethics, 32*(3), 191–204.

Economist. (2011). Africans are asking whether China is making their lunch or eating it. Retrieved August 15, 2014, from http://www.economist.com/node/18586448

Gold, T., Guthrie, D., & Wank, D. (Eds.). (2002). *Social connections in China*. Cambridge: Cambridge University Press.

Granovetter, M. (2004). *The social construction of corruption*. Presented at the conference The Norms, Beliefs and Institutions of 21st Century Capitalism, 8–9 October, Cornell University.

Gupta, A. (1995). Blurred boundaries: The discourse of corruption, the culture of politics, and the imagined state. *American Ethnologist, 22*(2), 375–402.

Gupta, A. (2005). Narratives of corruption: Anthropological and fictional accounts of the Indian State. *Ethnography, 6*, 5–34.

Haller, D., & Shore, C. (2005). *Corruption: Anthropological perspectives*. London and Ann Arbor, MI: Pluto Press.

Hsing, Y.-t. (1998). *Making capitalism in China: The Taiwan connection*. Oxford: Oxford University Press.

Jeffrey, C. (2002). Caste, class, and clientelism: A political economy of everyday corruption in rural North India. *Economic Geography, 78*(1), 21–41.

Joseph, R. (1987). *Democracy and prebendal politics in Nigeria: The rise and fall of the Second Republic*. Cambridge: Cambridge University Press.

Lampert, B., & Mohan, G. (2014). Sino-African encounters in Ghana and Nigeria: From conflict to conviviality and mutual benefit. *Journal of Current Chinese Affairs, 43*(1), 9–39.

Larmour, P. (2008). Corruption and the concept of 'Culture': Evidence from the Pacific Islands. *Crime, Law and Social Change, 49*, 225–239.

Li, A. (李安山). (2000). 非洲华人华侨史 (*A history of Chinese overseas in Africa*). 北京:中国华侨出版社 (Beijing: Chinese Overseas Publishing House).

Li, C. (2012). The end of the CCP's resilient authoritarianism? A tripartite assessment of shifting power in China. *The China Quarterly, 211*, 595–623.

Li, L. (2011). Performing bribery in China: Guanxi-practice, corruption with a human face. *Journal of Contemporary China, 20*(68), 1–20.

Lu, X. (2000). *Cadres and corruption: The organizational involution of the Chinese communist party*. Stanford, CA: Stanford University Press.

Luo, Y. (2011). The changing Chinese culture and business behavior: The perspective of intertwinement between guanxi and corruption. *International Business Review, 17*, 188–193.

Meng, Q. (2013). *Corruption in transitional China: A 33-year study*. Oisterwijk, The Netherlands: Wolf Legal Publishers.

Nonini, D. (2005). Towards a (proper) postwar history of Southeast Asian petty capitalism: Predation, the state, and Chinese small business capital in Malaysia. In A. Smart & J. Smart (Eds.), *Petty capitalists and globalization: Flexibility, entrepreneurship, and economic development* (pp. 167–200). Albany, NY: State University of New York Press.

Nwankwo, A. A. (1999). *Nigeria: The stolen billions*. Enugu, Nigeria: Fourth Dimension Publishing Co., Ltd..

Obiorah, N., Kew, D., & Tanko, Y. (2009). Peaceful rise and human rights: China's expanding relations with Nigeria. In R. I. Rotberg (Ed.), *China into Africa: Trade, aid, and influence* (pp. 272–295). Washington, DC: Brookings Institution Press.

Ochonu, M. E. (2011). Corruption and political culture in Africa: History, meaning, and the problem of naming. *The Law and Development Review, 4*(3), 27–58.

Okoosi-Simbine, A. T. (2011). Corruption in Nigeria. In S. Adejumobi (Ed.), *State, economy, and society in post-military Nigeria* (pp. 157–180). New York: Palgrave Macmillan.

Perera-Mubarak, K. N. (2012). Reading 'stories' of corruption: Practices and perceptions of everyday corruption in post-tsunami Sri Lanka. *Political Geography, 31*, 368–378.

Pierce, S. (2016). *Moral economies of corruption: State formation and political culture in Nigeria*. Durham: Duke University Press.

Portes, A., & DeWind, J. (2004). A cross-atlantic dialogue: The progress of research and theory in the study of international migration. *International Migration Review, 38*(3), 828–851.

Renmin Daily. 李克强牵挂中国在尼公民安全 (Keqiang Li cares about Chinese citizens' safety in Nigeria). Retrieved August 15, 2014., from http://politics.people.com.cn/n/2014/0508/c1001-24989237.html

Smart, A. (1993). Gifts, bribes, and guanxi: A reconsideration of Bourdieu's social capital. *Cultural Anthropology, 8*(3), 388–408.

Smart, A., & Hsu, C. L. (2007). Corruption or social capital? Tact and the performance of guanxi in market socialist China. In Nuijten & Anders (Eds.), *Corruption and the secret of law: A legal anthropological perspective* (pp. 167–190). Hampshire: Ashgate.

Smith, D. J. (2007). *A culture of corruption: Everyday deception and popular discontent in Nigeria*. Princeton, NJ: Princeton University Press.

Sun, I. (2013). Chinese businesses and corruption in Nigeria. Retrieved August 15, 2014, from http://www.chinaafricarealstory.com/2014/01/irene-sun-chinese-businesses-and.html

Sun, Y. (2004). *Corruption and market in contemporary China*. Ithaca, NY: Cornell University Press.

Taylor, I., & Williams, P. D. (2008). Political culture, state elites and regional security in West Africa. *Journal of Contemporary African Studies, 26*(2), 137–149.

Transparency International. (2014). Nigeria: Corruption and insecurity. Retrieved August 15, 2014, from http://www.transparency.org/news/feature/nigeria_corruption_and_insecurity

Umez, B. N. (1999). *The tragedy of a value system in Nigeria: Theories and solutions*. Bethesda, MD: International Scholars Publications.

Uslaner, E. (2008). *Corruption, inequality, and the rule of law: The bulging pocket makes the easy life*. New York: Cambridge University Press.
Waldinger, R. (2008). Between "here" and "there": Immigrant cross-border activities and loyalties. *International Migration Review, 42*(1), 3–29.
Wilson, S. (2002). Face, norms and instrumentality. In T. Gold, D. Guthrie, & D. Wank (Eds.), *Social connections in China* (pp. 163–178). Cambridge: Cambridge University Press.
Xiao, A. H. (2015). In the shadow of the states: The informalities of Chinese petty entrepreneurship in Nigeria. *Journal of Current Chinese Affairs, 44*(1), 75–105.
Yan, Y. (1996). *The flow of gifts: Reciprocity and social networks in a Chinese village*. Stanford: Stanford University Press.
Yang, M. M.-h. (1994). *Gifts, favors, and banquets: The art of social relationships in China*. Ithaca, NY: Cornell University Press.
Zhong, S. (2011). 拿出客观审视中非关系的勇气 (Objectively observe China-Africa relations). Retrieved August 15, 2014, from http://cpc.people.com.cn/GB/64093/64094/15595067.html

Allen Hai Xiao is a PhD candidate in the Department of Geography at the University of Wisconsin–Madison. He was trained in anthropology at the Chinese University of Hong Kong and has done extensive fieldwork on Chinese migration to Nigeria. He has published articles in *Politique Africaine* and the *Journal of Current Chinese Affairs*.

10

Emerging Governance Crises in Twenty-First Century Nigeria

Samuel Chukwudi Agunyai

Introduction

Since Nigeria's return to democracy in 1999, the country has witnessed five general elections (1999, 2003, 2007, 2011 and 2015). In 2015, Nigerians were eager to replace the ruling party, the People's Democratic Party (PDP) because of rampant bad governance, corruption, poor management of resources, underperformance, electoral fraud, and so on. All the indicators that shape good governance have never been observed and put into practice in Nigeria by its political leaders. Since 1999, the country has been plagued with cases of high profile corruption, poor service delivery, lack of accountability, the absence of the rule of law, fraudulent electoral process and unprecedented electoral and political violence. All these have been obvious symptoms of the governance crisis in the country. Nigeria, with its massive oil resources, has remained a seemingly

S.C. Agunyai (✉)
Department of Political Science, Obafemi Awolowo University,
Ile-Ife, Nigeria

A. Adeniran, L. Ikuteyijo (eds.), *Africa Now!*,
https://doi.org/10.1007/978-3-319-62443-3_10

failed nation due to bad governance. For instance, in spite of its huge resources, Nigerians are yet to see the efficient delivery of services in power, security, election administration, primary health care, justice and infrastructure. The accountability of leaders or of those who hold public office is non-existent. Also, there has been disobedience to the rule of law, in terms of non-compliance with court judgements by the political leaders, including the President. No doubt the President has an unmatchable measure of integrity with the political will to fight corruption, but this has to be done within the confines of the laws of the land. Any act or action that is not guided by the law is a threat to good governance. The rejection of court judgement in respect of the release of El-Zazaki, the leader of the Shiite sect in Nigeria, and Col. Sambo Dasuki is clear violation of the rule of law. The unlawful detaining of Nnamdi Kanu, the leader of Indigenous People of Biafra, despite an order by a competent court that he should be released, is unlawful. All these tend to reduce the trust that Nigerians place in the government.

Also, the electoral process has never been free of controversy and malpractice, while the appointment of incompetent public officials has repeatedly characterised the ministerial recruitment process in Nigeria. The fight against corruption has never been carried out with the intent to rid the country of this dreaded plague, but to punish a select few corrupt individuals. Political parties, on the other hand, have not fared better with regard to internal democracy. Activities of most parties in Nigeria have severely damaged good governance. Parties that win elections naturally use the state's resources to compensate their members via political appointments, irrespective of competence and qualifications. This accounts for the low performance, the poor service delivery and, in the long run, the governance crisis. Crisis in governance is inevitable where resources that are meant for service delivery are diverted to appease a selected few, or where the ruling government willingly receives defecting corrupt opposition party members into its folds for 'protection' from prosecution. As an example, the ruling All Progressives Congress (APC) has witnessed the cross-carpeting of former opposition state governors and notable politicians who were alleged to be mismanaging public money. The cases of Orji Uzor Kalu, the former Governor of Abia State, Joshua Dariye, the former Governor of Plateau State and Ken Namani, the former Senate President from Enugu State were among the numerous instances of self-interested defections to the ruling APC. Since gaining control of government, President Buhari and his party have failed to provide the country with good governance. The people voted in the last presidential election with greater hope that Nigeria's sordid socio-economic situation would change for the better, yet there has been nothing to indicate that good governance is being pursued.

Governance seems to be good when it endorses the promotion of the public good, elevates national interests over selfish motives and harmoniously controls the pressure from the demands of the cronies of public office holders. On the other hand, governance seems to be bad and crisis-driven when it strives only to protect the interests of a few people at the expense of the majority in a country. In Nigeria, there are obvious cases of bad governance because of the persistence craving for selfish interests over civic motives; this in turn has led to the stagnation of economic growth. Corroborating this view, Dahida and Akangbe (2013) aver that despite its enormous potential for development and its huge resources, the Nigerian economy has still witnessed a period of stagnant economic growth, and this has been partly blamed on corruption and the gross mismanagement of the country's vast resources.

Comparatively, there is a sharp difference between developed democracies and developing democracies in terms of the nature of governance. In developed democracies such as Britain and Canada, governance seems to be driven by the promotion of the public good, accountability, the rule of law and competency. Leaders or public office holders in these democracies have friends, loyalists and associates but it seems that they have devised a means of checking the excesses of their cronies. The strict application of the law, public asset declarations by public office holders, selflessness, a regard for competency and meritocracy as well as a strengthening of the law enforcement agencies and anti-corruption institutions are some of the measures used in checking the excesses or illicit actions of politicians' cronies in these countries.

In developing democracies like Nigeria, however, governance is commonly driven by placing self-interested motives over the national interest. The view of Odeh (cited in Nnamani and Illoh 2014) on governance was more revealing when he states that Nigerian society is configured by the interests of a limited number of elites, ethnic groups and friends—and not for the populace. Leaders and politicians disregard the rule of law, competency and merit for selfish reasons. In Nigeria, contracts are awarded on the basis of efforts contributed during elections and monies spent sponsoring politicians, with a total disregard for competency, integrity and merit. This has resulted in the governance crisis in Nigeria. In the same vein, the actions of cronies of politicians in terms manipulating elections results through electoral violence, intimidation of electorates and obvious cases of the snatching of voting materials, all in an attempt to favour their leaders in general elections, seem to aggravate bad governance. The consequence of this is that these leaders, when they gain control of government, seem to satisfy the interests of their cronies before meeting those of the public. This, in the view of Yahaya (1993), is classified as political corruption: bad governance seems to be caused as a result of electoral fraud, the rigging of election results, the awarding of false contracts, wealth acquisition through financial impropriety by political office holders, misuse and abuse of political or bureaucratic office, nepotism and tribalism. The foregoing notwithstanding, few intellectual resources have been deployed to investigating why the change mantra of the APC has not been able to provide Nigerians with good governance.

Theoretical and Methodological Framework

Neopatrimonialism states that there is a strong interdependence between the clients (cronies of politicians) and the patrons (politicians or public office holders) in that the patron depends on the support and effort of his clients to stay in power. Neopatrimonialism presents a two-way approach wherein the sponsors of public office holders and the sponsored political functionaries are in a symbiotic relationship. It is a political concept that reflects interdependence of funders of political campaigns and the beneficiaries of their sponsorship. Clients, in turn, obtain material benefits and protection. The theory further holds the view that resources are distributed through direct transfers in the form of the disbursement of cash, gifts and favours from politicians to their constituents. This theory sees the government or the politicians as a transfer pump: the government collects resources and distributes them to its supporters. These transfers benefit only particular groups connected to the politicians through patronage networks, at the cost of the rest of the constituents.

This theory is based on three basic structures: the 'ins', 'outs' and the government/public office holders. The nature of the interaction between the in-group political actors and the out-group political benefactors is basic to appropriate a peaceful political space, but necessary to progressive transformation of the larger society. They are of the view that the government derives its support by providing patronage to the 'ins' (such as clients and cronies) and funds this by taxing the 'outs'. While the 'ins' constitute the cronies of politicians, the 'outs' constitute the general public. This literally means that the politicians spend taxpayers' monies on providing patronage for their cronies (the 'ins'). The distribution of resources in neopatrimonial state seems to be primarily motivated by personal relationships or ethnic/tribal loyalties. In such cases, distribution can take the form of personal favours, such as appointing relatives or people from the ruler's ethnic/tribal group to important government positions. It is important to note that this phenomenon is predominant not only in Nigeria but in some African states. African leaders tend to deploy the patron–client model of neopatrimonialism to prolong their period in power and this inevitably degenerates into a governance crisis. This perspective could be further broadened by projecting the African political space as a conglomeration of personality-driven interests. Neopatrimonialism is a recurring feature of African politics. Accountability in neopatrimonialism regimes operates in two principal ways—through formal democratic mechanisms (such as elections, the rule of law, access to information etc.) and through informal pressures. Clients/constituents have the option of voting out the rulers if they fail to deliver on their campaign promises. It has been found that if the 'bargain' between leader and his followers is actually carried through then it is normal for the recipients to continue to support him. This theory is very relevant in explaining the emerging governance crisis in Nigeria.

Applying this theory to the governance crisis in Nigeria, the patrons, that is, the politicians/public office holders, naturally want to remain in power indefinitely by deploying means of resisting opposition from other party or politicians. For instance, in Nigeria, under President Obasanjo, politicians from the opposition party who were against his ambition for a third term were cruelly punished. Even those within his own party who were also against his ambition were threatened with anti-corruption agencies. Governance was in crisis during this period in Nigeria as public money meant for service delivery and good governance was looted to bribe parliamentarians to support Obasanjo's ambition, and money was collected but not refunded even after the ambition was discarded by lawmakers in the National Assembly. It also took the form of paying cronies in the form of compensating their efforts for success at the poll. The ministerial appointments of President Buhari were done in a bid to compensate those who sponsored his emergence as President. This accounts for the low performance, poor service delivery and low focus on good governance. At the state level, governors during gubernatorial election in their states loot taxpayers' money to galvanise their clients to vote them into power.

After gaining control of government, rather than appointing competent hands to handle public offices, they appoint their 'ins' (clients). This no doubt has been the cause of crisis in governance at the state level. Today, most state governors (for example in Osun, Oyo, Ondo, Ekiti, Imo and Bayelsa) in Nigeria cannot pay staff salaries because they have abandoned governance in favour of looting and personal aggrandisement. However, the theory stipulates elections, the rule of law and access to information as ways of making the patrons accountable. First, by election, the outs (public) can vote out patrons in government when there is crisis in governance. Secondly, the outs can depend on the rule of law to check the excesses of patrons and make them focus on good governance. Also, access to information can help inform the public about how the country's resources are managed. It will make them raise the alarm when they notice bad governance. Informal pressure from friends, religious groups or the public will make the patrons (politicians) focus. A good example was the voting out of the incumbent President Jonathan from office in 2015 by Nigerians who were tired his governance. Also, some concerned Nigerians went to court to sue the former President and some of his ministers over corrupt practices and abuse of office. Specific examples were the Ministers for Petroleum, Alison Maduekwe, and for Aviation, Stella Oduah. They were sued in the law courts by aggrieved Nigerians who felt that the ministers abused their office with corrupt practices. Apart from this, they were also invited to appear before lawmakers for questioning over alleged cases of abuse of office and corruption.

Primary data was collected among academics in professorial cadres in the field of governance, history, politics and public administration in universities of Ibadan and Ahmadu Bello, Nigeria. Data was collected through in-depth interviews (IDIs). A total of six male IDIs and six female IDIs were carried out for the study. The studies were conducted between 13 August 2016 and 26 March 2017. Information from secondary data complemented the data collected, which was analysed using content analysis.

Literature Analysis

Self-Interest/National Interest

In understanding what self-interest denotes in this context, it is important to begin with the understanding of the meaning of national interest. This will go a long way towards revealing the meaning of self-interest here as both concepts are direct opposites. National interests are the core values of a country that perpetually seek to promote the overall well-being and common welfare of the entire people, as against a few in a country. Mbah (2014), in the same vein, describes the national interest as that which covers a whole range of subjects from the sovereignty, unity and security of the state to communal amity, economic stability and standards of decency and morality. And, in the same process, Nwangwu and Ononogbu (2014) contend that the national interest includes factors such as ensuring the efficient and effective running of public services and an interest in ensuring the administration of justice. It is reasonable to argue that self-interest is motivated by a narrow concern with wealth, fame and power (Nwangwu and Ononogbu 2014).

In the same vein, this is a situation in which procedures governing the operation of state offices are used as fig leaves behind which state resources are appropriated; under the confiscatory system, government officials simply seize public assets without even attempting to camouflage their behaviour behind rules and procedures (Joseph 1999). It can be inferred that this view depicts three striking traits of governance in Nigeria. First, there has been a pervasive misappropriation of 'public' resources for personal or parochial gain. Second, such allocations have been patterned along ethnic identities and patron–client networks. Third, the distributive arena has been largely regionalised, and clientelist relations are diffuse. By the same token, Nwangwu and Ononogbu (2014) opine that if there is conflict between the national interest and self-interest interests among government decision-makers, the national interest suffers. It is interesting to note that good governance seems to be in play in a country where national interest is central, and where both leaders and the citizenry work towards achieving the national interest, which is their common objective. But when leaders and their cronies pursue their own interests, resources reserved for the public interest are diverted to private use, consequently leading to crisis-driven governance because there will be limited resources to meet public goods, which will in turn improve the quality of life of the entire citizenry.

Corruption

Within the disciplinary bounds of social sciences, the concept of corruption, like other concepts, has not been free from definitional disputations. The etymology of 'corruption' is derivable from Latin word '*corrumpo*', which means 'I break' (Oladele 2004). Corruption is an encompassing word which stands not only for institutional decadence but also includes moral and personal decadence such as bribery, favouritism, nepotism, cheating, sexual gratification and other unfair means adopted by any individual to extract some socially and legally prohibited favour (Dwivendi 1967; Akinbowale 2000). The famous international organisation on issues of public corruption, Transparency International, sees it as the misuse of entrusted power for private gain. The World Bank and its agencies describe corruption as the abuse of public office by a person or group of persons for private gain.

In the same vein, Lipset and Lenz (2000) have described corruption as the 'effort to secure wealth or power through illegal means—private gain at public expense, or a misuse of public power for private benefit' (p. 4). Because of the massive involvement of public office holders and their cronies in corruption, it has been described by Nje (cited in Ogoma 2014) as behaviour which deviates from the formal duties of a public role. It is a behaviour which violates the rules against the exercise of certain types of action for private gain, in particular regarding influence (Ogoma 2014). Elsewhere, the word connotes the lowering of one's integrity in the performance of a duty, especially in relation to the state or state-owned enterprises. It involves the act of an official or member of the judiciary who unlawfully uses his/her station or quality to procure some good for himself/herself or for another person, contrary to duty and the freedom of others. Perhaps indisputable is the fact that corruption, like any other social deviance or crime, necessarily involves a violation of some existing order or norm. Out of the numerous socio-economic and political ills and challenges that successive administrations have encountered in Nigeria's fourth republic, corruption seems to attract the greatest attention and concern (Yagboyaju 2010). This has probably grown out of Obasanjo's (cited in Yagboyaju 2010) classical conceptualisation of corruption in Nigeria as:

> The greatness single bane of (the Nigerian) society today will be tackled head-on at all levels. Corruption is incipient in all human societies and in most activities. But it must not be condoned. This is why laws are made and enforced to check corruption, so that society would survive and develop in an orderly, reasonable and predictable way. No society can achieve anything near its full potential if it allows corruption to become the full-blown cancer it has become in Nigeria. (p. 5)

It can be inferred from Obasanjo's view above that corruption, though it exists in all political systems and human activities, it constitutes the greatest cause of bad governance in Nigeria. It is also noted that the provision of law can be used to check and control it. By the same token, former Economic and Financial Crime Commissioner Farida Waziri in a conference proceeding shows forms of corruption and how it is manifest in Nigeria. She reveals that corruption takes any of these forms, although it is perceived differently from one geographical location to another. The following behaviours indicate or are red flags for corruption: embezzlement, conflict of interests, the awarding of contracts by public office holders to cronies and personally held companies, bribery, fraud, political corruption, nepotism or favouritism, ethnicity, the rigging of elections, the misappropriation and conversion of public funds for personal gain, bureaucratic corruption, extortion, the manipulation of procurement processes by over-inflation of contracts, leaking of tender information to friends and relations, and misappropriation of corporate funds through manipulation or falsification of financial records. The stories of Cadbury Plc and the Vaswani Brothers locally and Enron, Anderson and Anderson internationally are all clear examples of payment for favourable judicial decisions (judicial corruption) (Waziri 2010).

Given this discussion, it could be argued that corruption in Nigeria, whichever way one looks at it, is caused by greed and materialism, cultural and moral decay, a lack of contentment, poor law enforcement to prosecute corrupt persons, a lack of political will to punish corrupt persons or selective justice in the handling of corrupt cases. In line with these claims, the 4th Economic Summit in Nigeria in 1997 recognised the following as the causes of corruption in Nigeria: low income levels; ineffective control measures; and the organisational structure within Nigeria's civil service.

Governance

'Governance', just like the concept of national interest/self-interest and corruption analysed above, is used today in many different senses. According to Olowu, Williams and Soremekun (cited in Yagboyaju 2010), some of these include its application as:

a term to describe regime types, or the nature of the relationships between state and society; as a set of norms to appraise governmental systems and prescribe appropriate or acceptable practice of the manner in which power is utilised. (p. 18)

In recent times, 'governance' is also used as a synonym for 'democracy'. In this sense, it is used to denote whether a society is democratic in nature or not (Yagboyaju 2010). For example, a democratic society is one 'which possess a wide variety of state and society institutions which interact with one another to ensure that state institutions are efficient, responsive, accountable, transparent and bounded by the rule of law' (Olowu *et al.* 1999) (p. 5). If the foregoing view on governance is applied to the process of democratisation, then it means this process can be depicted as 'the movement from the dominance of state-society relations by one institution (usually the executive branch), "monocracy", to a polycentric structured society' (Yagboyaju 2010) (p. 20). This is important because of the huge cases of corruption, the subjection of the national interest for private motives and illicit actions by politicians' cronies manifested in electioneering fraud in terms of the manipulation of elections results, electoral violence and the unending craving for private benefits, especially in the fourth republic. This view was aptly corroborated when Bayart *et al.* (1999) aver that at the height of the absurdity in its operation and the exploitative acts of its own officials, the state in Africa was rapidly 'criminalized, while its officials were merely involved in the politics of the belly'. In other words, it means that the rapid increase in such negative practices as the privatisation of public institutions, maintaining private armies and the connection of national operators with global criminal networks are all features that the state itself was becoming a 'vehicle for organized criminal activity' (Bayart *et al.* 1999).

In the same vein, Collier (1982) had earlier described most of these views in the foregoing analysis as being different from the traditional opinion, which essentially links governance to the activities of 'specific occupants of public office who are in apposition to make binding decisions at any given time'. Governance in this sense presupposes that the totality of what public officials do or what they refuse to do is for the improvement of the generality of the people. In the light of this, it is not surprising that governance is often a conflict between the ideal and the actual facts on the ground (Yagboyaju 2010).

Bad governance is characterised by a lack of accountability, widespread corruption and rent-seeking, arbitrary policy-making, avoidable bureaucratic bottlenecks, inequitable and unenforced legal systems, the abuse of executive power, and an apathetic and uninterested civil society distanced from public life. In contrast, good governance features transparent, predictable, open and enlightened policy-making, a bureaucracy imbued with professional ethos acting for the furtherance of the public good, the rule of law, transparent processes, a respect for human beings and their individual rights, and a strong and engaged civil society participating in public affairs.

On a general note, the various analyses presented above should be sufficient to draw attention to the importance of the state in contemporary African politics. Indeed, in the early 1980s, there was a profound agreement among analysts regarding the importance of the state in Africa 'as the focus for the institutionalization of central power' (Chazan and Rothchild 1992). Incidentally, the opinion on the centrality of the state in Africa is also at the core of the debate that has alternately depicted the post-colonial state as 'weak and powerful', 'repressive and feeble', 'fragile and absolutist', 'dependent and autonomous', 'expanding and collapsing' (Chazan and Rothchild 1992) (p. 1). All of these are the bedrock of the perennial governance crisis confronting Nigeria and other parts of the African continent (Yagboyaju 2010). Although some elements of these challenges exist in every system across the globe, there is evidence of safety mechanisms which prevent them from fatally injuring the developed and advanced democracies (Rose-Acherman 1999). The effects and consequences of all these variables can better be understood in subsequent discussions, especially in the section on the effect of bad governance on democratic consolidation. This view is in line with that of Hilary Benn, a former British Secretary of State for International Development, who contended that 'corruption is both a cause and a consequence of poor and outright bad governance'.

Having analytically, though in a synoptic fashion, perused what constitutes self-interest/national interest motives, corruption and governance, it is appropriate to examine the effect of the governance crisis in

Nigeria. This is what constitutes the core of the discussion in the next section.

Outcomes of the Governance Crisis in Contemporary Nigeria

It is obvious that Nigeria faces a sustainable development crisis located at the nexus of governance and undue influence of the actions of politicians' cronies in terms of the looting of public funds for private interest. To tackle this crisis, first it is necessary to diagnose the actions of politicians' cronies that aggravate crisis-driven governance properly and then to confront this fully. In trying to trace the problem of bad governance and corruption, Odinkalu (2010) opined that:

> A major source of difficulty is the fact that the transition from intrusive, atomised communities to the anonymous society built on the Westphalian model of the nation-state took place all over the continent without creating mechanisms to infuse or enforce these richly rooted African values in governance or its institutions and processes. Thus we have evolved a bi-furcated system of public ethics that appears to subvert our community values at their point of contact with the post-colonial African State and, in so doing, denudes the State of both relevance to the people and the will to protect their best interests. This is not a cultural problem, however. Rather it is a problem of institution and state-building. (p. 16)

A critical look at the views of Odinkalu suggests that bad governance and corruption in Africa, and particularly in Nigeria, became more visible following colonialism. This is partly because there were no mechanisms to inculcate African values and ethics or morals into governance and available institutions, thereby making the people prone to all forms of corrupt practices. In explaining the effect of the actions of politicians' cronies on bad governance, three cases of their actions were used. These include bad governance resulting from the pervasive misappropriation of state resources for individual gains, resources allocation based on ethnic identities and regionalised distributive arenas.

The Governance Crisis Resulting from the Pervasive Misappropriation of State Resources in Nigeria

How has a country so richly endowed blown its opportunities for itself and its generations yet unborn so spectacularly? In addressing this question, cases of crises emanating from the misapplication of state resources in Nigeria are presented in this section.

Godfatherism and Governance Crisis in Nigeria

Following the argument of the neopatrimonialism theory, which holds the view that a governance crisis or bad governance is inevitable when the actions of actors, including the cronies of politicians, are based on 'give and take syndrome'. The case at hand to buttress this view is that of 'godfathers' and their sponsored politicians in Anambra and Oyo States. For instance, in Anambra, Chief Chris Uba was the godfather who sponsored Chris Ngige who won the 2003 gubernatorial election in the state. Uba provided the cash and the means through which Ngige, his godson, got to the governorship position:

> [F]or so many months, Ngige, the then governor, abandoned governance for battle over who becomes the 'superior' in the state with Uba. While we waited patiently after voting to see service delivery in terms of security of lives, job creation, primary health care, rule of law and accountability, what we see then in Anambra was threat to our lives through repeated gun shots from political thugs of Ngige and Uba. (IDI, male, professor of politics, 55 years, University of Nigeria, 25 March 2017)

The relationship between these two people (Uba and Ngige) later cracked and this led to a governance crisis in the state. This means that there could be breakdown in the relationship as any one of them, especially the godsons (politicians), could fail to fulfil an agreed decision. Amanbra State under the leadership of Chris Ngige went through a turbulent period, which almost led to the collapse of governance in the state, when Ngige refused to give Uba the chance of appointing commissioners and paying him the money that he (Uba) claimed he had spent on making Ngige the Governor of the State. The crisis got worst when the sitting Governor, Ngige, was abducted by Uba using the Nigerian police. This view was aptly corroborated when Onwumere contends that:

> A self-confessed godfather, Uba employed thugs and Nigerian police to abduct his godson, Chris Ngige, who was the elected governor of Anambra State. Ngige's sin was his refusal to allow Uba to nominate all political appointees, take the largest share of state's allocation, and instantly pay him a sum of N2.5 billion; the claimed cost of installing Ngige as governor. (p. 13)

A critical look at the above assertion, it can be deduced that Uba gave (paid) all the money needed for the ascension of Ngige to the post of governor of Anambra State, having calculated the benefits and gains he stood to receive in return for sponsoring Ngige. He had calculated that he would be given in return for his money the opportunity of appointing commissioners and getting the lion's share of the state's monthly allocation from the federation account, but the Governor refused to fulfil that promise. This however, led to a governance crisis in the state, because the Governor was abducted for some days and this in the long run adversely affected governance in the state, such that lives were lost and properties were burnt to the ground. This view is expressed in the submission of the following female IDI:

> I lost my car to the war between Ngige and Uba…I also got injured. The insecurity was much in my state then, I was ordered by hoodlums to vacate my car and was beaten seriously while on my way to Nssuka. It actually terrified my entire family who thought I was already dead. (IDI, female professor of governance, 57 years, University of Nigeria, 25 March 2017)

Prevalence of godfatherism in politics is usually at the detrimental of well-being of the larger society. It was not a case of two individuals fighting, but a desperate godfather (Uba and his 'troops') consuming within their reach when it became clear that his investment had been wasted. The only solution the federal government proffered was the threat to declare a state of emergency in the state.

The foregoing scenario is similar to the case in Oyo State where the self-acclaimed godfather, the late Alhaji Adedibu, fought and masterminded the impeachment of his godson Ladoja for refusing to allow him to appoint commissioners, and share in the state monthly allocation and security vote of the state. This action led to a continual governance crisis in the state. For instance, it led to hostility between the Governor and his deputy, Alao Akala, who later became the Governor after the ousting of the Governor by the late Baba Adedibu. This enmity continued between the Governor and his deputy even when the court quashed the impeachment and returned Ladoja as the Governor.

> [T]he crisis between late Adedibu and Ladoja was over the sharing of security vote. Baba Adedibu felt governor Ladoja was not remitting enough to him from the state's security vote and he planned for his removal from office. During this period, governance was jettisoned for excessive politicking. Governance was already threatened by the diversion of state's fund to appease Baba Adedibu, yet Baba was still making unbearable demands on Ladoja. This obviously threatened the ability of the governor to provide service, respect the rule of law and be accountable to the people. Citizens were harassed, intimidated and severely beaten by political thugs of both godfather and godson in Oyo State. There was a feeling of animosity between Adedibu and Ladoja, Ladoja and Akala, the deputy governor. (IDI, female lecturer, Political Science, University of Ibadan, 54 years, 27 March 2017)

The analysis above shows that a governance crisis is inevitable where politicians and their sponsors fall out over issues relating to political appointments, their share of the state's treasury and control of state apparatus. These, in the long run, create tension, insecurity, a breakdown of law and order, and abuses of human rights. Godfatherism is capable of thwarting the democratic growth of Nigeria.

From the foregoing analyses, it is evident that godfatherism threatens good governance in the states (Anambra and Oyo) considered. This, however, means that a governance crisis is inevitable in a state where money meant for the entire people is spent on the demands of godfathers, which in turn adversely affects accountability, service provision to the people, rule of law etc. The rule of law is thwarted when godsons abandon the relevant laws guiding their roles and position as Governor in granting the demands of their godfathers or when the godson (politicians) fail to account for the state's resources. This in the long run affects the living conditions of Nigerian citizens.

Electoral Malpractice and Governance Crisis

In the unfolding fourth republic in Nigeria, five general elections have been conducted: specifically, in 1999, 2003, 2007, 2011 and 2015. Numerous political functionaries, especially incumbent politicians, have been guilty of squandering the state's treasury in maintaining and sustaining political thugs and party loyalists across the country. Considering that neopatrimonialism, which is predicated on the 'give and take' syndrome, holds the view that interdependence between the cronies and politicians is not always one-directional, this means that there can be a breakdown in the relationship between the politicians and their cronies. And when this is the case, a crisis in governance is inevitable. It can be inferred that the cronies, in exchange for juicy political appointments from the politicians, will give their time and energy to the snatching of ballot boxes, the intimidation of voters and the wanton killing of political opponents for the success of their leader at the poll. On one hand, this in itself constitutes a governance crisis—this is because when the politicians attain power, they tend to loot public funds to satisfy their cronies' demands. On the other hand, if there is breakdown in the relationship, for instance, if the politicians fail to fulfil their promises, that is, if they fail to give while they (politicians) take positions of power, it will still result in a crisis in governance.

> At the state level, governance crisis is worst. Most state governors rigged their way into government through the despicable activities of their cronies

> during election. And when they are declared the winner of the election, the governor is bent at satisfying these cronies at the expense of masses. They loot treasury to meet the unending demands of their cronies, they consider their cronies first for any available appointive position regardless of competency and they unduly influence the police to shield their cronies from prosecution. All these prompt some state governors to spend tax payers' money in sustaining these cronies. This specifically undermines good governance. (IDI, male, Professor of Public Administration, Ahmadu Bello University, 60 years, 26 March 2017)

Most politicians, especially incumbent ones, routinely recruit thugs during elections and squander the state's resources in the process, all in an attempt to return as President, Governor or Local Government Chairman. They tend to give these thugs whatever they demand provided that they retain their seat of power. Buttressing this view, Yagboyaju (2010) gives examples of political assassinations that are related to electoral violence in Nigeria: Chief Layi Balogun, one-time presidential aspirant, who was killed in 2000 while preparing to make submissions before the Oputa Panel on human rights abuses in Nigeria; Chief Funsho Williams and Dr. Ayo Daramola of PDP in Lagos and Ekiti States, respectively, were assassinated for their political views. During the 2007 general election, the actions of the cronies of some politicians did cause electoral violence. For instance, in Kano, the actions of some political contenders led to the burning of the Panshekara police post, in the Kumbotso Local Government Area, and the killing of the Divisional Police Officer, his wife and eleven others, due to delay in the announcement of the results of the state governorship election.

During the 2015 general elections, there were cases involving incumbent governors, who misappropriated the state's resources on cronies in a bid to retain their seats of power. This has in no little measure caused electoral upheavals in Kaduna and Bauchi States. This position is further buttressed by the submission of the following IDI:

> [W]hat we had in Kaduna state during 2015 gubernatorial election was war between cronies of the then Governor Mukhtar Ramaian Yero and supporters of Nasiru Ahmed El-Rufai. During this period, there was no focus on governance but politicking. The state's treasury was looted to

maintain political thugs who disrupted the election and hijacked election materials. They were resisted by El-Rufai's supporters, but the truth is that we were terrified and some of us cannot vote because of fear of being intimated by these hoodlums. There was total breakdown of law and order and insecurity in the state. This adversely affected governance in the state. (IDI, female, professor of policy science, Ahmadu Bello University, Zaria, 65 years, 26, March, 2017)

Appointive Positions and the Governance Crisis

Good governance thrives in a society where the rule of law is respected, public office holders are accountable, efficient service delivery is prioritised, a transparent electoral process is in place, there is an independent judiciary and public office appointments are based on competency, expertise and integrity. However, when the appointment of ministers and cabinet members is done on a compensatory basis, then the mismanagement of government ministries and agencies will become inevitable. In Nigeria, successive governments and administrations have continually recruited people, some of whom are grossly incompetent to handle the posts they have been appointed to. For instance, ministerial appointments were mainly done on the basis of political compensation, not merit.

The administration of President Buhari has not been spared the compensatory recruitment of public office holders. Ministerial appointments under his administration have been largely criticised for being used as compensation for party members. This, however, accounts for the low performance of the ministries in terms of service delivery to the people. At the lower level of governance, especially at the state level, the recruitment of incompetent party members to head important government offices has undermined good governance. As an example, most state governments tend to compensate those who support them at the poll with caretaker committee appointments. Such committees are made up of politicians' cronies who had given support to the government at the polls. The case of states like Osun, Anambra and Imo are quite apt here. The majority of these states had jettisoned the constitutional provisions for having local governments in place. This development has been corroborated by various observations and evidence from the literature, which reflects how various state governments have blatantly refused to conduct local government elections. Instead, they have resorted to the use of caretaker committees, which are mainly composed of party loyalists and sometimes members of their own ethnic groups. This has been responsible for the lack of participation in politics and poor governance at the grassroots level. One of the ingredients of good governance is participation—how can there be good governance when the people are not allowed to elect their representatives?

> The existence of the appointed caretaker committees does not promote good governance as they are merely in place to satisfy the governors, who must have given them the job, and not the people. (IDI, male, senior lecturer, Peace and Conflict Studies, 50 years, University of Ibadan, Nigeria, March 2017)

Resource Allocation Based on Ethnic Identity

This is a situation where politicians or public office holders give preferential treatment to people from their own ethnic group, specifically in terms of appointed positions and allocation of resources. This has often been due to massive numbers of votes or support such politicians receive from such people/cronies during the general elections. Neopatrimonialism aptly explains this since the cronies, who are from the same ethnic group as the politician, seem to get government attention, appointed positions and other sought-after jobs irrespective of their competence and qualifications, and at the expense of more qualified citizens.

> [I]f you look at the first political appointments made by President Buhari, it was based on ethnic sentiment. This attracted serious criticism of his administration. A country like Nigeria with a multi-ethnic configuration that practices a federal character needs to spread political appoints across those ethnic lines, but must appoint the competent ones and not just based

> on compensatory motive. There is nothing wrong appointing ministers, service chiefs, and special advisers from different ethnic groups, but the point is that they must be competent. Since we practice a federal character, it must be followed during appointment in order to enhance national integration and a sense of belonging, a key ingredient of good governance. (IDI, male, Professor of History, 51 years, University of Nigeria, March 2017)

For instance, the first set of appointments made by President Buhari was largely criticised on the grounds that it favoured the northern part of the country to the detriment of the south.

Regionalised Distributive Arenas

In Nigeria, the regions (states or geo-political regions) form the basis of sharing resources. This has brought to light the federal character principle, which was introduced in 1979. This principle was introduced to give a semblance of order to the distribution of privileges and benefits among the federating units in such a way that no group has predominant and exclusive access to the opportunities and benefits inherent in the exercise of state powers (Ekeh 1989). Today, this principle seems to be used to promote ethnicity rather than solve the problem of ethnic identity. Thus, according to Usman (1979), federal character has deepened the reach of ethnicity and extended its ravages. In effect, federal character is the solution that has deepened the problem it was devised to tackle (Akindele *et al.* 2002). The foregoing has shown that resources in Nigeria seems not to be distributed on the basis of national intentions but for regionalised motives. It has become a situation where states or geo-political regions perpetually strive by different means to gain a considerable share of the country's resources.

> [T]he intention of the federal character was to foster a sense of belonging by distributing resources based on equity, fairness and competency. Even when we based our recruitment on the principle, competent people from each ethnic group should be reached out to, not mediocre. If this is done,

> it will go a long way to enhance good governance. If our ministries, commissions and boards are manned by competent hands, there will be efficiency in service delivery, accountability, and transparency in electoral process. For instance, look at when Prof. Jega was the chairman of INEC [Independent National Electoral Commission], the commission was very capable in handling elections, but today the stories we are hearing about elections at state level sincerely put the competency of the new chairman in question. (IDI, male, Professor of History, Ahmadu Bello University, 43 years, March 2017)

This seems to be the bane of good governance in Nigeria because the application of ethnic sentiment in the distribution of societal resources, rather than national considerations, will amount to the substitution of competency, qualification and merit for mediocrity. Of course, the following IDI's submission sheds more light on this:

> [G]overnance in Nigeria, especially at state level and local level, is in a serious mess. Governors and local government chairmen have shifted their focus to politicking and have abandoned governance. Governance represents respect for law, efficient service delivery, accountability, transparency and competency. How many of our governors or chairmen are accountable to the people who voted them into power? Why are we still experiencing poor service delivery? In some cases, there is no service delivery. It is in Nigeria that people pay for electricity that they do not make use of, our hospitals are not well equipped, roads are not in proper shape. Look at the Abuja airport, it now that the government remembers to fix it, when it is long overdue for repair. The government is just wasting our taxpayer money on excessive politicking. Now Buhari is already talking of the 2019 election, when he has not delivered good governance to the people. It is really a pathetic situation. (IDI, male, professor of history, 65 years, University of Ibadan, 27 March 2017)

In the same vein, Akindele *et al.* (2002) note that when the resources of a country are distributed on the basis of 'regionalised arenas', such a process destroys national loyalty and collective consciousness. The idea of unity and one Nigeria, which post-colonial political leaders like Obafemi Awolowo, Aminu Kano, Joseph Tarka, Okoi Arikpo, Anthony Enahoro and so on had nurtured has vanished into thin air. Thus, it has led to prevalent fragmentation of loyalties, elitism and regional allegiances in the country (Akindele *et al.* 2002).

Policy Redress for Bad Governance in Nigeria

There is need to deploy an imperative control mechanism as presented by the theory of neopatrimonialism. Elections as one of the control mechanisms can be used to vote out politicians squandering the state's resources for their cronies. In Nigeria, during the 2015 Presidential election, the incumbent President was voted out of office as a result of numerous allegations of corruption, abuses of office, looting of funds and placing cronies' interest over the national interest. At the state level, some states, such as Bauchi, Jigawa and Kaduna, had their sitting governors removed from office through a vote. The point being made here is that, as suggested by the theory of neopatrimonialism, Nigerians can rely on elections to remove leaders who loot public funds. This will serve as a deterrent to other politicians who may want to loot public in the future.

> [I]f you look at the 2015 Presidential election, you will notice that people were tired of the governance of President Jonathan. He was voted out of office because he was not able to address cases of corruption levied against some of his cabinet members and ministers. For so many times, the Ministers of Petroleum and Aviation were accused of corruption, but Jonathan did not act decisively on the matter. In fact, he was also accused of saying that a bribe is not corruption. Besides, Nigerians are tired of PDP government which has dominated seat of power for 16 years. (IDI, Female, Professor of Political Science, 59 years, University of Ibadan, March 2017)

Also, neopatrimonialism advocates for the prevalence of the rule of law as one of the control mechanisms that can check the excesses of politicians and their cronies. The independence of the judiciary is central to the efficacy of the rule of law. The judiciary has a role to play in ensuring good governance in Nigeria by strictly upholding the sanctity of law, without fear or favour. More importantly, there is need to re-orientate and educate the entire citizenry and the public office holders/politicians, in particular, to the principle of promoting the country's national interest, upholding strong ethical values, merit and non-partisanship when it comes to addressing issues of resources, appointments, wealth distribution and political benefits in Nigeria. This view was aptly supported by Waziri, who opined that:

> The war against corruption like terrorism is a special kind of war. It admits of no conventional methods. It is a war against human selfishness and greed. It is a war against rapid and senseless primitive capital accumulation. It is a war against decadence of mind, ethics and morals. Because of these special characteristics of the war, it requires a strong and uncompromising political will. It must be approached holistically. Casual and superficial approaches will not work. Rhetoric must match concrete action. Like all wars on salvation and restoration, friends will be hurt; families and associates will equally be hurt. And above all, politics has no place in the war. (Waziri 2010: 3)

Instead of looting the state's treasury in order satisfy their cronies, Nigerian political elites should empower them with the requisite skills and knowledge that will make them employable and move them away from committing electoral crimes and violence. This will reduce the constant tendency of towards political patronage. It is also important that politicians understand that squandering the state's funds on their cronies will not end poverty, which can be reduced only by creating new businesses. This will reduce the utilisation of cronies as political thugs by politicians.

> [P]oliticians should rather design programmes like empowerment programme that can minimally serve as temporary means of job for their cronies rather than placing them on job they do not have competency to do. This can help reduce pressure on governors and local government chairmen. And this in the long run will make them focus on governance than

politicking. (IDI, Senior Lecturer, Department of Public Administration, 42 years, University of Nigeria, March, 2017)

In addition, the cronies of politicians should understand that their interests should not supersede the country's national interest. And politicians should grant the request of their cronies only when it is within the confines of rule of law. This will enhance their accountability to the public and will save them from the pressures of having to loot the public treasury to satisfy their cronies.

Further, the policy of asset declaration by politicians in the country will reduce the looting of state's funds. It will show the financial strength of politicians and this will in turn inform their cronies of the financial status of their principals (politicians), who are mostly believed to be 'super rich'. Leadership by example can be helpful in addressing the problems of the actions of cronies of politicians in Nigeria. Countries such the United States, Canada, Uruguay under Mujica, South Africa under the late Nelson Mandela, Hong Kong, Singapore and so on, have achieved good governance under leaders who ruled by example while occupying public office. Curbing the actions of politicians' cronies over self-interested sentiments requires strong political will—and having political will that governs without fear or favour requires a good leader with strong character. The Nigerian political class should borrow a leaf from leaders like Mujica, the late Nelson Mandela, Barack Obama, and even from the country's own late leaders, such as Obafemi Awolowo, Nnamdi Azikwe, Anthony Enahoro and Aminu Kano, who unrelentingly strived for national unity, service and the promotion of national interests. These leaders were selfless and national in their thinking and outlook. It is suggested that if our leaders in this political dispensation could emulate such former leaders, the country would be galvanised towards good governance.

It should be noted that Nigeria's problem with bad governance is closely related to the citizens' collective inability to take up active roles, such as ensuring the declaration of accurate votes during elections and during the population census. Corroborating this view, Odinkalu (2010) reveals that three processes are essential in ensuring the effective functioning of a country. These are: the processes of legitimising public power (elections); the processes of quantifying the demographic coverage of the country (census); and the processes of estimating and distributing the commonwealth (public accounts, including revenues and appropriations). These three inter-related processes—elections, demography and public accounts—rely on the basic skills and institutions of honest policy numeracy. During elections, this involves the counting of votes and the conferment of a mandate, usually on the persons with the highest number of valid votes. During a census, the people are counted, which in turn helps to determine the bases for allocation of representation, social services, revenue and sundry public goods. In the management of public accounts, we quantify in numerical terms the size of the commonwealth, so as to know exactly the pool of resources that those who have the legitimate mandate through elections can distribute for the benefit of those who have been counted (Odinkalu 2010). Nigeria's multiple crises of governance exist because these three processes do not operate fully or effectively.

Concluding Comments

This chapter has put the concept of governance and its related variables into perspective, clarifying the unfolding Nigerian case. Various challenges are posed by the illicit actions of politicians' cronies, as shown by their inordinate demands for self-interested benefits to the detriment of the national interest. In the ensuing process, it has been argued that corruption and the undue actions of politicians' cronies is borne out of poverty, greediness, selfishness, pride and laziness, among others. The study has contended that by establishing small-scale businesses, educating the cronies of politicians, leading by example, formulating empowerment policies for cronies and by ensuring disciplined habits and upholding strong ethical behaviour by both cronies and politicians, as well as voting out corrupt politicians from public offices, it is possible to reduce the trend to loot the state's resources by the politicians.

It has also been argued that good governance as is seen in other developed democracies is presently lacking in Nigeria. This has been partly due to a lack of accountability on the part of politicians, a lack of access to information by the people, corruption, poor governmental policies, youth unemployment and the untenably wide gap between the rich and the poor, as well as greed, which have militated against the promotion of the national interest in most cases. The governance crisis would not have surfaced if the autonomy and functionality of the modern state and its agencies had been strengthened in contemporary Nigeria. Worryingly, not only is democracy endangered in the country, but Nigeria's corporate existence is also imperilled by the damaging activities of most politicians, who seem to be above the law. Just as most of these elected and appointed public office holders, as well as their cronies, escape reproach for different offences, so also many ordinary citizens who increasingly perpetrate other forms of crime that further incapacitate the state also escape the law.

References

Akinbowale, J. O. (2000, February). How to fight corruption. *The Tribune*, p. 12.

Akindele, S. T., Olaopa, O. R., & Obiyan, A. S. (2002). The theory of public administration and its relevance to Nigerian administrative ecology. *Journal of Social Sciences, 6*(4), 247–256.

Bayart, J. F., Ellis, S., & Hibou, B. (1999). *The criminalization of the state in Africa*. Oxford, UK: James Curry.

Chazan, N., & Rothchild, D. (1992). *Politics and society in contemporary Africa*. Boulder, CO: Lynne Rienner.

Collier, R. (1982). *Regimes in tropical Africa*. Berkeley: University of California Press.

Dahida, D. P., & Akangbe, O. M. (2013). Corruption as a bane for underdevelopment in Nigeria: Issues and challenges. *International Affairs and Global Strategy, 15*, 76–83.

Dwivendi, O. P. (1967). Bureaucratic corruption in Developing Countries. *Asia Review, 7*, 245–253.

Ekeh, P. P. (1989). The structure and meaning of federal character in Nigerian Political System. In P. P. Ekeh (Ed.), *Federal character and federalism in Nigeria* (p. 21). Nigeria: Heinemann Educational Books.

Idowu, A. A. (2013). Revisiting the 2007 elections in Nigeria: Road to justice for aggrieved parties. *Ife Journal of the Humanities and Social Studies, 1*(Maiden Issue), 29–40.

Joseph, R. A. (1999, July). Nigeria and the challenge of leadership. *Tell.*

Lipset, S. M., & Lenz, G. S. (2000). Corruption, culture and markets. In L. E. Harrison & S. P. Huntington (Eds.), *Culture matters*. New York: Basic Books.

Mbah, P. (2014). Demystifying intra-legislative conflicts in Nigeria: The senate and the house of representatives, 2007–2011. *Mediterranean Journal of Social Sciences, 5*(1), 677–690.

Nnamani, D. O., & Iloh, J. O. (2014). Good governance and national value: Where does Nigeria stand at 53? *International Journal of Democratic and Development Studies, 2*(1), 26–35.

Nwangwu, C., & Ononogbu, O. A. (2014). The pursuit of material interest and proliferation of political parties in Nigeria, 1999–2013. *Global Journal of Art Humanities and Social Sciences, 2*(6), 64–76.

Odinkalu, C. A. (2010). Corruption and governance in Africa: How do we break the cycle in Nigeria? Paper presented at the *National Conference on Corruption and Governance Challenge in Nigeria*, held at De Renaissance Hotel, Lagos, January 2010.

Ogoma, D. E. (2014). Leadership, corruption and governance in Nigeria. *International Journal of Politics and Good Governance, 5*(5.3) Quarter III, 1–19.

Oladele, O. O. (2004). Policy initiatives towards efficient functioning of the corporate Affairs Commission. *Ibadan Bar Journal, 3*(1), 20–29.

Olowu, D., Williams, A., & Soremekun, K. (Eds.). (1999). *Governance and democratization in Nigeria*. Ibadan: Spectrum.

Rose-Ackerman, S. (1999). *Corruption and government: Causes, consequences and reform*. London: Cambridge University Press.

Usman, B. (1979). As quoted in P. P. Ekeh (ed.) 1989, *op cit.*

Waziri, F. (2010). Corruption and governance challenge in Nigeria. *Cleen Foundation Monograph Series* No. 7, 1–7.

Yagboyaju, D. A. (2010). *The state and governance crisis in Nigeria: A comparative approach*. Ibadan: College Press and Publishers Limited.

Yahaya, I. (1993). The economic crisis, resources scarcity and decline in ethical standard in public agencies. In Olowu (Ed.), *Ethics and accountability in African Public Service*. Ethiopia: A. A. PAM.

Samuel Chukwudi Agunyai is an Assistant Lecturer/Researcher in the Department of Political Science, Obafemi Awolowo University, Nigeria, where he is also an interdisciplinary doctoral student. His research fields include public policy analysis, public administration, public governance and democratic studies.

Part IV

Transnationalism, Education and Human Development

11

Migration of Nigerians to Canada for Undergraduate Education: Implications of the Brain Drain for Development

Charles T. Adeyanju

Introduction

As an immigrant in Canada and a university instructor of many years' standing in Canadian universities, I have noticed the growing number of Nigerians in undergraduate programmes across Canada in recent years. The growing desire of Nigerians for undergraduate education in Canada is a recent development. In the mid-1990s, when I enrolled as an adult student in an undergraduate programme in a Canadian university, in one of the most cosmopolitan cities in the world, Toronto, there were hardly any parent-sponsored undergraduate students from Nigeria. Today, Nigeria is among the top six countries with international students in Canada. It is this personal observation that has made me research the migration of Nigerians for undergraduate education in Canada.

C.T. Adeyanju (✉)
Department of Sociology & Anthropology, University of Prince Edward Island, Charlottetown, PE, Canada

A. Adeniran, L. Ikuteyijo (eds.), *Africa Now!*,
https://doi.org/10.1007/978-3-319-62443-3_11

African migration is often framed in dystopian narratives (Simmons 2010); that is, African migration is driven by immiseration, poverty and violence on the continent of Africa. Falola and Afolayan (2007) have pointed out that African immigrants in the West face 'nightmarish realities'. Falola and Afolayan (2007: 3) note:

> Essentially, the human cost of migration is expected to rise proportionately to the capital gain which, often times, is the arena of only a select few who are usually ashamed to confess the nightmarish sacrifices and compromises they are often compelled to make, in order to achieve and sustain the ever glorified Western dream.

However, Flahaux and De Haas (2016: 2) state that the notions that African migration is connected to a myriad of socially caused problems in the continent 'are based on assumption, selective observation or journalistic impressions rather than on sound empirical evidence'. Flahaux and De Haas (2016: 2) add, 'The focus of media, policy and research on irregular migration, smuggling, trafficking and the high death toll among trans-Mediterranean "boat migrants" reinforce the impression that African migration is essentially directed towards Europe and driven by despair.' Their approach is a critique of the determinist materialist explanation of African migration, undermining agency and subjective understanding of the complex nature of the migration phenomenon. The condition of African migration reminds one of the 'Buddhist parable' of the blind scholars and their encounter with the elephant recounted by Steger (2009: 11). This parable suggests that divergent views on African migration by scholars are normal. According to the parable, the scholars did not know what an elephant looked like because of their limitations; therefore, they decided to gain a mental image of the animal by touching it. One felt the trunk and 'argued that the elephant was like a lively snake' (Steger 2009: 11). Another rubbed along its enormous leg and 'likened the animal to a rough column of massive proportions' (Steger 2009: 11). The third blind scholar held to its tail and 'insisted that the elephant resembled a large, flexible brush' (Steger 2009: 11), and the last one 'felt its sharp tusks and declared it to be like a great spear' (Steger 2009: 11). Each of the blind scholars was convinced that his interpretation of an elephant was right and so held firmly to his feeling and end up arguing with the others. African migration is studied from different perspectives and the experiences of African migrants are subject to diverse interpretations (Falola and Afolabi 2007). Therefore, the divergence of views on African migration to the West is not an anomaly.

Alan Simmons (2010) uses two ideal types to explain migration phenomena. The first one is a 'dystopian' form of migration, which he describes as 'founded on colonial and authoritarian principles, in which a few actors gain and many lose' (Simmons 2010: 4). The second type is the 'Utopian migration system', which is 'a "win-win" situation in which the goals of many actors are realized at least in part within the context of supportive political-economic and cultural processes' (Simmons 2010: 4). This chapter examines migration for higher education to Canada by Nigerian youths of higher socio-economic backgrounds. The chapter uses the case of these Nigerian students in undergraduate programmes in universities in different Canadian provinces to challenge the prevailing view in the West that African migration to the West is consequentially dystopian. Those involved in this mode of migration are young, urban and of high social-economic status. Their profile deviates from the particular social types of African migrants (e.g. refugees and asylum seekers) which often feature in the Western media. Nevertheless, this type of migration can contribute to the brain drain from Nigeria, and undermine the socio-economic development of the country if the students choose not to return home after completing their studies.

The chapter has four main sections. The first deals with the methods employed to collect data for the study. The second accounts for factors that motivate Africans, including Nigerians, to pursue higher education in the West (especially Europe and North America). The section 'Migration to Canada for Undergraduate Education by Nigerians' specifically discusses the migration of Nigerians to Canada for undergraduate education and its implications for Nigerian development. The final section concludes and highlights the key points of the chapter.

Methods

This study makes use of two sources of evidence. The first is the primary evidence comprising semi-structured interviews with participants in Canada and Nigeria, and participant observations in Nigeria in 2015 and 2016. Interviews with participants in Nigeria were conducted in the two cosmopolitan cities of Lagos and Port Harcourt. Specifically, research participants for the study comprise an unrepresentative sample of Nigerian parents who have sent their children to Canada for an undergraduate programme (*N* = 8); parent-sponsored Nigerian undergraduate students in Canada (*N* = 10); one prospective undergraduate student, who was in the process of applying for a Canadian student visa in Nigeria; international student recruitment agents (*N* = 2) who recruit students for Canadian universities; international student advisors; and one private high school teacher.

The sponsored Nigerian students interviewed for the study were studying in Canada at the time of the interviews, and were in the following universities: the University of Windsor, in Windsor, Ontario (*N* = 2); the University of New Brunswick, in Saint John, New Brunswick (*N* = 1); the University of Prince Edward Island, in Charlottetown, Prince Edward Island (*N* = 4); Brock University, in St. Catharines, Ontario (*N* = 1); the University of Alberta, in Edmonton, Alberta (*N* = 1); and Simon Fraser University, in Burnaby, British Columbia (*N* = 1). To minimise travel costs across Canada, interviews with students in jurisdictions other than Prince Edward Island were conducted via Skype (see Hulme, Thomson, Hulme and Doughty 2014[1]). I should add that I did face-to-face interviews in Nigeria because of the poor communication networks, which make the use of audio-visual technologies unreliable. In addition, I observed an interview between an international student recruitment agent in Port Harcourt and a Nigerian interested in travelling overseas for higher education; and I attended a high school graduation ceremony in a private high school in Port Harcourt, where representatives of international student recruitment agencies were present to advertise their companies and attract the interest of graduating students, parents and teachers in overseas education. All the students interviewed for this study were products of private high schools in Nigeria.

The secondary source of evidence is the literature on African migrations. This source comprises published academic books, refereed journal articles, and book chapters produced by other scholars.

Why Migration for Higher Education?

The question, 'why migration for higher education?' ought to be preceded by another—'why do people migrate?' Different theoretical approaches have been taken to provide an answer (Dako-Gyeke 2016; Massey *et al.* 1994). Theories of migration can be divided into two strands: the individual and the structural (Gogia and Slade 2011). The individual approach claims that individuals are rational beings who evaluate opportunities in other places and relocate to maximise their employment and wage opportunities. This approach is also known as 'human capital theory' (Gogia and Slade 2011: 11). The shortcoming of this approach is that the decision to migrate is rarely made by individuals. Migration decisions are often influenced by family members and networks of contacts in the origin and destination countries. In addition, the focus of this theory on money/wages/employment is reductionist. It overlooks the non-materialist motives of migration.

The second approach, identified by Gogia and Slade (2011), is the structural approach, also referred to as 'push-pull' theory. This perspective argues that factors associated with under-development, especially poverty, unemployment and conflict, 'push' people, while favourable conditions of employment opportunities and socio-political stability attract migration. This perspective has been criticised by scholars (see Massey *et al.* 1994; Simmons 2010; Adeyanju and Oriola 2011) for over-simplifying the element of human intention. In their critique of 'push-pull' views of migration, Flahaux and De Haas (2016: 4) note that migration is motivated by 'ambitions' and 'resources'. Therefore, migration is a function of 'aspirations' and 'capabilities'. Access to information and exposure to lifestyles elsewhere 'conveyed through education, media and advertising tend to change people's perceptions of the "good life" alongside increasing material aspirations and a growing appetite for consumer goods' (Flahaux and De Haas 2016: 4).

A number of factors have been adduced for African migration to the West for higher education. It is often believed that under-development in Africa induces international migration, including migration for higher education. However, Flahaux and De Haas (2016: 4) argue that development increases rather than decreases migration. Theirs is a critique of neoclassical or push-pull theories of migration that claim that poverty and material scarcity predict migration or that migration decreases as a society develops. In their words, 'In reality most migrants do not move from the poorest to the wealthiest countries, and the poorest countries tend to have lower levels of emigration than middle-income and wealthier countries' (Flahaux and De Haas 2016: 4). They claim:

> [T]o understand why development is generally associated to more migration, it is important to move beyond views of (African) migrants as objects which are passively pushed around by external 'push' factors such as poverty, demographic pressure, violent conflict or environmental degradation, analogous to the way physical objects are attracted or repelled by gravitational or electromagnetic forces…Development processes typically expand people's access to material resources, social networks, education, media and knowledge. At the same time, improvements in infrastructure and transportation, which usually accompany development, make travel less costly and risky, enabling migration over increasing distances. (Flahaux and De Haas 2016: 4)

Those who migrate for higher education in Nigeria are mostly members of the upper high socio-economic class, which has access to social and cultural capital (see Bourdieu 1986).

Globalisation plays a pivotal role in migration for higher education. Migration for higher education by Africans is one of the 'flows' generated by globalisation (Ritzer 2011). The notion of 'time-space distanciation', as articulated by Giddens (1990: 141), means that in the global age communication technologies connect us with 'events, with actions, and with the visible appearance of physical settings thousands of miles away from where we happen to live'. Prospective students are able to access information about schools of higher learning in faraway countries and receive adequate information instantaneously while universities in the West are able to reach educational markets or recruit students in distant places.

In addition to the influence of globalisation on ease of transport and access to information about other places, Kritz (2011: 157–158) lists a number of factors that account for the growing demand in Africa for tertiary education overseas. These include: first, increased demand in tertiary education in many African countries due to the growth in secondary school education completion. Olutayo (2016) has also noted this factor in regards to Nigeria, where existing universities are not able to absorb university admission applicants; and second, faculty shortages, increases in class sizes, heavy teaching loads, and the circulation of teachers with weak credentials also account for the desire for international education. Relatedly, many professors and researchers have left the African region for the West due to the parlous state of the education sector in many African countries. Nigeria is among the countries in Africa that have experienced a loss of professionals, including university instructors and researchers, as a result of the myriad number of problems confronting the public education institution (Olutayo 2016).

Overall, one recurring factor regarding the motivation to pursue higher education overseas in the literature and public discourse in many African countries, especially Nigeria, is the poor quality of public education (Adepoju and van der Wiel 2010). For example, Kritz (2011: 159) quotes the World Bank for corroboration:

> Tertiary education fares no better [than secondary education] in terms of quality. Part of its curriculum is often dedicated to remedial efforts to make up for shortcomings at the lower levels of education. In a certain sense, the price of African education for its impressive enrolment expansion has been a loss of educational quality.

Apart from the objective indicators of the poor quality of higher education in African countries, many Africans are attracted by tertiary institutions in the West based on their perception, rather than the reality, that living and studying in the West is prestigious. This perception is not unconnected to the colonial discourse on Africa, in which the West is held to be better than Africa (Adeyanju and Oriola 2011).

Migration for Higher Education: Retention and Brain Drain

Kritz (2011) notes that a significant number of Africans who receive tertiary degrees abroad do not return. Although there are no reliable estimates of skilled professionals who initially left as students and remain in their destination countries after completing their education, it is safe to assume that some of them choose to remain after completing their studies due to the social, economic and political conditions of Africa (Kritz 2011: 167; Sako 2002: 25). Therefore, non-returning African students are a constituent part of many categories of the African diaspora (Oucho 2007: 55). The Canadian government's immigration policies have been described as neoliberal because of its focus on highly skilled immigrants, those with post-secondary education, and funds for settlement (Simmons 2010). International students, by default, fit this category upon graduation. Canada has already created programmes to capture international students with Canadian work experience and university education.[2]

The retention of international students of African origin in the West can contribute to brain drain in Africa, even though the emigration of Africa's professionals and skilled workers to Europe and North America for employment purposes is the primary cause of this brain drain. Aginam (2007) points out that many African health professionals who now live and work in Europe, North America and the Gulf states contribute to the brain drain that affects the human material condition of Nigerians and a larger portion of the African continent. For example, it is estimated that one third of the most qualified African nationals live in North America and Western Europe (Sako 2002: 25). It is noted that factors contributing to brain drain in Africa are political instability, poor infrastructure, the high rate of employment, poor working conditions, corruption, and so on. (Sako 2002). Sako (2002) discusses three schools of thought on brain drain. The first views brain drain as having detrimental effects on the sending society and beneficial ones on the recipient country, i.e. brain drain is a zero-sum game. In other words, brain drain allows the receiving countries to reap the benefits of professional immigrants without having to pay the cost of developing them, which is borne by their country of origin. The second position on brain drain argues that the migration of professionals is beneficial to the sending country because it enables the professionals to avoid unemployment, underemployment and the under-utilisation of their skills and professions in their home country. It is also pointed out that the investments made in professional migrants by the sending country are partly recouped through remittances. The third perspective, according to Sako (2002), claims that brain drain benefits neither the source nor the receiving country. It is argued that skilled and unskilled immigrants are poorly treated in Europe and the Americas. Professionals among them do less congenial work than their native counterparts. They become underemployed, working in occupations that are of no relevance to their training. Over time, they lose their skills and are unable to contribute to either the host society or the home country.

Nevertheless, it is noted that a brain *gain* emerges from international migration. Brain gain has often occurred in Africa in the form of knowledge transfer by returning immigrants who share with their home societies the knowledge they have acquired abroad. The Carnegie African Diaspora Fellowship Program is considered an exemplary example of brain drain reversal. According to Foulds and Zeleza (2014: 17), 'The goal of this programme and model [is] to ensure that African institutions are the driving forces in identifying needs and opportunities for engagement, as well as providing to diaspora scholars and African institutions the space to build and expand their scholarly alliances.' Remittances are another means by which immigrants enrich their countries of birth. Remittances worldwide in 2015 are estimated to amount to $582 billion (Torkington 2016).

Migration to Canada for Undergraduate Education by Nigerians

Nigeria, which is the most populous country in Africa, has the largest number of students abroad among African countries (Kritz 2011: 168). Simmons (2010) predicts that Nigeria and South Africa will be major sources of future immigration to Canada for three reasons. First, Nigeria has highly educated citizens confronted with unemployment stemming from the parlous state of its economy. Second, Nigeria, as the most populous country in Africa, has an urban population that is willing to relocate. Third, Nigeria's lingua franca is English, and is a member of the Commonwealth, as is Canada.

Over the past 15 years or so, Canada has been a destination for Nigerians for undergraduate education. Nigeria was among the top six countries with international students in Canada in 2015.[3] Nigerians are 'pushed' to Western countries for post-secondary education because of the poor state of the education system in their own country, inadequate opportunities for the burgeoning Nigerian youth population, and a rising affluent class. They are 'pulled' by Canada's post-industrial, neoliberal immigration policy, which targets 'self-made' immigrants with capital and skills (Simmons 2010). Most of these undergraduate students are sponsored by their affluent parents, who pay foreign tuition fees (Olutayo 2016) two to three times higher than those for native students. Their age range is 17–22. These students are in post-secondary institutions across Canada. These Nigerians are from a high socio-economic class. Anecdotal evidence has shown that some Nigerian parents buy homes (outright) for their children in university towns across Canada. Given the invidious political economy of inequality in Nigeria, that deprives *most* Nigerians of their basic needs, it is apt to evoke Stuart Mills' quote that the parents of these international students 'grow rich in their sleep'. There is curiosity surrounding their enormous wealth in the midst of grinding poverty confronting the vast majority of the Nigerian population.

The recurring themes concerning the desire of Nigerians to undertake post-secondary education in Canada are: recognition of foreign credentials; and the poor quality of education in Nigeria. These two factors are the subject of the next section.

Recognition of Foreign Credentials and Quality of Education

Two interrelated factors that pull Nigerians towards Canadian higher institutions are employers' preference for foreign credentials and the poor quality of education in Nigeria. It is not difficult to make a connection between the poor quality of public education in Nigeria and employers' interest in credentials obtained in developed countries. An international student recruitment agent based in Lagos, Nigeria, notes:

> If a student wants a certificate that is well acceptable in this country, such a student should study in the UK, Canada or US; and because of the school fees and then the standard of education in Canada, so the certificate is also acceptable here in Nigeria. So if a student is able to study a degree course in Canada, if he or she is able to come back to Nigeria with that certificate, he or she will get a job here, because of the quality of education in Nigeria.

Allied to the above viewpoint is the notion that Canada delivers high-quality education that interests Nigerian employers. The international student recruitment agent adds:

> Canada is considered a G8 country, so once you school in Canada, you are getting one of the top certificates you can get anywhere in the world. Also, the universities are all top rated. You know, there is a body that rates universities, McLean's.ca, so I usually advise parents to go there, so when they check out the ratings of Canadian institutions, so they are getting one of the best educations anywhere in the world.

This recruitment agent refers parents to the internet for the ratings of Canadian universities. This shows the significance of the internet in instantaneously connecting people to information in faraway places.

Another viewpoint on the quality of Canadian university education is expressed by a student studying in a Canadian university. This student was a medical student in a private university in Nigeria, but gave up the programme for a non-professional degree programme in a Canadian university. The following is his perspective on the recognition of foreign credentials in Nigeria:

> A lot of Nigerians have that viewpoint, that it is better to get a degree from one of the schools abroad than to get a degree from the schools in Nigeria, because…I mean (pause)…I don't know about qualification, but I know a lot of the stuff taught here…it's like…it's almost like the same stuff, but I think here they go into more detail and you become very good in it. So that's what I think. So I think they were like more interested in the reputation you get when you bring a degree from outside the country.

This student's view on the quality of education and migration for higher education is more nuanced than that of the recruitment agent. He believes, to some extent, that the poor quality of education in Nigeria vis-à-vis the West might be based on 'reputation'.

One student simply responds to the question regarding the attraction of Canadian education as follows: 'The way they teach. The teachers give more examples, and explain better.' A parent with two international students in a Canadian university is more elaborate:

> [J]ust like my husband has said, they have the privilege of doing presentations, you know, before all the class members and the professors; so it builds confidence in them. Yes, and then asking them to do research and all that, you know, makes them understand not just head knowledge of what they are being taught, when they get involved in getting the knowledge from the internet, from books, and all that, they get to understand the course more. Yes, that's why we say that that education over there is superior to the one we have here, and we are happy that they are being exposed to a tougher system.

This same parent attributes the preference for foreign credentials to the poor quality of post-secondary education in Nigeria. In her words:

> [O]f recent we found that most graduates from Nigerian universities are not easily employed in the labour market. And when you interview them, you find they cannot defend the course they have studied. My husband said when they began to recruit people who studied abroad, yes, they prefer their certificates, and so we thought our children too could benefit from that if we could afford to send them overseas to study. We should give them the chance of getting the best of education so that eventually if they come back and have to work here, they would be able to secure employment. It won't be difficult for them.

One of the deficiencies of the Nigerian post-secondary education system, participants note, is that the system is heavily theoretical. A parent, whose daughter is studying at a university in the Maritimes notes, 'they have access to library books, e-learning materials…and for the courses they are studying, especially Public Health, they do a lot of travel, they interact, so that would also help and equip them, they do practical stuff.'

Student Brain Drain and Implications for Development

Olutayo (2016) refers to migration for higher education as 'students' brain drain'. Although the contribution of international students of Nigerian origin to Canada's economy has not been quantified and documented, given the higher tuition rates for international students, compared with the rates for domestic students, it must be enormous in Canada. International students contribute over $8 billion to the Canadian economy annually (Enright 2016). Besides tuition, their expenses on accommodation and living costs boost the local economy. Students who stay with host families pay for their room and board and the care they receive from the latter, who serve in loco parentis. Therefore, it goes without saying that one outcome of migration for higher education is a loss of foreign exchange for Nigeria and other developing countries.

From interviews with parents and Nigerians in Canadian universities, three ideal type outcomes emerged concerning decisions after graduation. These outcomes have implications for development. First, a minority of the students either intend to return, or their parents wish them to return. The following is a sample of their responses: a Business Administration student at a University in Ontario is certain about returning to Nigeria after completing his studies. He says, 'I think there are more opportunities in Nigeria than here [Canada]. There are more business opportunities in Nigerian than Canada...because I found that bankers make $20,000 a year or $30,000 a year [here]; if you go to Nigeria, some bankers are making that in two months to three months.' It is evident from his response that he believes there are more opportunities for him in Nigeria than in Canada. His intention is driven by the rewards he will receive for his investment in human capital. Another student of sociology at a university in the province of Alberta intends to return to Nigeria, even though 'not immediately', to contribute to the development of Nigeria. In her words: 'I feel like the kind of education I have acquired here, I would be able to help in a little way, help like in the development of the country in a minute way...but, I feel like I want to go back to the country because I want to help, I want to be part of the development, and part of the change that the country needs.' Unlike the male student who wishes to return to reap the rewards from his investment in human capital, this student wants to contribute to the development of Nigerian society.

A parent with two children, one of whom he claims is a medical student, studying at universities in the provinces of Ontario and Alberta asserts that his children will return to Nigeria upon the completion of their studies. In his words:

> I want them to come back home when they finish, to do their housemanship here in Nigeria so that if they want to work in Nigeria, they will be recognised. They can work in any government hospital. They can also go back and work there. But I want them to be more of home than abroad…I want to get close to them. I want to be seeing them. Seeing their performance and how they are faring is very important, that's why I am training them. I am not training them to go and send them away to foreign countries, no.

In this particular case, it is the parent that intends to dictate the students' future relationship with Canada and Nigeria after their graduation. After all, he is the one bearing the costs of their education in their entirety.

The second ideal outcome is the desire to stay in Canada after graduation, with intermittent visits to Nigeria. One of the students who intends to stay in Canada for a protracted length of time (because she intends to go to medical school) shares her intention as follows:

> No, I don't intend to go back to Nigeria, but things can happen…because I still have to go to Med School, after my Biology…getting a Biology degree here. So I might not stay here, but not going back to Nigeria, probably go to another country…because I don't know, maybe because I have already mixed with other people, so I don't want to go to the same place, I guess, where I am coming from [Nigeria].

She elaborates on her point of view regarding her unwillingness to return to Nigeria:

> And the thinking is different in Nigeria…Yes, in Nigeria, you have to go with the flow, like…everybody believes you should do this, and you have to do it, if you don't do what people think you should do, it becomes a problem. Like everybody just tries to control your life. You can't really be independent, I think that's the word…but here [Canada] everybody has an open mind. If you tell them this is what you believe, they won't question you, like they give you that ability to just be yourself, and mind yourself. But in Nigeria, you just have to be in a way pretend because they expect you to follow the culture.

Apart from the time it will take for this student to complete degrees in Biology and Medicine, she prefers the Western culture of independence and autonomy to the 'controlling' Nigerian culture.

The third ideal type outcome is the plan to gain a Canadian residency permit, which will enable the students to live without uncertainties around their status upon graduation. German sociologist Ulrich Beck (1992) has argued that we now live in an era of uncertainty and insecurity. A sense of insecurity and uncertainty, both real and perceived, pervades all areas of life in a 'risk society', particularly with respect to labour markets and employment. The students (in cahoots with their parents) desire an opportunity to have a choice over where to live, especially between Nigeria and Canada. They also wish to have unhindered travel to places, which the Canadian travel documents make possible. In the words of a parent, who is of Igbo ethnicity, interviewed in Port Harcourt, Nigeria: 'the plan is that he stays there, and comes here on vacation to learn more about his roots, and know how to associate with people in his village, to be able to behave well and respect his culture and the people and village he comes from.' Another parent, interviewed with his son, on holiday in Nigeria at the time of the interview, says:

> Even my son said 'Mama I want to stay, you know what it means to stay there before you get your citizenship.' I said okay if it's your citizenship, stay there and get your residency, you know, start working there, give them some years, work there, go into the system so that you know you can always come back there to get a job, to work, if you need to. So that's why we encourage him to do his Master's there and get a job.

There is conflict between the desire of the students and the wish of parents regarding future plans after graduation. In one instance, it is the wish of a parent that her children studying at a university in the Maritimes return after graduation to enable them to contribute to the development of Nigerian society. One of the children has a different intention from his mother. In the parent's words:

> [W]hen they were going, we were asking them…me personally wanted them to come back, yes because I remember the day we sat together and we were talking…he said he does not think he would come back. I just started crying. Yes, but my husband is of opinion that anywhere in the world anyone feels comfortable to stay, that the person can stay, you understand; but me I wanted them to come back. One is that I would have loved a situation where these children will go over there, learn their education and come back here to give it back to us here in Nigeria. It will make this country a better country. You understand? And then, we have just two of them. If we have two of them over there, I don't know how…it would be too lonely here.

Some of the participants want to return to contribute to the development of their home country, many intend to stay permanently, and most of them are unsure but desire to have some certainty in the form of a permanent residency permit for the host society. It is noteworthy that most, if not all, participants, regardless of the intention to stay or return, want a Canadian permanent residency permit.

The decision by Western-educated youths to return or remain in the country of study is important because of the implications for the future development of the sending country. Most of the participants express their desire to hold permanent residency status upon graduation because it will enable them to not only to work legally and live permanently in Canada, but to also cross national boundaries unencumbered. After all, one of the attractions of Nigerians to Canada for higher education is the opportunity to work and study while in Canada, and to obtain Canadian permanent residence upon graduation. A travel agent describes the benefits of migration to Canada for higher education for Nigerians as follows:

> [T]he benefits are that when they come in, number one, for their first year they can apply, after six months in the programme, for a student work permit, you know, that doesn't apply in many other countries. So a student begins to apply what you study, you can start applying it in the first year of schooling in Canada. And you are able to work throughout the whole... four years, of course, during holidays and during...and if it's during school, it is 20 hours per week and during holidays you are able to take on a full-time job, and then of course once you graduate and you know that you are not going directly into a Master's programme, you can apply for a graduate work permit, you know, after graduate work permit now, once you do more than two-year programme, whatever programme you do, once it is more than two years, you are offered a three-year graduate work permit. And that's a huge advantage, because the Canadian government doesn't want to chase students out, they want to retain them. And so the students have opportunities of staying and working and they can be easily be absorbed into the economic system.

One of the attractions of Canada to many international students is the opportunity for them to be future members of the Canadian nation. Most of the Nigerian students, with the support of their parents, want to have an opportunity to have access to multiple countries. They want to live in a 'transnational world' (Simmons 2010). The Canadian permanent residence or citizenship could make this wish a reality. Back in the 1990s, Basch *et al.* (1994: 7) conceived the notion of transnationalism and defined it 'as the processes by which immigrants forge and sustain multi-stranded social relations that link together their societies of origin and settlement.' Basch *et al.* (1994: 7) refer to immigrants 'who develop and maintain multiple relationships-familial, economic, social, organizational, religious, and political-that span borders' as 'transmigrants'.

Deriving from the legacy of Basch *et al.* (1994), transnationalism does not indelibly transform the host society; rather, it reinforces the existing inequalities there. In their seminal work, the authors compare immigrants from the Caribbean and those from the Philippines in the United States. In spite of these groups being of different racial backgrounds and from different regions of the world, they face a similar situation in the United States because they are non-whites. Although these groups are on the margins of US society, Basch *et al.* (1994: 261) note, '[I]t is striking that, in both the Philippines and the Caribbean, many transnational practices have been motivated by a concern for maintaining or improving the migrants' class and social position.' That is, the immigrants are primarily concerned about perpetuating their class location in their home country or advancing their social-economic status in their country of origin. For Basch *et al.* (1994), transformation of unequal social relations in their home country is not on their agenda. For most of the parents of Nigerian students in Canada and their wards, their main interest is not to uplift Nigeria developmentally. Their main concern is sustaining the existing political economy of Nigeria or improving their position in Nigerian society. The intention of most of those who intend to return after graduation is to occupy the available few well-paying jobs in the private and public organisations. They will combine their foreign education with their cultural capital to gain access to opportunities in Nigerian society. They will also use their social capital, that is, the social connections they have acquired from attending elite private elementary and high schools and from sojourning in the West, to gain access to the corridors of power. Those who intend to stay behind in Canada will enjoy the deference and prestige that transnational migration confers on transmigrants in Nigerian society (Adeyanju and Oriola 2011).

International students who return to their home society on the completion of their studies may contribute to the development of their society or reverse the brain drain, if they share the knowledge and experiences they have acquired in the West. This is exactly what some of the Nigerian students intend to do upon return. At the ideational level, however, Olutayo (2016) is averse to the notion of brain gain in the form of diasporas teaching or influencing their African societies because he believes their intervention erodes the local cultures of their country of origin or sending country. Olutayo (2016)'s point is unconvincing simply because he fails to denote and define the notion of 'local culture' in the global age of cultural ambiguity and cultural hybridity.

Nigeria is a highly stratified society. Nigerian society is also a social laboratory for testing the relationship between formal education and the reproduction of social inequality in the larger society. Over the past 20 years, there has been a preponderance of private educational institutions, from elementary to university. They now exist alongside public education, which is poorly funded and poorly managed (Olutayo 2016). The public system draws its students from the bottom stratum of society. The private system, albeit with different gradations, is patronised by the elites. The public system is being gradually abandoned by the elites, who now send their children to private schools from elementary level to university. Due to the decline in the quality of higher education, most products of the public institutions have become unemployable (Adepoju and van der Wiel 2010). Employers prefer the graduates of private universities or those with foreign credentials from countries in the West. Apart from poor infrastructure in post-secondary public institutions, there is a plethora of anti-social behaviour in these schools, including gangsterism, hooliganism, armed robbery, prostitution, etc. These social problems are symptoms of a larger malaise in the wider society and they are signs of rebellion by restive youths abandoned by the politico-economic class. There are also allegations of sexual harassment of female students by male lecturers, both proven and unproven.[4] Overall, there is a feeling of insecurity, physically and psychologically, in the public school system in Nigeria, which is in part responsible for why members of the advantaged class create an alternative in private schools.

Therefore, it is no hyperbole to say that a system of caste/apartheid is in place, or being put in place, in Nigeria through the education system. In addition to the private and public systems, a subset of the advantaged class has opted for higher education abroad, especially in the West. If the students sent abroad for higher education return to Nigeria after completing their education, they receive preferential treatment in the Nigerian labour market and deference in the larger society. If they become transnationals, they are equally recognised by their society.

Concluding Remarks

This chapter has lent further credence to the notion that African migration is not primarily induced by war, conflict, poverty or deprivation, as Western media and some academic literature tend to portray it. Migration for higher education, in the case of Nigerian migration to Canada for post-secondary education, is an indication that Africans migrate extra-continentally for varying reasons and motivations. They participate in global capitalism when they move funds to Western capitals for their school tuition and other expenses related to higher education. Although there is no empirical evidence in this study to support the claim that migration for higher education leads to brain drain in Nigeria or developing countries, there is proof that developed economies maintain policies towards the retention of international students[5] because they are a resource to their society and not a liability or burden. There is also evidence that many international students of Nigerian origin are interested in Canadian permanent residence due to employment opportunities in Canada and opportunities to cross national borders that Canadian permanent residence and citizenship offer.

Retention of students by Canada or developed countries does not foreclose development for the sending countries as it is possible for overseas residents to transfer knowledge through various means in this age of globalisation. For example, the Carnegie African Diaspora Fellowship Program is one of the ways that African scholars in North America are reversing the brain drain. According to Foulds and Zeleza (2014: 17), 'The goal of this program and model [is] to ensure that African institutions are the driving forces in identifying needs and opportunities for engagement, as well as providing to diaspora scholars and African institutions the space to build and expand their scholarly alliances.'[6] However, findings from this study have shown that part of the reason for patronising higher education in private schools and overseas is poor funding of public education and government neglect of higher education in the public system. Concomitantly, the education system in Nigeria has accentuated social inequality rather than attenuating it, depriving those from the disadvantaged class of the experience of social mobility.

As has been pointed in this chapter, most of the products of the public system are not employable. However, from personal observation, the private system alternative in Nigeria, especially the more highly priced schools, are not lower in standard than some of the higher institutions Nigerians patronise in Western Europe and North America. In other words, many private universities in Nigeria have equal or higher standards than some of the degree-awarding universities in the West. Therefore, I do not think it is the higher institutions in Western Europe or North America that are rated higher by many Nigerian parents and their wards studying in the West; it is the West and its 'civilisation', often associated with 'whiteness' by some, that many Nigerian elites and non-elites adore and thereby rate higher than Nigerian/African culture.

Notes

1. Hulme et al. (2014) use a mix method of audio-recorded interviews conducted via telephone, Skype video and face-to-face conversations to collect data for their study of the role of agents in international student recruitment in British institutions.
2. http://www.cic.gc.ca/english/immigrate/apply.asp
3. http://www.iie.org/Services/Project-Atlas/Canada/International-Students-In-Canada#.V2q7C_krIdU
4. These social problems within publicly funded higher institutions of learning in Nigeria are presented anecdotally here because there is dearth of literature on the subject. Nevertheless, social media and traditional media sources have copious information on malfeasance in the Nigerian public education system.
5. See www.cic.ca
6. The website provides more details: http://www.iie.org/Programs/Carnegie-African-Diaspora-Fellows-Program#.WM8MWtQrL4Y

References

Adepoju, A., & van der Wiel, A. (2010). *Seeking greener pastures abroad: A migration profile of Nigeria*. Onireke, Ibadan: Safari Books.

Adeyanju, C., & Oriola, T. (2011). Colonialism and contemporary African migration: A phenomenological approach. *Journal of Black Studies, 42*(6), 943–967.

Aginam, O. (2007). "Predatory Globalization?": The World Trade Organization, general agreement on trade in services, and migration of African health professionals to the West. In T. Falola & N. Afolabi (Eds.), *The human cost of African migrations* (pp. 65–77). New York: Routledge.

Basch, L., Schiller, N. G., & Szanton Blanc, C. (1994). *Nations unbound: Transnational projects, postcolonial predicaments, and deterritorialized nation-states*. Amsterdam: Gordon and Breach Science Publishers.

Beck, U. (1992). *Risk society: Towards a new modernity*. New York: Sage.

Bourdieu, P. (1986). The forms of capital. In J. C. Richardson (Ed.), *Handbook of theory and research in for the sociology of education*. New York: Greenwood Press.

Dako-Gyeke, M. (2016). Exploring the migration intentions of Ghanaian youth: A qualitative study. *International Migration and Integration, 17*, 723–744.

Enright, M. (2016). Hefty fees from foreign students are bailing out Canadian Universities and that's not good. *CBC: The Sunday Edition*. Retrieved from http://www.cbc.ca/radio/thesundayedition/kellie-leitch-wants-all-immigrants-screened-for-canadian-values-1.3854030/hefty-fees-from-foreign-students-are-bailing-out-canadian-universities-and-that-s-not-good-1.3854036

Falola, T., & Afolabi, N. (2007). Introduction: Migration fantasies, nightmarish realities. In T. Falola & N. Afolabi (Eds.), *The human cost of African migrations* (pp. 1–15). New York: Routledge.

Flahaux, M.-L., & De Haas, H. (2016). African migration: Trends, patterns, drivers. *Comparative Migration Studies, 4*(1), 1–25.

Foulds, K., & Zeleza, P. T. (2014). The African academic diaspora and African higher education. *International Higher Education, 76*(Summer), 16–17.

Giddens, A. (1990). *The consequences of modernity*. Cambridge: Polity Press.

Gogia, N., & Slade, B. (2011). *About Canada: Immigration*. Halifax: Fernwood.

Hulme, M., Thomson, A., Hulme, R., & Doughty, G. (2014). Trading places: The role of agents in international student recruitment from Africa. *Journal of Further and Higher Education, 38*(5), 674–689.

Kritz, M. M. (2011). African international student mobility: Country differences in numbers and destinations. In A. Adepoju & J. O. Oucho (Eds.), *Migration in the service of African development: Essays in honour of professor Aderanti Adepoju* (pp. 157–205). Onireke, Ibadan: Safari Books.

Massey, D., Arango, J., Hugo, G., Kouaouci, A., Pellegrino, A., & Taylor, E. (1994). An evaluation of international migration theory: The North American case. *Population and Development Review, 20*(4), 699–751.

Olutayo, A. O. (2016). Democracy, funding of higher education and students' 'Brain Drain' in Nigeria: Development implications for now and the future. *Journal of Management and Social Sciences, 5*(2), 1–14.

Oucho, J. (2007). African brain drain and gain, diaspora and remittances: More rhetoric than action. In A. Adepoju, T. Naerssen, & A. Zoomers (Eds.), *International migration and national development in Sub-Saharan Africa: Viewpoints and policy initiatives in the countries of origin* (pp. 49–69). Lieden: Brill.

Ritzer, G. (2011). *Globalization: The essentials*. Malden, MA: Wiley-Blackwell.
Sako, S. (2002). Brain drain and Africa's development: A reflection. *African Issues, 30*(1), 25–30.
Simmons, A. (2010). *Immigration and Canada: Global and transnational perspectives*. Toronto: Canadian Scholars' Press.
Steger, B. M. (2009). *Globalization: A very short introduction*. New York: Oxford University Press.
Torkington, S. (2016). For a number of families, remittances is their main source of income. *World Economic Forum*. Retrieved from https://www.weforum.org/agenda/2016/11/migrants-sent-half-a-trillion-dollars-home-in-2015-these-maps-show-where-the-money-went

Charles T. Adeyanju is Associate Professor of Sociology in the Department of Sociology and Anthropology at the University of Prince Edward Island. He holds a PhD in Sociology from McMaster University. Adeyanju's research interests lie in the areas of immigration, social inequality, media, and race and ethnicity. He has researched a range of social issues, including African migration, media representations of race and taser use by the Canadian police. His publications have appeared in leading peer-reviewed journals, including the *Journal of Social Identities*, *Journal of Black Studies* and the *Canadian Journal of Criminology and Criminal Justice*. Charles Adeyanju is the author of *Deadly Fever: Racism, Disease and A Media Panic*, published by Fernwood in 2010.

12

Imperialism and the African Indigenous Knowledge System

Taiwo Olabode Kolawole

Introduction

Development is an imperceptible process; no society can attain significant development within a short time frame and without adequate planning. The implication is that all societies must have gone through both the Palaeolithic and the Neolithic processes in attaining appreciable progress. These two processes were eras when societies were close to nature and human beings used stone and wood as major materials to take adequate control of their environs. During these periods, there was no sophisticated technology and the main occupation was pure agriculture. It is very interesting that human beings survived these periods. As time went by and human beings continued to develop their control of their environment, other occupations, requiring seemingly advanced technology, began. Africa, at this stage, was growing apace with global standards.,

T.O. Kolawole (✉)
Department of Sociology, Federal University Oye Ekiti, Oye Ekiti, Nigeria

A. Adeniran, L. Ikuteyijo (eds.), *Africa Now!*,
https://doi.org/10.1007/978-3-319-62443-3_12

Unfortunately, however, these processes were interrupted by imperialism, culminating in the thwarting of socio-economic, political and cultural development in Africa. Of this, Rodney (1982: 51) says, '[i]t has been shown that, using comparative standards, Africa today is underdeveloped in relation to Western Europe and a few other parts of the world; and that the present position has been arrived at, not by the separate evolution of Africa on the one hand and Europe on the other, but exploitation.' As is well known, Africa has had prolonged and extensive contact with Europe, and one has to bear in mind that contact between different societies changes their respective rates of development. To set the record straight, four operations are required:

(a) Reconstruction of the nature of development in Africa before the coming of Europeans;
(b) Reconstruction of the nature of development which took place in Europe before expansion abroad;
(c) Analysis of Africa's contribution to Europe's present 'developed' state; and,
(d) Analysis of Europe's contribution to Africa's present 'underdeveloped' state.

Moreover, it is very important to note that 'culture is a total way of life'. It embraces what people eat and what they wear; the way they walk and the way they talk; the manner in which they treat death and greet the new-born. Obviously, unique features come into existence in virtually every locality with regard to all social details. The continent of Africa south of the Sahara Desert formed a broad community where resemblances were clearly discernible. For example, music and dance had key roles in 'uncontaminated' African society. They were ever present at birth, initiation, marriage, death, etc., as well as appearing at times of recreation. Africa is the continent of drums and percussion. African peoples have reached a pinnacle of achievement in that sphere.

With imperialism, Europe developed at the expense of Africa. To date, Africa has not substantially benefited from Europe's socio-political and economic well-being. The Western superpowers seemingly knowingly inhibit the potential and capability of Africa and impede opportunities that may engender rapid development. This is particularly obvious in the 'digital divide' between the Global North and the Global South (Arnold 1985), which simply means the gap between demographics and regions that have access to modern ICT, and those that do not have access at all or have restricted access to technology such as the Global System for Mobile Communications, television, personal computers and the internet, to mention but a few. This chapter assesses the effects of imperialism on ICT, especially in the area of its application and use on the African continent and examines whether the indigenous knowledge system can proffer solutions to the problem of ICT in Africa.

An Overview of Imperialism in Africa

Colonialism is the direct and overall domination of one country by another on the basis of state power being in the hands of that foreign power. An example is the direct and overall domination of Nigeria by Britain between 1900 and 1960. It is also a form of temporally extended domination by people over other people and, as such part, of the historical universe of forms of intergroup domination, subjugation, oppression and exploitation (cf. Horvath 1972). The first objective of colonialism is political domination. Its second objective is to make possible the exploitation of the colonised country. Colonialism in Africa refers to an event which took place between 1800 and the 1960s. It is a phenomenon which is part and parcel of another phenomenon called imperialism. In fact, colonialism is a direct form of imperialism. This is why it is often said that all colonialism is imperialism, but not all imperialism is colonialism.

Colonialism began as a result of changes in the mode of production in Europe (for example, the emergence of the Industrial Revolution). The Industrial Revolution ushered in a new process of production, in place of the earlier slave-based economy. The Industrial Revolution was a revolutionary trend in the history of mankind. The problem of how to lubricate machineries came up with the emergence of the Industrial Revolution. The slave trade and slavery had by this time fulfilled their basic function of providing the primitive capital. The quest for the investment of the accumulated capital and the need for raw materials led to the colonisation of Africa. This process snuffed out the growth of the indigenous knowledge system which started in Africa during the hunter-gatherer era as Africans interacted daily with their environments.

From a world system perspective, much of the history of the capitalist world economy is a history of colonialism, consisting of repeated and more or less successful attempts by the core to create a periphery, and to control it politically in order to exploit it economically (cf. Sanderson 2005: 186). Both the capitalist and pre-capitalist world systems have had colonial empires (Chase-Dunn and Hall 1997: 210). However, we are more specifically interested in the impact of European, American and Japanese colonialism in their heyday between the mid-nineteenth and mid-twentieth centuries, what Bergesen and Schoenberg (1980) have identified as the second wave of colonial expansion and contraction (1826–1969). This is the period of extension and intensification of colonial domination during which 'colonial economic development took a new direction. The extensive penetration of Western commodities, organization, and control ushered in the era of the export economy, during which colonialism reached its peak' (Birnberg and Resnick 1975: 3). Our sample is broadly defined as the parts of the modern world system which were under colonial control in the nineteenth and twentieth centuries. It consists, as in the previous research mentioned above, of 83 countries of Africa and Asia, which contained around 90 % of the population under colonial rule in 1920.

Colonial Causes of Underdevelopment in Africa

This section is an interrogation of whether or not colonialism can be regarded as a cause of underdevelopment in Africa. Some of the measures taken by colonisers did in fact bring benefits to African countries whereas some did not. At that time, technology and innovation were not as developed as they are nowadays and economists were unable to clearly observe all the challenges Africa was facing, such as the existence of various diseases. This major point should be taken into account. However, the problems are now identified, and the developed countries should aid Africa in reaching the same welfare level as the rest of the nations.

During the twentieth century, Africa suffered from the consequences of colonialism and neocolonialism, and misrule and mismanagement by the members of the political African class. The new generations of Africans, born after the 1960s, blame the Africans in power for the bad situation in Africa due to bad leadership, corruption and a 'get rich culture' that ignores the values of hard work and creativity (Tangie 2006). Based on this, it appears as though Africans contributed to the destruction of Africa. The African population should realise that poor countries from Africa are supporting the well-being of Europeans by providing cheap labour, raw materials, markets, etc. African labour played a key role in building up the 'Atlantic system', and was an important factor in American and European development.

Slavery was also practised in African societies, and it was African leaders and intermediaries who brought most of the slaves to the coast. Colonial rule can be regarded as a phase in Africa's integration into the international system (Simensen 1999). *The Stanford Encyclopedia of Philosophy* defines 'colonialism' as the process of European settlement and political control over the rest of the world. Therefore, colonialism is considered to be a relationship between the indigenous majority and a small number of foreign invaders in which decisions made by colonisers directly affected the lives of the colonised people. In most cases, the colonisers rejected cultural compromises with the colonised population and tried to impose their own rules, maintain order and show superiority.

The manner in which Africa is portrayed nowadays is derived from an encounter with the West, the intervention of the post-1885 'Scramble for Africa', and the partition of the continent between European powers. The map of Africa has changed since then; the only countries that escaped from colonisation were Liberia and Ethiopia, although they were never completely free. Colonial rule lasted between 60 and 80 years for the most part. The changes that appeared were deep, touching various aspects, including cultures and institutions. The exploitative connection between Africa and the global economy changed the way the whole world perceived Africa. Colonial conquest was just a means of showing racial arrogance, humiliating African leaders and their peoples (Toyin 2005).

Colonialism encouraged competition and conflict between ethnic groups. The question is: to what extent can Africa self-develop? Africa was self-developing before Europeans came to colonise the continent. With violence, colonialism created new frontiers, and developed new political and economic objectives. After the end of colonialism, Africa started developing on its own, though in Western terms. The main achievement of colonial rule is state-building. The colonisers regarded themselves as agents of change. They developed modern export systems, and the infrastructure and education facilities that were necessary to make the whole colonisation venture profitable. The commercialisation of land, labour and products gave a specific monetary value to activities that had been previously used in social matters. In pre-colonial Africa, farmers and peasants were producing for either their own use or for trade, though the concept of production for the global market did not exist for them.

Later on, the concept of rent-seeking was introduced in Africa. This is a type of behaviour associated with interest groups in heavily regulated economies. In economics, rent-seeking is an attempt to derive economic rents by manipulating the social or political environment in which economic activities occur, rather than adding value. It involves costs, such as expenditures, in getting laws passed by government (Toyin 2005). Europeans brought two institutions to Africa: plantation agriculture and private property rights. Plantation agriculture required more land than was utilised by the average African farmer and increased the demand for labour services. On the other hand, private property rights replaced the communal ownership systems (ibid.). Colonisers divided the land and set up property rights. The colonial police provided the coercive force needed to prevent the Africans from reclaiming their rights over the land. Europeans suffered a labour shortage and they came up with the following solutions to solve the problem (ibid.):

(a) Recruit Africans to work under European supervision;
(b) Import immigrants from Europe as contract workers;
(c) Import labour from Asia; and
(d) Establish a capitalistic wage labour system.

Europeans invested in resource extraction. The private side focused on exporting the resources, while the public side chose to invest in education, health and infrastructure. The building of roads and railways to carry goods to Europe, as well as other measures, were considered beneficial for Africa's development. However, this is not quite the truth. Europe conquered Africa primarily to benefit itself, not to uplift the African continent. Europe's goal was to accelerate its own development at the expense of others (Mentan 2002).

In order to understand better what occurred, it is best to study the colonial past and understand the legacy of that era on the present. Colonialism was a big step in the incorporation of Africa into a world capitalist system. In the end, colonialism was a period of international commerce, starting with palm produce, cocoa, rubber and peanuts, which attracted many European merchants and trading organisations. The Portuguese established a 'triangular trade' system in 1446 when they arrived in West Africa. Manufactured articles flowed from Europe to Africa; slaves from West Africa were sent to America; and sugar, gold, silver, indigo, tobacco and other products flowed from America to Europe (Mentan 2002). Fifteen million people were taken across the Atlantic and sold into slavery. Approximately 14 % of slaves sent to the New World were children under the age of 14; 56 % were male adults; and 30 % were female adults. Guns and textiles were sent out of Europe and traded in Africa for slaves (Mentan 2002).

The transatlantic slave trade has advantages and disadvantages on both sides. Mainly, slavery gave value to colonies. However, it negatively impacted African countries in several ways. It resulted in an increase in violence, a drain of people and an economy itself relying on slavery. The consequences were deeper than simply the massive depletion of African people, and were manifested in the massive destruction of the entire fabric of African society, the pillaging of goods under the pretext of international trade and so on (ibid.). Most African nations are at present reliant on the export of one or two primary commodities or unpolished goods. Their aim is to sell these exports at the market prices as the global market dictates. As a direct result of colonialism, African economies cannot rely on indigenous manufacturing or industry; in fact, there are very few domestic industries or companies, excluding multinational corporations. As a consequence, African economies are vulnerable to the ups and downs of the global market. On the other hand, most of the countries produce and export a similar commodity, which reduces the ability to bargain and gain favourable terms of trade. African economies have become import- and export-dependent as one of the effects of colonialism in Africa. The import of cars, machines and all types of food is necessary for economic survival.

In 1880, less than 10 % of the continent was under European control, which produced significant changes in the social order and a political landscape of many African nations (Mentan 2002). African territory was divided with no respect for pre-existing divisions. France and England owned large areas within Africa, while small shares went to Belgium and Italy. Germany had a temporary influence on the colonies of Togo, Cameroon, Namibia and Tanganyika. Europeans settled in poor, uninhabited regions where they could invest. Moreover, Africans say that exploitation was the main objective of Europeans. They developed an economy based on cash crops. Farmers and traders received larger incomes from cash crops, though the cost of purchasing food was higher. Furthermore, in the event that the colony lacked resources, Europeans used the African people as a 'labour basket' (Toyin 2005).

After the exploration of the territories, the Europeans agreed that capitalism must 'nestle everywhere' (Duignan and Gann 1975). It took them a short time to shift from the 'civilising mission' to profit and power. The European powers felt that it was their duty to bring Western civilisation to the uneducated population. In the beginning, Europeans implemented capitalism as the economic system. Capitalism allows every person to work for their own profit and to compete, which leads to an improvement in the welfare of society. This assumes that economic growth is created by profit and the value added by a wide range of small businesses. However, capitalism does not take into account a series of characteristics that these societies have.

The absence of sustained, long-term growth in the volume of production or trade—combined with the lack of technology and capital accumulation—constitutes an obstacle to development (Sender and Smith 1986). Supply and demand are both affected. Supply is restricted by the inadequacy of roads and the attempts in some areas to maintain the balance of power by limiting the access to trade. Demand is influenced by the impact of trading relations with advanced capitalist economies. The rate of growth of demand for tradable goods was seriously limited by the technical achievements of pre-capitalist trading partners. Even the most developed African economies were subject to serious limitations concerning economic specialisation. The rapid development of Europe in the nineteenth century and the formation of economic and political forces in some of the countries created a new vision of Africa in terms of investment and exchange of commodities. The flow of goods implied not only their production, but also included transport, storage, packaging materials, markets and lands prepared for the increasing demand (Table 12.1).

Table 12.1 Trade of Africa in 1897

	Imports £(000)	Exports £(000)
Southern Africa	23,915	20,467
Lagos	771	811
Niger Protectorate	640	750
Sierra Leone	457	401
Gambia	140	164
	2,919	*2,984*
British Somaliland	348	350
Egyptian Sudan	91	66
British East Africa	198	73
Zanzibar and Pemba	144	81
British Central Africa	*86*	*27*
Portuguese East Africa	4,350	273
Angola	1,031	1,195
Guinea	8	3
Total Portuguese	*5,389*	*1,471*
Guinea	306	269
Senegal	1,167	843
Ivory Coast	188	189
Dahomey	330	231
French Congo	143	211
Total French	*2,134*	*1,743*
German East Africa	479	262
South West Africa	244	62
Togoland	99	39
Kamerun	295	185
Total German	*1,117*	*548*
Congo Free State	950	537
Grand Total	*37,391*	*28,347*

Source: S. Herbert Frankel, *The tyranny of economic paternalism in Africa: a study of frontier mentality 1860–1960* (Johannesburg 1960: 17)

By 1897, for example, only 4.3 % of British imports derived from Africa, and only 10.4 % of British exports went to Africa. As shown in Table 12.2 above, the most developed African countries were located in the European settlement areas of Southern Africa (Duignan and Gann 1975). Due to the fact that African societies could not deal with the capitalist system, colonisers had no choice but to change their approach to one of imperialism. The transformation, according to Lenin (1999), included:

1. The concentration of production and capital developed to such a high stage that it created monopolies which govern economic life;
2. The merging of bank capital with industrial capital, which formed a 'finance capital' of a financial oligarchy;
3. The export of capital was more important than the export of commodities;
4. The formation of international monopolist capitalist associations which were ruling the world; and
5. The territorial division of the world among the biggest capitalist powers.

The main purpose of imperialism is to change quantity into quality, by transforming free competition into a monopoly. Colonial imperialism added to the characteristics mentioned above (Gino 2007):

(a) A new system of administration;
(b) A process of exploitation; and
(c) A production system often geared towards the creation of capitalist relations of domination-subordination, and the economic and socio-cultural enhancement of the coloniser.

Table 12.2 Investment in Africa (in million US dollars)

	Years	Public aid	Private investment
Great Britain	1956–1963	2,867	2,791
France	1956–1963	6,811	2,853
Belgium	1956–1980	444	523

Source: *African Economic Outlook*

Governments tightened up their control over European companies and became more concerned with African needs and interests. They forced colonies to pay and encouraged bankers, investors, traders and business groups. Tax systems were created in order to attract investors. The British expected each colony to pay its debts, while the Germans and French invested money in their colonies. In order to run the colonies, the new rulers levied import duties and imposed hut taxes (Phillips 1989). Companies were obliged to pay taxes in Europe, so the colonial administrators were never in possession of these. The export duties imposed were used to funnel goods to the metropolitan markets in order not to raise revenue. The policy of each colony was to attract and serve European enterprises and to encourage farmers to export their crops. The colonial administration determined laws on land and on labour class, and on economic planning and control. Usually, the first sector of the economy to be regulated was commerce (ibid.). British Africa was a free-trade area for most of the colonial period. The British did not control the colonial exchange economies for the benefit of the metropolis and they did not set up a monopoly on the purchase of colonial products. Instead, they encouraged the exchange of goods between other colonies.

Conversely, France attempted to protect its colonies and push them to serve the metropolis's interests. The colonies followed an open-door policy only when international treaties forced them to do so. Colonial exports to France were favoured and exempted from taxes. In addition, French ships did not pay certain charges in some colonial ports. As shown in Table 12.2, France dominated the export-import trade of its African colonies to a far greater extent than the others colonisers.

Between 1890 and 1939 over 32,000 miles of railroad were built in Africa. New cities appeared and over 400,000 miles of roads were created. Plantations, and light and service industries served mines, towns and farms. Cars and trucks were brought to Africa to make some regions more accessible for commerce. So, the overall effects of foreign trade indicates a considerable influx of new commodities. Moreover, people developed new skills and found better job opportunities (Easterly and Levine 2002).

The speed of economic development differed between the different regions in Africa, mainly because it was not attractive enough to European investors. During the colonial period, merely 9.3 % of French foreign investment and only 15 % of British capital benefited Africa. As shown in Table 12.2, government investment exceeded private investment in African colonies. Prior to 1945, capital extended to where it could receive the greatest return. What was lacking from Africa was not the capital, but the educated workforce that could easily adapt to a modern society. French and Belgian investments were larger after 1945, when governments decided to expand the scope of colonial development to include research and social services. A serious number of voices criticised the colonial system, not only the colonisers, but also outsiders such as philanthropic non-governmental organisations and African associations from abroad (for example, the International African Service Bureau) (ibid.).

Colonial Impact and its Post-Colonial Reverberations

The colonial powers used brutal policies and devious methods to subjugate the African people in order to acquire full control over their lands and resources. These policies and methods included the consistent inferiorisation of indigenous cultures, and concerted efforts to erase existing systems of knowledge, replacing these with Western-driven belief and knowledge systems. Such pre-mediated policies successfully culminated in, on one hand, the absolute submission of the communities and stigmatisation of their knowledge systems. This consequently led to the trapping of most of the communities in the perpetuating of their own subjugation (Progler 1999) through Western education, Christianisation and the degeneration of relatively self-sufficient economies into dependent consumers (Eyong 2005). On the other hand, the colonial design succeeded in producing the economic imbalances essential for the growth and domination of European capitalism and imperialism.

This project of domination continued to prosper during the post-colonial (also called neocolonial) period, with articulations in the economic and political domains (Progler 1999) and the current systems of production and dissemination of academic knowledge in Africa, which reflects Western hegemony (Zeleza 2006). As Zeleza states, 'the provisions of intellectual exchange are distinctly unequal; as the African Studies in Europe constitute a marginal part of the academy while the European epistemology remains central in African Studies' (ibid.: 197).

Therefore, the Western domination of knowledge and the marginalisation of African systems of knowledge continue to be an academic challenge that calls for a comprehensive evaluation, rigorous planning and watchful implementation of policies that ensure the recognition and provisions of space for these in the existing political, economic, cultural and pedagogical domains. It is only then that African indigenous knowledge systems may successfully be established and gradually contextualised within these domains.

Conceptualisation of Indigenous Peoples

The term 'indigenous peoples' is in itself a contested category of people, so too is indigenous knowledge. The former refers to 'culturally distinct ethnic groups with a different identity from the national society, drawing an existence from local resources and are politically non-dominant' (Melchias 2001: 35). The World Bank (1991) adds a development perspective by stating that indigenous peoples are 'social groups with a social and cultural identity distinct from the dominant society that makes vulnerability to being disadvantaged by the development process'.

The UN has no universally accepted definition but states that 'indigenous communities, peoples and nations are those which, having a historical continuity with pre-invasion and pre-colonial societies that developed on their territories, consider themselves distinct from other sectors of the societies now prevailing in those territories, or parts of them' (Cobo 1987). There are some common grounds from these definitions. Indigenous peoples are people living in an area within a nation state, prior to the formation of a nation state, but may identify with it; they have maintained a great part of their distinct linguistic, cultural, social and organisational characteristics. They can be differentiated, to some degree, from the surrounding populations and dominant culture of the nation state. It would make sense to note that, for this claim to be valid, others must see such groups as indigenous. Found mainly in areas where they have lived for thousands of years, indigenous people inhabit nearly a fifth of the planet. Contemporary discourse on who is an indigene has led to the talk of 'autochthones', indigenous inhabitants, and 'non-indigenes', migrant settlers.

However, this distinction is politically laden, especially in the Cameroonian context. The cloudy atmosphere surrounding the definition of indigenous peoples also explains the variance in estimates of their numbers. Estimates show that between 300 and 500 million indigenous peoples speak a vast majority of the world's languages and represent the majority of cultural diversity (Melchias 2001). These groups and their cultures must be preserved for posterity.

Indigenous Knowledge: Definitions and Scope

Indigenous knowledge refers to what indigenous people know and do, and what they have known and done for generations—practices that evolved through trial and error and proved flexible enough to cope with change (Melchias 2001). This definition draws our attention to the racist colonial idea that indigenous knowledge is characterised by trial and error while Western (modern) knowledge is scientific, characterised by experimentation. Hence while the former is presumed to be concrete and inaccurate, the latter is painted as intangible, correct and imbued with universal reasoning. Indigenous knowledge systems were also developed by experimentations, though these experiments were not documented and the knowledge systems were legitimised and fortified under suitable institutional frameworks, culture and practices. They have been passed on to other generations (though discriminatorily) and have enabled indigenous people to survive, manage their natural resources and the ecosystems surrounding them, such as animals, plants, rivers, seas, natural environment, economic, cultural and political organization.

The knowledge of these elements forms a set of interacting units known as 'indigenous coping systems'. Here, indigenous knowledge systems refer to the set of interactions between the economic, ecological, political and social environments within a group or groups with a strong identity, drawing its existence from local resources through patterned behaviours that are transmitted from generation to generation to cope with change. These patterns are sustained by micro-level institutional arrangements vested with differentiated responsibilities that ensure the group's continuous survival. Unfortunately, these systems are fast eroding due to colonialism, commercialisation, globalisation and modernisation, lack of efficient codification, breakdown of the traditional family structure and function (the institution that helps in the socialisation of tacit knowledge), developmentally induced human displacements, the decline in the practitioner base and many other factors.

Technological Stagnation and Distortion of African Society

It has already been indicated that in the fifteenth century European technology was not totally superior to that of other parts of the world. There were certain specific features which were highly advantageous to Europe—such as shipping and (to a lesser extent) guns. Europeans trading to Africa had to make use of Asian and African consumer goods, showing that their system of production was not absolutely superior. It is particularly striking that in the early centuries of trade, Europeans relied heavily on Indian cloth for resale in Africa, and they also purchased cloth in several parts of the West African coast for resale elsewhere. Morocco, Mauretania, Senegambia, Ivory Coast, Benin, Yorubaland and Loango were all exporters to other parts of Africa—through European middlemen. Yet by the time Africa entered the colonial era, it was concentrating almost entirely on the export of raw cotton and the import of manufactured cotton cloth. This remarkable reversal is tied to technological advance in Europe and to the stagnation of technology in Africa, owing to trade with Europe.

Europe benefited technologically from its external trade contacts, while Africa either failed to benefit or actually lost out. Vital inventions and innovations appeared in England in the late eighteenth century, after profits from external trade had been re-invested. Indeed, the new machinery represented the investment of primary capital accumulated from trading and from slavery. African and Indian trade strengthened British industry, which in turn crushed whatever industry existed in areas that are now called 'underdeveloped' countries.

There are many social factors which combine to determine when a society makes a breakthrough from small-scale craft technology to equipment designed to harness nature so that labour becomes more effective. One of the major factors is the existence of a demand for more products than can be made by hand, so that technology is asked to respond to a definite social need—such as that for clothes. When European cloth became dominant on the African market, it meant that African producers were cut off from the increasing demand. The craft producers either abandoned their tasks in the face of cheap available European cloth, or they continued on the same small hand-worked instruments to create styles and pieces for localised markets. Therefore, there was what can be called 'technological arrest' or stagnation, and in some instances actual regression, since people forgot even the simplest techniques handed down by their forefathers. The abandonment of traditional iron smelting in most parts of Africa is probably the most important instance of technological regression.

In Africa, the trading groups could make no contribution to technological improvement because their role and preoccupation took their minds and energies away from production. Apart from inventiveness, we must also consider the borrowing of technology. When a society, for whatever reason, finds itself technologically trailing behind others, it catches up not so much by independent inventions but by borrowing. Indeed, very few of humanity's major scientific discoveries have been separately discovered in different places by different people. Once a principle or a tool is known, it spreads or diffuses to other peoples. Why then did European technology failed to make its way into Africa during the many centuries of contact between the two continents? The basic reason is that the very nature of Afro-European trade was highly unfavourable to the movement of positive ideas and techniques from the European capitalist system to the African pre-capitalist (communal, feudal and prefeudal) system of production.

Japan was already a highly developed feudal society progressing towards its own capitalist forms in the nineteenth century. Its people were neither enslaved nor colonised by Europe, and its foreign trade relations were advantageous. For instance, Japanese textile manufacturers had the stimulus of their own growing internal market and some abroad in Asia and Europe. Under those circumstances, the new Japanese capitalist class (including many former feudalist landowners) borrowed technology from Europe and successfully domesticated it before the end of the nineteenth century. The use of this example from outside of Africa is meant to emphasise that for Africa to have received European technology the demand would have had to come from inside Africa—and most probably from a class or group who saw profit in the new technology. There had to be both a willingness on the part of Europeans to transfer technology and African socio-economic structures capable of making use of that technology and internalising it.

Hunting for elephants or captives did not usually induce in Africa a demand for any technology other than for firearms. The lines of economic activity attached to foreign trade were either destructive, as slavery was, or at best purely extractive, like ivory hunting and cutting camwood trees. Therefore, there was no reason for wanting to call upon European skills. The African economies would have had little room for such skills unless negative types of exports were completely stopped. A remarkable fact that is seldom brought to light is that several African rulers in different parts of the continent saw the situation clearly, and sought European technology for internal development, intended to replace the trade in slaves.

In the early nineteenth century, one king of Calabar (in Eastern Nigeria) wrote to the British asking for a sugar refinery, while around 1804 king Adandozan of Dahomey was bold enough to ask for a firearms factory. By that date, many parts of West Africa were going to war using European firearms and gunpowder. There grew up a saying in Dahomey that 'he who makes the powder wins the war', which was a far-sighted recognition that Africans were bound to fail before the superiority of Europeans in the field of arms technology. Of course, Europeans were also fully aware that their arms technology was decisive, and there was not the slightest chance that they would have agreed to teach Africans to make firearms and ammunition. As seen so far, capitalism introduced into Africa only such limited aspects of its material culture as were essential to more efficient exploitation, but the general tendency has been for capitalism to restrict African access to technology.

Indigenous Knowledge in Technological Use in Africa

The term 'indigenous knowledge' is used to describe the knowledge systems developed by a community as opposed to the scientific knowledge that is generally referred to as 'modern' knowledge (Ajibade 2003). Indigenous knowledge is the basis for local-level decision-making in many rural communities. It has value not only for the culture from which it evolves, but also for scientists and planners striving to improve conditions in rural localities (Robinson and Herbert 2001).

Warren (1991) and Flavier et al. (1995) present typical definitions by suggesting that indigenous knowledge is the local knowledge—knowledge that is unique to a given culture or society. Indigenous knowledge contrasts with the international knowledge system generated by universities, research institutions and private firms. It is the basis for local-level decision-making in agriculture, health care, food preparation, education, natural-resource management, even technology and a host of other activities in rural communities (Warren 1991). Indigenous knowledge is the information base for a society, and it facilitates communication and decision-making. Indigenous information systems are dynamic, and are continually influenced by internal creativity and experimentation as well as by contact with external systems (Flavier et al. 1995: 479).

Indigenous knowledge systems refer to intricate knowledge systems acquired over generations by communities as they interact with their environments. They encompass technological, economic, philosophical, learning and governance systems. They refer to a body of empirical knowledge and beliefs handed down through generations, by cultural transmission, of long-time inhabitants of a specific locale about the relationship of living beings with each other and their environment (Warren 1991; Johnson 1992: 4). These systems are about reopening crucial functional templates that were closed in the process of modernisation and development in which the cultural, socio-economic life of indigenous peoples was negatively impacted.

Indigenous knowledge encompasses spiritual relationships, relationships with the natural environment and the use of natural resources, relationships between people, and is reflected in language, social organisation, values, institutions and laws. These knowledge systems are usually embedded in naturalistic epistemologies and belief systems, which differ radically from those of scientific systems (IUCN 1997). However, indigenous knowledge has value not only for the culture in which it evolves but also for researchers who are interested in improving conditions in rural localities (Warren 1991). The recognition of indigenous knowledge systems is, therefore, crucial for the economic and cultural empowerment of indigenous people in particular, and the world in general.

Indigenous people improve their livelihood through the use of simple technology as applicable to all areas of their lives in order to make life better and comfortable for all and sundry in the community. Many of these communities work on finding and developing technologies that improve the lives of the poorest and provide openings for small local enterprises. Such ventures include bee-keeping, making soap from local materials, planting weeds that repulse animal pests from the fields, and traditional crop varieties that grow best in local conditions. The work of Gibbons (cited in Appiah-Opoku 1999: 217) in Sudan, for instance, suggests that by relying on local informants and drawing on their technical knowledge, it may be possible to undertake a perfectly satisfactory soil survey and mapping in a few days even though a formal scientific approach could take several months. Similarly, pastoralists in Mali who noticed that drinking a lot of tea made people nervous and irritable wondered as to whether the tea or sugar was the causal agent. To identify the cause, they poured sugar and water on a liver from a freshly slaughtered animal and there was no visible reaction. A similar experiment with tea and sugar gave a visible reaction. They concluded, rightly, that tea was the causal agent (Chambers 1983). Similarly, Ghanaian farmers are able to identify precursors of climatic and weather conditions. There is always a strong possibility of a prolonged dry season when a vulture is seen incubating at a time when rainfall is widely expected. Farmers can tell the time accurately by observing the range of their shadow under the tropical sun. They also have a good knowledge of the characteristics of soils in the local environment. Indicators commonly used by the farmers to determine soil fertility include the presence of clay, a low proportion of sand and moisture, decayed organic matter, and earthworms and certain plant species (Asaman-ntorowa or solanum torvum) (Appiah-Opoku 1999: 221).

In Nigeria, the simple traditional technology explored by the goldsmiths has been totally overtaken by the introduction of modern technology. The development of science and technology does not benefit Africa in any form. For instance, a former Federal Capital Territory (FCT) Minister in Nigeria quizzed the management of multinationals in Nigeria on why Nigerians were not found in decision-making cadres in their organisations? They simply replied that Nigerians cannot operate the kind of modern technology they use in discharging their specified or specialised duties in the workplace. The question is, does that mean Africans (Nigerians) are mentally incapable of operating modern technology? Was there any effort to train them on how to use or operate any of these ICT facilities before reaching such an irrational conclusion? This also explains the challenges and problems Africa is facing in the application or adoption and use of ICT facilities in different sectors. For instance, the studies conducted by Richard and Richard (2001) in Botswana and Kolawole (2005) in Nigeria on the adoption and use of ICT in selected manufacturing companies showed certain problems in ICT application and use, such as low capital investment, a lack of trained personnel in ICT use and development, the excessively high cost of ICT facilities and the maintenance cost of these, to mention but few.

It is noteworthy that when one travels to any of the developed nations, things are well organised and functional such that if care is not taken, one would exhaust their savings without handling cash. That is why Africa is encouraged to develop its own technology and improve on it until its indigenous technology can favourably compete with that of the developed nations of the world. There is still room for the development and integration of indigenous technology in Nigeria as there are lots of organisations in the country, be it in the service or manufacturing sector, that are yet to comply with the policy of adopting and using relevant home-grown ICT facilities. To this end, the indigenous knowledge in the area of technology can be developed within the culture of every community in order to boost meaningful growth and development in our own way, rather than leveraging on technology that does not improve the state of the country.

Conclusion

In conclusion, this chapter considered that the nations of Africa have been totally ostracised from the deployment and use of relevant modern-day ICT facilities because of the influence of imperialism, which has widened the digital divide between developed and underdeveloped nations. Africa is now at receiving end. Instead of hoping for help or assistance that will never come, Africa should boldly return to its abandoned indigenous knowledge systems and internalise it in this modern day to suit work processes, structures and designs. For instance, in the Nigeria banking sector, one of the sectors that has adopted and used ICT facilities because banking operations and services have been built round such facilities, customers in this sector are not in any way satisfied at all because ICT does not provide adequate service and operations to the customers. Customers are often disappointed with the use of ICT facilities to carry out any banking operation. A customer once said, '*Nba za su yi ha ba, ba dole ba ne, mu ko ma inda mu ke yi da,*' meaning, 'it isn't by force, if we cannot do it using ICT facilities, let us go back to the traditional way of doing banking operation and services.' Therefore, if Africa must develop to compete with world powers, the new perspective and emerging alternatives Africa must embrace fully is the indigenous knowledge systems in the area of technology since nothing can be done without (ICT) technology.

References

Appiah-Opoku, S. (1999). Indigenous economic institutions and ecological knowledge: A Ghanaian case study. *The Environment, Issues, 19*, 217–227.

Arnold, G. (1985). *Aid and the Third World: The North/South divide*. London: Robert Royce Ltd.

Bergesen, A., & Schoenberg, R. (1980). Long waves of colonial expansion and contraction. In A. Bergesen (Ed.), *Studies of the modern world-system* (pp. 231–277). New York, NY: Academic Press.

Chambers, R. (1983). *Rural development: Putting the last first*. London: Longman.

Chase-Dunn, C., & Hall, T. D. (1997). *Rise and demise. Comparing world-systems*. Boulder, CO: Westview Press.

CIA World Fact Book. (2009). *Equatorial Guinea*. Retrieved from https://www.cia.gov/library/publications/the-world-factbook/geos/ek.html

Cobo, M. J. (1987). *Study of the problem of discrimination against indigenous populations*. New York: United Nations Working Group on Indigenous Populations.

Duignan, P., & Gann, L. (1975). *Colonialism in Africa 1870–1960* (Vol. 4). London: Cambridge University Press.

Easterly, W., & Levine, R. (2002). Tropics, germs and crops: How endowments influence economic development. *NBER Working Paper 9106*, National Bureau of Economic Research, Cambridge, MA.

Eyong, C. T. (2005). Indigenous knowledge and sustainable development in Africa: Case study on Central Africa. In K. B. Emmanuel & H. Luc (Eds.), *Indigenous knowledge systems and sustainable development: Relevance for Africa*. Published in *Tribes and Tribals*. Special Volume No. 1. (pp. 121–139).

Flavier, J. M., et al. (1995). The regional program for the promotion of indigenous knowledge in Asia. In D. M. Warren, L. J. Slikkerveer, & D. Brokensha (Eds.), *The cultural dimension of development: Indigenous knowledge systems* (pp. 479–487). London: Intermediate Technology Publications.

Gino, R. (2007). *French colonialism unmasked: The Vichy years in French West Africa*. London, UK: University of Nebraska Press.

IUCN. (1997). *Indigenous peoples and sustainability: Cases and actions*. Amsterdam: IUCN.

Johnson, M. (1992). *Lore: Capturing traditional environmental knowledge*. Ottawa: Dene Cultural Institute/IDRC.

Kolawole, T. O. (2005). *Adoption and use of ICT in selected manufacturing industries in Ibadan city*. Unpublished M.Sc thesis dissertation.
Lenin, V. (1999). *Imperialism: The highest stage of capitalism*. Sydney: Resistance books.
Lynch, C. (2010, April 29). *UNESCO's strongman prize for life sciences*. Retrieved from http://turtlebay.foreignpolicy.com/posts/2010/04/29/unescos_strongman_prize_for_life_sciences
Melchias, G. (2001). *Biodiversity and conservation*. Enfield: Science Publishers, Inc.
Mentan, T. (2002). Exiting the whirlpool? Pan-Africanism caught in the cross-fire of identity and globalization. *African Journal of International Affairs, 5*, 1–2.
Ocheni, S., & Nwanko, B. C. (2012). Analysis of colonialism and its impact in Africa. *Cross-Cultural Communication Issues, 8*(3), 46–54.
Pambazuka News. (2010, February 26). *UNESCO-Obiang Nguema Mbasogo International Prize for Research in the Life Sciences*. Retrieved May 20, 2010, from http://www.pambazuka.org/en/category/action/62590
Phillips, A. (1989). *The enigma of colonialism. British policy in West Africa*. London, UK: Currey.
Progler, Y. (1999, December). Towards defining Western research on Indigenous Knowledge. *Muslimedia*, pp. 1–15.
Richard, D., & Richard, H. (2001). *Information and communication technologies and small enterprises in Africa*. Published by Institute for Development Policy and Management.
Rodney, W. (1982). *How Europe Underdeveloped Africa*. London and Dar-Es-Salaam: Bogle-L'Ouverture Publications and Tanzanian Publishing House.
Sanderson, S. K. (2005). World-systems analysis after thirty years. *International Journal of Comparative Sociology Issues, 46*, 179–213.
Sender, J., & Smith, S. (1986). *The development of capitalism in Africa*. London: Methuen.
Simensen, J. (1999). *Africa: The causes of under-development and the challenges of globalization*. Oslo: University of Oslo.
Tangie, N. F. (2006). The state and development in Africa. Retrieved from https://www.codesria.org/IMG/pdf/tangie.pdf?1422/
Toyin, F. (2005). *The dark webs: Perspectives on colonialism in Africa*. Durham, NC: Carolina Academic Press.
Warren, D. M. (1991). *The role of indigenous knowledge in facilitating the agricultural extension process*. Paper presented at International Workshop on Agricultural Knowledge Systems and the Role of Extension, Bad Boll, Germany.

World Bank. (1991). *Indigenous people. The World Bank operational manual, operational directive*. Washington, DC: World Bank.
Zeleza, P. T. (2006). The disciplinary, interdisciplinary and global dimensions of African studies. *International Journal of African Renaissance Studies, 1*(2), 195–220.

Taiwo Olabode Kolawole holds a doctorate in sociology from Obafemi Awolowo University, Ile-ife, Nigeria. He currently teaches sociology at the Federal University, Oye-Ekiti, Nigeria. He has published in notable local and international journals.

Part V

Crime, Conflicts and Regional Dynamics

13

North Africa: Can Common Security Challenges Promote Regional Integration?

Rawia Tawfik

Introduction

In the last two decades, security started to feature clearly on the agenda of regional organisations in Africa. In West Africa, the Economic Community of West African States (ECOWAS) Ceasefire Monitoring Group (ECOMOG) played an active role in peacekeeping and conflict management in Liberia, Sierra Leone, Côte d'Ivoire and Guinea Bissau. The Southern African Development Community (SADC) created its body on Politics, Defence and Security in 1996 based on its declaration's statement that security and stability are prerequisites for development. In the same year, the Inter-Governmental Authority on Drought and Development (IGAD), a regional community established in 1986 and focused initially on drought and famine in the Horn of Africa, expanded its mandate to cover peace and security. Subsequently, the organisation

R. Tawfik (✉)
Faculty of Economics and Political Science, Cairo University,
Orman, Giza, Egypt

A. Adeniran, L. Ikuteyijo (eds.), *Africa Now!*,
https://doi.org/10.1007/978-3-319-62443-3_13

has played an active role in facilitating peace agreements in Somalia and Sudan.

Compared to other regions, North Africa has lagged behind in terms of security cooperation and regional integration. The Arab Maghreb Union (AMU), set up in 1989 to include five Arab-African countries—Algeria, Libya, Mauritania, Morocco and Tunisia—has been almost frozen since the mid-1990s. The security role of the Community of the Sahel and Saharan States (CEN-SAD), which comprises more than 20 countries from North Africa and the Sahel, rests more on individual member states than on the organisation itself.

In the last few years, the North Africa uprisings have shaped a new regional environment. The fall of the Muammar Gaddafi regime in Libya has fuelled arms smuggling not only in North Africa, but also south to the Sahel. His overthrow left a power vacuum that was filled by competing militias, and more recently by forces affiliated with the Islamic State (IS). Extremists based in different Libyan cities have executed operations inside Tunisia and targeted Egyptian workers in Libya. In Algeria, a country that had its share of civil strife throughout the 1990s, Western interests have been targeted, with extremists holding hostages in Ain Amenas in January 2013 to revenge the Algerian decision to open its air space to French planes bombing northern Mali. But can these new security challenges provide a basis for cooperation and regional integration in North Africa? This question is the focus of this chapter.

The chapter presents a central argument. In spite of common security threats, the regional and international factors that have contributed to fostering security cooperation and regional integration in other parts of the continent do not yet exist in North Africa. According to the Regional Security Complexes (RSCs) theory, the absense of security interdependence between states sharing the same region can be attributed to the intervention of international powers to suppress regional dynamics, and/or the lack of capabilities of the states in the region to act beyond their borders. Given their interests in the region, international powers, which tended to delegate the responsibility of conflict resolution to regional powers and organisations in other parts of the continent, are eager to intervene to manage conflicts in North Africa. The intervention of the North Atlantic Treaty Organisation (NATO) in Libya in 2011, and the subsequent deployment of French secret special forces and the launching of American airstrikes against IS in Libya illustrate this interest. Equally important, there is no clear regional hegemon that is ready to lead the process of transforming regional organisations, carry a disproportionate share of maintaining them and work to achieve regional consensus. Regional politics is still dominated by the historical rivalry between Algeria and Morocco. In spite of recent efforts of revitalisation, the AMU remains dysfunctional and the CEN-SAD lacks the capacity to manage regional conflicts.

The chapter is divided into six sections. The first section outlines the main arguments of the Regional Security Complexes (RSCs) theory that explain the prescence or absence of security interdependence at the regional level. The second section illustrates how the agenda of regional economic communities in Africa has developed to include peace and security and how this agenda links peace and development. The third section evaluates the record of regional integration in North Africa, illustrating why the region has lagged behind in security and development cooperation compared to other parts of the continent. The fourth section illustrates the role of international powers in the region and its impact on stability and development. The fifth section explains the efforts made towards reviving regional integration after the North Africa uprisings and the limitations of these efforts. The last section discusses and summarises the findings of the chapter.

Regional Security Complexes: From a Traditional to a New Security Agenda

The concept of RSCs emerged in the mid-1980s to refer to security interdependence and cooperation between states belonging to the same region. RSCs refer to 'a set of states whose major security perceptions and concerns are so interlinked that their national security problems cannot reasonably be analyzed or resolved apart from one another' (Buzan 2003: 141). Consequently, security interdependence is more intense between states inside such RSCs than with states outside it (Buzan 1983: 105–106; 2003: 141; Buzan and Waever 2003: 47–48). The links that tie the regional complex together may be geographical, political, strategic, historical, economic, or cultural (Buzan 1983: 106).

This definition of RSCs indicates that the existence of objective security threats is not a sufficient condition for fostering closer cooperation between states. These states should have shared perceptions on the impact of these threats on their stability and/or survival and the necessity of coordination to face them. This does not necessarily mean that these states have always enjoyed peaceful and cooperative relations. Even rival states can engage in security cooperation to reduce conflicts and contain mutual threats (Buzan 2003: 142).

However, RSCs may be absent in regions that face common security challenges and whose member states perceive these challenges as bases for collaboration. According to the RSCs theory, two conditions can explain the absence of a regional security complex. The first condition is the lack of states' capacity to act beyond their own borders. A related aspect identified in other writings is the absence of a regional hegemon that is able to lead the cooperation process, provide the largest share of the costs of this cooperation, and achieve consensus between different members of the complex (Clapham 2001: 67). The second condition is the strong presence of outside powers in the region in a way that suppresses and marginalizes the regional security dynamics between states. This presence, or overlay as labeled by the RSCs theory, transcends the level of penetration to subordinate the regional interactions to the larger patterns of major power strategies and competition (Buzan 2003: 142–143).

In addition to the explanation of the presence and absence of security complexes in particular regions, the RSCs theory has been interested in the changes in the structures of these complexes. These changes may occur due to internal transformations in the distribution of power among member states and/or the patterns of amity and enmity between them. They may also result from external transformations that add or exclude major states that may affect power relations or patterns of conflict and cooperation within the regional complex. International powers may also intervene in the regional complex affecting its security dynamics (Buzan 2003: 143–144).

Although the RSCs theory advocates the regional level, it actually links four levels of analysis: the domestic level by looking at the state's definition of its security interests and threats; the regional level defined by the intense interactions between states; the region's interactions with neighboring regions; and the global level by examining the role of international powers in the region. In spite of the importance of the global level, it is the regional pattern of conflict and cooperation that determines the engagement of international powers in the region (Buzan and Waever 2003: 51–52).

With the end of the Cold War, the concept of security was redefined to include aspects that are not purely political or military, but that are not totally devoid of security characters. This invited a revised definition of regional complexes. Economic and environmental security are examples of these new aspects that encouraged states to agree on regional arrangements and regimes to face common concerns. Although these arrangements may appear at face value to be driven by economic and environmental motives, they also represent political and cultural projects in which countries come together to defend a set of values and/or rules of management. Societal security is another new dimension linked to protecting the identity of nations whose geographical expansion may correspond with the borders of the states or transcend them. According to the RSCs theory, an approach that acknowledges the linkages between these different elements of security and examines how they are tackled in policy-making at the national and regional levels by different state and non-state actors is necessary to capture the complexity of current RSCs (Buzan 2003: 152–159; Buzan and Waever 2003: 70–71).

The rest of the chapter applies the concept of security complexes, in its revised definition, on Africa in general and North Africa in particular. It illustrates the reasons for the absence of an RSC in North Africa compared to other regions on the continent. Following the indicators and conditions proposed by the RSCs theory, it examines the states' perceptions of security interdependence in the region, the capabilities of regional powers to act beyond their borders, the role of international powers in the region, and the transformations of these elements after the popular uprisings in 2010–2011.

Regional Integration in Africa: The Security–Development Nexus

Since independence, African governments have regarded regional integration as a driver for development in the continent. The Lagos Plan of Action for the Economic Development in Africa adopted by the Organisation of African Unity (OAU) in 1980 proposed a 'regional approach based primarily on collective self-reliance'. The plan reaffirmed Africa's commitment to set up an African Economic Community by 2000 to ensure the economic, social and cultural integration of the continent (OAU 1980). In 1991, African countries adopted the Abuja Treaty for the establishment of an African Economic Community. The treaty details measures of integration and the gradual harmonisation of policies in the fields of agriculture, industry, transport, trade, education and culture. It follows a market integration approach that starts with strengthening the existing Regional Economic Communities, then moves to the establishment of a free trade area followed by a customs union, and ends with a common market, monetary union and a single currency (OAU 1991).

Since the beginning of the 1990s, there has been a growing emphasis on security in the integration agenda of the OAU and African regional organisations. The end of the Cold War, and thus the retreat of superpower rivalries contributed to the emergence of 'a more regionalized international security order' in Africa and elsewhere (Buzan 2003: 145). International powers were more reluctant to intervene directly in African conflicts. This happened at a time of proliferating civil conflicts on the continent and increasing recognition by African governments and international institutions of the impact of conflicts on both development and regional integration. Equally important, in some regions, there has been a regional hegemon that is ready to lead the process of transforming regional organizations to maintain security, carry a 'disproportionate share' of the costs of maintaining the grouping, and constantly work to achieve consensus within it (Clapham 2001: 62–67; Fawcett and Gandois 2010: 620–630). At the continental level, this emphasis on the security role of regional organizations was reflected in a number of documents. The Cairo Declaration establishing a Mechanism for Conflict Prevention, Management and Resolution was adopted in 1993. The Declaration admitted the precarious state of socio-economic development in Africa and noted that 'no single internal factor has contributed more to the present socio-economic problems in the continent than the scourge of conflicts in and among our countries' (OAU 1993, article 9). It mandated the new mechanism to undertake peace-making and peace-building functions in order to facilitate the resolution of emerging conflicts (OAU 1993).

The emphasis on security as a condition for development and cooperation was reiterated in the Solemn Declaration of 2000. The declaration adopted a concept of human, rather than merely state, security. It noted that 'the concept of security must embrace all aspects of society including economic, political, and social and environmental dimensions of the individual, family, and community, local and national life. The security of a nation must be based on the security of the life of the individual citizens to live in peace and to satisfy basic needs while being able to participate fully in societal affairs and enjoying freedom and fundamental human rights' (OAU 2000, article 10). A similar conception was adopted by the New Partnership for Africa's Development—the economic blueprint of the African Union (AU). In its framework document, the New Partnership defined peace, security and democratic governance as conditions for sustainable development (AU 2001, Section v). This is why the AU, as the successor of the OAU, has developed a peace and security architecture to promote the role of the continental body in conflict management and peace-building (Dersso 2014, 53–54).

At the regional level, security interdependence, and thus the sophisitication of the RSCs, has varied from one region to the other. Southern Africa represented a full-fledged RSC with South Africa as a regional power at its core (Buzan and Waever 2003, 233–235). In 1996, SADC created its body on Politics, Defense and Security. This was based on SADC's conception that peace, security and stability are prerequisites for development and for the improvement of the quality of life of the people. However, it was noted that SADC protocol on Politics, Defense and Security did not clearly define security, mixing elements of both traditional conceptions of state security and more recent ideas of human security. Furthermore, SADC's expansion into regional conflict management 'has not been based on a coherent policy approach' in linking security and development in practical terms. Rather, it has been largely driven by the diverse and conflicting interests of major regional partners (Francis 2006: 16–17). Security dynamics have been affected by regional rivalries, especially between Zimbabwe and South Africa, and by the occasional reluctance of the latter to provide regional leadership (Buzan and Waever 2003, 235).

Although referred to as a weak security regime (Buzan and Waever 2003, 239), ECOWAS was able under the leadership of Nigeria to intervene to restore peace to Liberia, Sierra Leone, Cote d'Ivoire and Guinea Bissau in the 1990s. In 1999, ECOWAS adopted a Mechanism for Conflict Prevention, Management, and Resolution. With the new peace and security architecture of the AU, ECOMOG has been transformed into the ECOWAS standby force (Ebo 2007: 5–6). But, as in the case of SADC, 'the efforts to respond to regional conflict and security challenges have not been based on constructive and long-term strategies' to link security and development. It was also noted that Nigeria used these interventions to achieve its strategic and security interests and that the countries of the region were not able to agree on a common security policy (Francis 2006: 15). Additionally, Nigeria as a regional power has exhibited symptoms of weakness and fragility given its regional, ethnic, and religious incohesiveness and tensions (Buzan and Waever 2003, 240).

In East Africa, decades of bilateral conflicts (Ethiopia and Somalia, Ethiopia and Eritrea) and support of insurgencies in other states (Sudanese support for rebel groups in Northern Uganda and in different Ethiopian regions, Ugandan and Ethiopian support for the Southern People's Liberation Army in Southern Sudan for example) have weakened security interdependence (Buzan and Waever 2003, 241–242). The regional organisation, IGAD expanded its mandate in 1996 to cover peace and security. The organisation played an active role in facilitating peace agreements in Somalia and Sudan. Less evident was its role in economic and developmental integration. As with the case of ECOWAS, IGAD members have not agreed on a common foreign and security policy (Francis 2006: 15–6; Al-Effendi 2009: 7–8).

In short, most regional economic communities in Africa have played a security role in the last two decades with different degrees of success and limited institutionalisation of the relationship between security and development. Scholars have explained this growing emphasis on peace and security in Africa by reference to a number of international and regional factors. Internationally, the end of the Cold War, and thus the retreat of superpower rivalries in Africa, contributed to the reluctance of great powers to intervene directly in African conflicts. This happened at a time of increasing conflicts and proxy wars on the continent and growing recognition of African governments and international institutions of the impact of conflicts on both development and regional integration.

At the regional level, there has been a regional hegemon that is ready to lead the process of transforming regional organisations to maintain security. An effective regional security operation often rests on leadership by a single state or a close group of states that are ready to bear a 'disproportionate share' of the costs of maintaining the grouping, but also constantly work towards achieving consensus within it (Clapham 2001: 62–67; Fawcett and Gandois 2010: 620–630).

These factors conformed to a global tendency to delegate responsibility for security to regional actors and the discussion about 'regional security complex' in regional integration literature (Al-Effendi 2009: 1–2). All these factors have contributed to the emphasis on the *military* role of African *economic* communities (Goldstein 2002: 21).

The next three sections examine the factors that have hindered cooperation and integration in North Africa and question whether the rising security threats after the North African uprisings have provided a new incentive for reviving this cooperation and developing a regional security complex.

Integration and Development in North Africa: Why Did the Region Lag Behind?

North Africa is the least economically integrated region in the world (Kolster *et al.* 2012: 20). Its main regional organisation, the AMU, has almost been frozen since 1996. Established in 1989, the Union brings together Morocco, Algeria, Libya, Mauritania and Tunisia. It has a broad integration agenda with objectives including maintaining the security of its members, coordinating their policies in various spheres and working towards the removal of any barriers to the movements of people, capital, goods and services between member states. The founding treaty of the organisation did not set any timetable for the achievement of these objectives. In the field of security, the treaty stated that aggression against any member state shall be considered as aggression against the other member states. But it did not identify collective security measures to be taken in the case of aggression against member states. It stated that member states are committed not to support any organisations that threaten the security, territorial integrity or the political system of other member states. Member states are also committed, according to the treaty, to abstaining from joining any military or political alliance directed against the political independence or territorial integrity of other member states (AMU 1989).

Common history, culture and economic challenges provided sound bases for cooperation between the five countries of the AMU. These countries shared the experience of European colonialism and anti-colonial solidarity, and the same origin of its Arab and Berber population. The union was also established to deal with the economic crisis faced by Algeria and Libya as a result of the fall in oil prices and the European Union (EU) enlargement to include Portugal and Spain, the traditional competitors of the Maghreb countries (Zoubir 2012: 85–86; Layachi 2016: 247–248).

However, security challenges have badly affected the march of integration in the region. The rise of fundamentalism in Algeria and domestic challenges in Morocco and Mauritania since the beginning of the 1990s distracted the members of the AMU from strengthening their newly established union. Border disputes between member states have also left a legacy of mistrust and suspicion. In the case of Algeria and Morocco, this dispute led to a full military confrontation in October 1963. In the 1970s and 1980s, AMU countries settled their border claims peacefully through bilateral agreements (between Algeria and Tunisia in March 1983, between Algeria and Mauritania in December the same year, and between Algeria and Morocco in 1972), but this did not end the tension in the region. Furthermore, the disagreement over Western Sahara remained a stumbling block towards cooperation, although Algeria and Morocco attempted to neutralise their relationships through the establishment of the AMU.

As noted by the RSCs theory, the domestic level of analysis is important in examining the regional security dynamics in general, and in the

African continent in particular. In this context, it was noted that the mistrust and suspicion between AMU countries was exacerbated by the different ideologies and political orientation of incumbent regimes. Leftist regimes in Libya and Algeria considered the Moroccan monarchy as 'reactionay and oppressive' (Layachi 2016, 248–250). Additionally, incumbent regimes of the AMU have often tended to use and escalate their tensions with their neighbours in the region to gain legitimacy at home or divert the attention away from domestic problems. In so doing, regimes in Maghreb countries focused on a narrow definition of security that gives priority to regime security over developmental regionalism contributing to human security. In this context, almost all the countries of the region accused other members of supporting opposition against incumbent regimes. For example, Algeria accused the other four members of the AMU of supporting Islamist opposition at the beginning of the 1990s. It supported, and invited Libya and Tunisia to support, the Popular Front for the Liberation of Saguia el-Hamra and Río de Oro—the movement fighting for the independence of Western Sahara. In return, Morocco attempted to neutralise Libya, which, in turn, aimed at breaking out its international isolation (Al-Adawy 2005: 207–216; Khafagy 2005: 106; Zoubir 2012: 84–86).

It was thus unsurprising that the organisation failed to deal with the threats facing member states, or to provide a supporting bloc for these states in crucial international confrontations. Its stance towards UN sanctions against Libya in the wake of the Lockerbie crisis in 1988 did not go beyond verbal support. The same applies to the organisation's stance towards Mauritania in its border dispute with Senegal (Bey 2005: 265, Abdellani 2005: 297–298).

It is also little wonder that the North Africa Regional Capability (NARC), established as a regional arm for the Africa Standby Force in 2007, is not managed by the AMU. It was primarily established to fill the regional vacuum that resulted from the dysfunctionality of the AMU. Morocco, a non-member of the AU, remains outside NARC. Egypt and Western Saharan, not members of the AMU, are members of NARC. The North Africa brigade headquarters is to be located in Cairo. An executive secretariat for NARC was set up in Tripoli in 2009. But until 2010, Egypt and Algeria were yet to sign the hosting agreement with Libya. Thus, the operationalisation of the regional structures of NARC has lagged behind compared to other regional brigades (AU 2010: 45–47).

The ineffectiveness of the AMU encouraged Libya to propose another regional organisation—CEN-SAD. Established in 1998, it extended its membership in a few years to more than 20 African countries, prioritising the 'establishment of a comprehensive economic union based on a strategy implemented in accordance with a developmental plan that would be integrated in the national development plans of the member states'. It gradually moved to include cooperation in fields of security, adopting a charter that committed its members to refrain from using its territories as bases for aggression on other member state and from providing military and/or political support to rebels in other states (United Nations Economic Commission for Africa n.d.).

In 2003, the community established a mechanism for conflict resolution. In 2004, member states signed another agreement for security cooperation, including cooperation in combating terrorism. The procedures in the agreement included sharing information to prevent all forms of crime, joint training of police and coordination of security policies through consultation and regular meetings between senior officials. By the end of the 2000s, issues of peace and security had topped the agenda of the summit meetings. The Community's summit in 2009, for instance, noted the deteriorating relations between Libya and Chad, expressed deep concerns about increasing violence in Somalia, condemned the assassination of President Joao Bernardo Vieira of Guinea Bissau, and denounced arms and drug trafficking in the Sahel region. As a response, the summit adopted a decision to hold joint military manoeuvres and exercises (*Tripoli Post*, 30 May 2009).

The organisation's role in conflict resolution has, however, been very limited. Mediation efforts to resolve conflicts in the region, such as the mediation of Libya, Nigeria and Chad in the Darfur conflict or Libya's efforts to resolve the succession crisis in Togo after the death of Gnassingbe Eyadema in 2005, came from individual member states rather than from the organisation (Hanafy 2005: 106–109). Besides, the Libyan role in establishing the CEN-SAD sparked controversy. Scholars debated whether it aimed at weakening the AMU, especially given that Algeria decided not to join the community. This may have contributed to the limited effectiveness of the organisation in spite of its potential in fields such as energy and combating desertification.

To sum up, regional factors, including enduring tensions between incumbent regimes in the region as a result of a legacy of border disputes, different ideological and foreign policy orientations and disagreement over the Western Sahara conflict, hindered regional cooperation, rendering AMU ineffective. This left a vacuum that CEN-SAD was not able to fill. The two groupings have not thus represented RSCs with a clear regional hegemon and increased security interdependence. Evidently also, domestic and regional dynamics increased the dependence of the countries of the region on external powers, thus contributing to inhibiting the development of a RSC. The links with these powers are illustrated in the next section.

External Forces: A Driver for, or Impediment to, Regional Integration?

The last section argued that the reluctance of Western powers to directly intervene in African conflicts in the last two decades and their willingness to support the security role of African regional organisations have promoted regional integration on the continent. In contrast, the heavy involvement of international powers, as maintained by the RSCs theory, may hinder the development of regional security interdependence. This section examines the impact of the policies of international powers on North Africa's security and development integration.

North Africa has long been one of the most vulnerable regions to external influence. It was subject to a continuous process of 'colonial subordination and externally enforced norms' by Western powers in the last two centuries, a process that shaped state building and affected the potentials of regional integration (Lustick 1997: 656–657). In the last few decades, the US and European powers have shared the objectives of combating radicalism, securing access to energy resources and to markets, and supporting Israel as a regional power and partner as far as the region is concerned (Cavatorta and Durac 2016: 4–5).

In the field of energy, the EU is heavily dependent on foreign supplies to meet its demand for oil and gas. Its Energy Security and Solidarity Action Plan made energy a major focus of the Union's international relations. Accordingly, the EU has sought to diversify these foreign suppliers, setting up new energy partnership with North and West Africa (EU Commission 2008: 9). By the end of 2000s, around 10 % of the EU's crude oil imports came from Libya and around 18 % of its gas imports came from Algeria, which is the third largest gas provider to Europe after Russia and Norway. Algeria's natural gas goes particularly to France, Italy and Spain through several pipelines. Libya has the largest oil reserves in Africa (more than 40 billion barrels) and comes fourth in terms of proven gas reserves (after Nigeria, Algeria and Egypt). Several French, British, Dutch, Spanish, Italian and Austrian oil companies have signed agreements with Libya for oil and gas exploration. Most of Libya's oil and gas exports have gone to Europe (Bahgat 2010: 39–48; Bahgat 2012: 509–511).

To achieve its interests in promoting free market ideology and ensuring political stability, the EU has signed partnership agreements with most North African countries. Unlike in other regions on the continent, North Africa countries negotiated with the EU individually rather than as a regional bloc. Tunisia was the first country in the region to establish an economic cooperation agreement with the EU (in 1995). The agreement provided significant funding to Tunis in return for a reduction in Tunisian tariffs. Other association agreements were signed with Morocco (1996), Egypt (2001) and Algeria (2001). Morocco became the largest beneficiary of EU's neighbourhood initiative, which aims at supporting EU's trade partners. In 2008, Morocco was the first country granted an 'advance status', a status that would be translated into closer political ties, the alignment of Morocco's legal system with the EU, the conclusion of a comprehensive Free Trade Agreement, and more cooperation in economic development (Friedman and Sharkey 2010: 35–43). The EU remains the top trading partner for all the countries of the region (see Table 13.1). In contrast, intra-regional trade ranged between 1 and 3% of the Maghreb's total trade (Layachi 2016: 252; Africa Research Bulletin 2014).

Table 13.1 Top trading partners of North African countries in 2015 (in million Euros)

	EU	US	China	Turkey	Brazil	Russia	Tunisia	Algeria	Egypt	Libya
Algeria	43,485	4623	8175	2417	2619	1981	1970	–	–	–
Egypt	25,655	4667	8580	3934	2361	3073	–	–	–	–
Libya	11,606	343	2654	1568	–	182	1190	–	627	–
Morocco	29,192	2993	3026	1985	1318	1587	–	910	567	–
Tunisia	20,456	1001	1379	930	353	348	–	2082		1152

As illustrated in Table (1), the EU remains the top trading partner for all the countries of the region, in spite of the growing volume of trade with rising powers, especially China, which replaced the United States as the second largest partner for all the countries of the region.

The United States expressed an interest in promoting a regional trade bloc in the Maghreb. In 1999, it launched the US North African Economic Program, which aimed at linking the US and the three Maghreb countries, Algeria, Morocco and Tunisia, more closely in terms of trade and investment, as well as reducing the barriers of integration between the three countries. During the Bush administration, the initiative was subsumed into the Greater Middle East Partnership Initiative. The United States has often urged the two regional powers, Algeria and Morocco, to resolve their dispute over the Western Sahara, and has preserved a balanced position to maintain its strategic relationship with the two powers. However, its political efforts towards resolving the conflict have been unsuccessful (Zoubir and Benabdallah-Gambier 2005: 186–194).

In the field of security, the 5 + 5 Dialogue Forum, which includes five European members (Spain, France, Portugal, Italy and Malta), and the five Maghreb countries was revived in May 2002 as a cooperation framework. European scholars and practitioners considered this forum as a means to build trust not only between northern and southern Mediterranean countries, but also among the countries of the Maghreb (Nunez 2012; Coustelliere 2006: 6). It has to be noted, however, that the policies of these European powers have contributed to rift between Algeria and Morocco over the Western Sahara. France has supported the Moroccan side in this conflict. Since 2002, more than quarter a century after it handed the administration of the Western Sahara to Morocco and Mauritania, Spain, the former colonial power in the region, declared its backing for a peaceful resolution of the conflict based on the UN resolutions. This meant that Spain supported the Algerian position on the conflict and opposed the view of Morocco, with which it has also been engaged in border disputes (Zoubir and Benabdallah-Gambier 2005: 183–184).

Algiers remains a key partner to the US and EU in counter-terrorism activities in North Africa and the Sahel. In the wake of the attacks of 11 September 2001, joint initiatives, intelligence missions and training sessions between subsequent American administrations and the Algerian government have taken place. Algeria was also incorporated into NATO's Mediterranean Dialogue in 2000. Since 2008, it has signed arms deals with France, Italy and Germany and joined navy and air forces military exercises with the US and the members of the 5 + 5 Dialogue (Ammour 2013: 5–6).

Morocco has a long tradition of cooperation with the US and NATO in military exercises in Mediterranean waters. During the Cold War, it was the second-largest recipient of American military aid in the Arab world (after Egypt) and has obtained more than one fifth of total American aid to the African continent (Zoubir and Benabdallah-Gambier 2005: 188). Like Algeria, it has collaborated with countries of the 5 + 5 Dialogue in the fight against drug smuggling, terrorism and illegal migration. It is the only country in North Africa with which the US has signed a Free Trade Agreement. Morocco is seen as a country which has managed to introduce successful economic and political reforms that have enabled it to avoid the costly changes that have afflicted other countries in the region. In 2012, the US signalled its interest in boosting bilateral relations by launching a Strategic Dialogue to strengthen cooperation in different fields (Moroccan American Center for Policy 2012).

With the end of international sanctions against Gaddafi's regime at the beginning of the last decade, Libya showed an interest in strengthening relations with Europe. Being excluded from the Barcelona process, which set the foundation for partnership between the European Union and the Southern Mediterranean Arab countries in addition to Israel, it hosted the 2002 Foreign Ministers Conference to revive the 5 + 5 Initiative. It signed agreements to establish military and industrial cooperation with France and closely cooperated with Italy in fighting illegal migration (Nunez 2012: 8–12; Zoubir 2012: 87–88).

In short, the countries of North Africa cooperated economically and politically more with Western powers, especially the EU, than with one another. These powers have often negotiated economic and security arrangements with individual North Africa countries. They may have an interest in promoting security cooperation between their partners in the region. However, given their security interests in the region, Western powers are more willing to intervene directly to protect these interests than to support regional intervention, as happened in other parts of the continent. NATO's intervention in Libya in March 2011 is a case in point. What was portrayed as a humanitarian intervention to protect civilians, proved—as revealed by e-mails recently released by the US State Department—to be motivated by Libya's oil wealth and French discontent with Gaddafi's increasing influence in Francophone Africa (Hoff 2016).

Unsurprisingly, North African countries held different positions in relation to this intervention. Algeria was at odds with other countries of the region in opposing NATO's operation, a position that seems to have been vindicated after the tremendous security challenges faced by Libya in the wake of the intervention (Boukhars 2013). At the same time, no regional hegemon has been ready to lead the efforts to deal with these security challenges. This is illustrated in some detail in the next section.

Increasing Security Threats and Regional Integration: The Missing Link

The popular uprisings in North Africa which succeeded in overthrowing Ben Ali's regime in Tunisia in December 2010 and gained momentum with the toppling of Mubarak in Egypt in February 2011 'put the legitimacy and effectiveness of regional organizations to the test' (Pinfari 2012: 135). The two sub-regional organizations, AMU and CEN-SAD, have taken no clear positions. Their member states took different stances towards the popular uprisings in North Africa in general and the intervention in Libya in particular. As noted earlier, Algeria strongly opposed NATO's intervention in Libya, considering it a Western plot to overthrow Gaddafi's regime, and refused to recognise the internationally recognised National Transitional Council. Mauritania called for a political solution and recognised the Council only in November 2011, nine months after its formation. As far as the revolution in Tunisia is concerned, Algeria expressed a neutral position, stating that it supported the Tunisian people, but did not go as far as to ask Ben Ali to step down. Expectedly, post-Ben Ali, Tunisia supported the revolution in Libya, but its recognition of the National Transitional Council came almost six months after the latter's formation, around the same time as Morocco (Zoubir 2012: 96).

The fall of Gaddafi's regime has had wide regional implications. It left a power vacuum that was filled by armed groups with different agendas and led to the proliferation of arms smuggling in North Africa and the Sahel. The killing of the US Ambassador in Benghazi in September 2012, the subsequent storming of the US embassy in Tunisia, the January 2013 hostage crisis in Ain Amenas, Algeria, and the kidnapping of the former Libyan Prime Minister Ali Zidan in October 2013 indicated the fragility of the new Libyan authorities and its impact on regional security and international interests.

These security threats were further exacerbated by the collapse of Libya into a new civil war (from February 2014) between an Islamist government installed in Tripoli by the elected, but expired, General National Council and another internationally recognised administration in Tobruk appointed by a parliament elected in June 2014. Military confrontations have not been confined to forces supporting the two factions. Most notably, Libya has become the first host of the extremist IS group outside Syria and Iraq. Advances achieved by IS across the country prompted the Libyan government in Tobruk to call for an Egypt-led regional intervention to fight the extremist group, which had slaughtered tens of Libyans and Egyptian migrant workers (Reeve 2015). Mediation efforts by the United Nations led to the signing of a political agreement in December 2015 and the formation of a Government of National Accord (GNA) in March 2016. But the new government hardly represents all the country's political groups and factions.

To face these challenges and secure their interests, international powers have further militarised their policies in the region. Although NATO ended its operations in October 2011, French and American air forces have continued their military and intelligence presence in Libya. France had to acknowledge that its forces are still operating in Libya when three of its special forces soldiers were killed in a helicopter crash in July 2016. The French intervention was slammed by the new GNA, which confirmed that the French troops operate without its knowledge or approval (*Libyan Prospect*, 22 July 2016). At the same time, the GNA requested the United States to launch air strikes against IS in Sirte (Reuters, August 1, 2016).

But to what extent have these challenges of radicalisation and foreign intervention contributed to the revitalisation of regional organisations and the strengthening of regional cooperation in North Africa? Following the North Africa uprisings, Maghreb countries indicated their willingness to revive the AMU. Their foreign ministers met in Rabat in February

2012, in the AMU's first high-level meeting since 1996, to boost political and economic cooperation. In the 5+5 Foreign Ministers' Ministerial Conference held in Rome in the same month, members expressed similar interest in revitalising the union. The ten countries discussed the Maghreb highway project, the solar energy plan, and encouraging the role of small and medium enterprises as drivers of development and cooperation in the area (Nunez 2012: 7).

Subsequently, King Muhammed VI called for a 'new Maghreb order' that takes account of the changes in Libya and Tunisia (Department of Communication of the Kingdom of Morocco 2014). Algeria encouraged drawing up a roadmap to revitalise the regional organisation. Its leader, Abdelaziz Bouteflika, underscored the need for a pan-Maghreban body able to defend the region's interests with global partners. Its foreign minister expressed his country's interest in fostering cooperation in the fields of countering terrorism and organised crime. Tunisia has shown enthusiasm towards the revival of the AMU as a way to face its domestic economic challenges. Its new president, Moncef Marzouki, held talks with Morocco, Algeria and Mauritania during his state visit in January 2012 to revive the long-stalled union (Zoubir 2012: 97; *Africa Research Bulletin*, 1 February, 2012a, 19151).

But the five member states have shown little more than verbal commitment. The foreign ministers of the five countries who met in Algiers in July 2012 highlighted the need to elaborate a common security strategy. But it is notable that the coordination has not taken place among defence ministers of the five countries, who tend to coordinate separately with the countries of the Sahel (Africa Research Bulletin, July 1, 2012b, 19333). In spite of hopes of holding an AMU summit before the end of 2012, no summit was held, with the Algerian foreign minister indicating that the conditions for holding a successful summit did not exist (*Echoroukonline*, 23 June 2013).

The second regional organisation, CEN-SAD, has also witnessed some efforts towards revitalisation. In the light of the absence of a rival inside the organisation after the fall of Qaddafi and its desire to counter Algeria's regional pre-eminence, Morocco led these efforts. Rabat hosted the organisation's summit in June 2012 and proposed the revision of the community's treaty and the restructuring its organs (Nickels 2013; Zoubir 2012: 94). The Moroccan proposals put more emphasis on the organisation's security role. They recommended reframing the organisation's activities to focus on a culture of peace and stability and sustainable development. To achieve this end, Rabat proposed establishing two committees for peace and security, and for sustainable development. The new organs are expected to contribute to collective security through increasing cooperation in countering terrorism, illegal trafficking and arms smuggling.

A number of experts have shown optimism about the revitalisation of CEN-SAD. According to the noted Mauritanian former diplomat Ahmedou ould Abdallah, 'given its diversified membership, with countries crossed by or bordering the Sahel-Saharan strip, the CEN-SAD appears to be a perfect diplomatic and institutional setting, for engaging in dialogue and action, in efforts to define and implement any relevant, inclusive and multidimensional response to the prevailing security challenges.' He invited international organisations and donors to support the organisation (Ould Abdallah 2013). There is, however, no evidence that international actors have an interest in contributing towards the revitalisation of the organisation. Given its limited role in meeting the challenges in the region, there is also no indication that the community has been transformed into a viable framework providing security coordination between the Sahel and Saharan countries.

In the light of these regional relations, there is no single state or group of states that are willing to play a leadership role, bear the costs of this role, and coordinate its policies with other countries in the region. North African countries have failed to coordinate their positions on the crisis in Libya, even after the success of IS to control Sirte and expand its control in Libyan territories. Morocco, backed by the UN, hosted rounds of talks between the Libyan rival factions in Skhirat that led to the signing of the political agreement in December 2015 (Stitou 2015), a role that put it in competition with Algeria. Algeria avoided openly criticising Egypt's air strikes against IS in February 2015 in retaliation for the killing of 21 Egyptian Christian migrant workers. However, it openly opposed Arab military intervention in Libya against IS, when this option was deliberated in the Arab League in August 2015, stressing the importance of a political solution to the crisis (*Ahram online*, 19 August 2015). Egypt launched retaliatory air strikes and proposed to lead an Arab military force, but was not seen as a credible neutral mediator on account of its support for the anti-Islamist forces and the Toruk government (Muhlberger 2016). Thus, Egypt and Algeria stood at odds in the Libyan conflict. Whereas Algeria supported a peaceful resolution of the conflict, opposing both regional and international intervention, Egypt supported one party to the conflict and was more willing to intervene to give weight to this party.

Algeria's non-interventionist military policy led scholars and observers to argue that Algiers was unwilling to assume the responsibilities of regional leadership and often adopted positions that isolated it regionally (Ammour 2013: 1; Boukhars 2013). Algeria's power struggles, its preoccupation with domestic economic challenges, and its fears of opening a new front against armed groups outside its borders may explain the government's conservative position (Ammour 2013: 2–3; Benlakehal 2015).

This situation does not negate the fact that some sort of security coordination has been taking place at the bilateral or trilateral level in the region. Looking at Tunisia as a potential buffer against instability in Libya, Algiers strengthened its security cooperation with Tunis in the areas of counter-terrorism, border security, organised crime and smuggling. On its part, incidents of political assassinations, suicide bombings and killings of security forces in northern and western Tunisia have encouraged Tunis to seek to learn from Algeria's experience in counter-terrorism. At the end of 2012, Algeria and Tunisia signed a border security agreement for facilitating joint patrols and operations. The two countries have established a joint intelligence unit and the military component of their bilateral commission has been active in the last few years (Boukhars 2013; Zoubir 2012: 89–90).

In the same vein, Algeria and Libya signed agreements on common security issues in 2012 and established a joint commission in 2013 for cooperation in different fields, including security. At the trilateral level, the prime ministers of Algeria, Tunisia and Libya held a meeting in January 2013 at which they agreed on joint teams to coordinate security along their borders, but the resurgence of civil war in Libya hindered its contribution to these teams (Bakrania 2014: 10).

Although Morocco and Algeria fostered their security cooperation by signing an agreement in July 2016, the disagreements between them remain unresolved. These will remain the stumbling blocks towards any effort to revitalise the regional grouping. This means that without resolving the sources of political tensions between Rabat and Algiers, the prospects of sustainable security and developmental cooperation in North Africa will remain bleak.

North Africa: Why a Regional Security Complex is Lacking?

The last three sections illustrated that North Africa represents a case of non-existent RSC. There have been objective security threats that could have encouraged the states in the region to increase their interdependence in traditional and new security sectors. Weak states and ungoverned spaces to the South have been expected to prompt states in North Africa to build strong security frameworks to protect their borders and contain the activities of violent extremist movements and militias, organized crime, and illegal smuggling. These activities of non-state actors have endangered not only the political and military security of states in the region, but also their economic security, a threat that increased after the fall of Gaddafi's regime in Libya. Targeting foreign nationals in the gas facility in Ain Amenas by Al-Qaeda affiliated terrorists, and oil infrastructure in Libya by militias are just two examples.

Equally important, the North African uprisings indicated that new dimensions of security are at stake. Some of the countries of the region have achieved a high level of economic growth (e.g. Egypt since 2004), and remarkable levels of human development (e.g. Tunisia), outperforming many other African countries. However, the lack of social equality and political freedom has given rise to popular protests that overthrew incumbent regimes in Tunisia, Egypt, and Libya.

New regimes in North Africa have voiced their recognition of the importance of security and developmental interdependence, especially after the popular uprisings. However, the region is no less fragmented compared to the past. Regional organizations expose this fragmentation and fail to represent a framework of effective cooperation. The AMU does not include Egypt that is considered part of the Arab Mashreq, while CEN-SAD excludes Algeria, the traditional rival of Morocco. North African countries cooperate more with other countries outside the region rather than among themselves, creating a web of overlapping international and regional arrangements that weaken, rather than strengthen, their cooperation. This means that regional transformations since 2010 have led to little change in terms of the states' perceptions of security and developmental cooperation and the patterns of amity and enmity in the region. The Algerian diplomatic effort to prevent the suspension of the SADR to allow for the readmission of Morocco to the AU is just one example on this lack of change. This position came in spite of the agreement between the two countries to boost their security and intelligence cooperation. In a nutshell, the conditions for transforming the region into a RSC continue to be lacking.

Conclusion

Popular uprisings since 2010 have raised security threats in North Africa. These threats have posed a challenge to new and existing regimes in the region, but have also represented an opportunity for reviving security, and subsequently economic, cooperation. The evidence, however, shows that bilateral and trilateral arrangements have been adopted to deal with the rising security threats, but no common strategy has been developed that could turn North Africa into a regional security community. Efforts towards revitalising regional organisations have been limited and more rhetorical than real. Members of these organisations are still cooperating more with international powers than among themselves. These powers have increased their military and/or intelligence presence in the region to protect their interests. Equally important, the largest military power in the Maghreb, Algeria, is unwilling to directly intervene in regional conflicts and/or bear the largest share of the costs of this intervention.

This means that the conditions that pushed African countries to strengthen the security role of regional organisations in other parts of the continent do not yet exist in North Africa. Resolving historical tensions between the two major regional powers, Algeria and Morocco, investing in a sub-regional organisation that could provide a framework for security and economic cooperation and supporting building state institutions in Libya are key to fostering security and developmental integration in the region.

References

Abdellani, A. (2005). The Arab Maghreb Union: From fragmentation to union. In M. Ashour & A. Salem (Eds.), *African regional integration: Perspectives and prospects*. Cairo: Cairo University, African Regional Integration Support Project.

Africa Research Bulletin. (2012a). Arab Maghreb Union: Algiers meeting, July 1, p. 19333.

Africa Research Bulletin. (2012b). Arab Maghreb Union: Renewed optimism, February 1, p. 19151.

Africa Research Bulletin. (2014). Arab Maghreb Union: Lethargic group, April 1, p. 20298.

African Union. (n.d.). *African Peace and Security Architecture: African Union's blueprint for the promotion of peace, security and stability in Africa*. Addis Ababa: AU.

African Union. (2010). *African Peace and Security Architecture: 2010 Assessment Study*. Addis Ababa: AU.

Ahram Online. (2015). Is the Arab League setting the stage for military intervention in Libya, August 19. Retrieved August 26, 2016, from http://english.ahram.org.eg/NewsContent/2/8/138271/World/Region/Is-the-Arab-League-setting-the-stage-for-military-.aspx

Al-Adawy, M. (2005). Regional integration between claims of national security and new concepts of security. In M. Ashour & A. Salem (Eds.), *African regional integration: Perspectives and prospects*. Cairo: Cairo University, African Regional Integration Support Project.

Al-Effendi, A. (2009). The perils of regionalism: Regional integration as a source of instability in the Horn of Africa? *Journal of Intervention and State Building, 3*(1), 1–19.

Ammour, L. (2013). Algeria's role in the Sahelian security crisis. *Stability: International Journal of Security and Development, 2*(2), 2–11.

Arab Maghreb Union. (1989). Treaty instituting the Arab Maghreb Union, Marrakesh, Morocco, February 17.

Bahgat, G. (2010). The geo-politics of energy: Europe and North Africa. *The Journal of North African Studies, 15*(1), 39–49.

Bahgat, G. (2012). The impact of the Arab spring on the oil and gas industry in North Africa: A preliminary assessment. *The Journal of North African Studies, 17*(3), 503–514.

Bakrania, S. (2014). *Libya: Border security and regional cooperation*. Birmingham: GSDRC.

Benlakehal, M. (2015, April 1). Algerian support for joint Arab military would be limited. *Al-Araby Al-Jadeed*.

Bey, A. (2005). Bases and obstacles for integration in the Arab Maghreb Union. In M. Ashour & A. Salem (Eds.), *African regional integration: Perspectives and prospects*. Cairo: Cairo University, African Regional Integration Support Project.

Boukhars, A. (2013). *Algerian foreign policy in the context of the Arab spring*. Op-ed. Washington, DC: Carnegie Endowment for International Peace.

Cavatorta, F., & Durac, V. (2016). Diverging or converging dynamics: EU and US policies in North Africa–an introduction. *The Journal of North African Studies, 14*(5), 1–9.

Clapham, C. (2001). The changing world of regional integration in Africa. In C. Clapham, G. Mills, A. Morner, & E. Sidiropoulos (Eds.), *Regional integration in Southern Africa: Comparative international perspectives*. Johannesburg: South African Institute of International Affairs.

Coustelliere, J. F. (2006). The 5+5 security and defense initiative. 5th International Seminar on *Security and Defence in the Mediterranean; Multi-Dimensional Security*. Barcelona: Barcelona Centre for International Affairs.

Ebo, A. (2007). Towards a common ECOWAS agenda on Security Sector Reform. *Policy Paper No. 23*. Geneva: The Geneva Centre for the Democratic Control of Armed Forces.

Echoroukonline. (2013). Security conditions for holding AMU Summit, June 23. Retrieved August 25, 2016, from http://www.echoroukonline.com/ara/articles/160723.html

European Commission. (2008). *Second strategic energy review: An EU energy security and solidarity action plan*. Brussels: Commission of the European Communities.

Fawcett, L., & Gandois, H. (2010). Regionalism in Africa and the Middle East: Implications for EU studies. *Journal of European Integration, 32*(6), 617–636.

Francis, D. (2006). Linking peace, security and developmental regionalism: Regional economic and security integration in Africa. *Journal of Peace Building and Development, 2*(3), 7–20.

Friedman, S., & Sharkey, H. (2010). A tale of two economic developments: Tunisia and Morocco. *College Undergraduate Research Electronic Journal*. The University of Pennsylvania, 1–69.

Goldstein, A. (2002). The new regionalism in Sub-Saharan Africa: More than meets the eye? *OECD Development Centre Policy Brief No. 20*.

Hanafy, K. (2005). An evaluation of the CEN-SAD integrative performance. In M. Ashour & A. Salem (Eds.), *African regional integration: Perspectives and prospects*. Cairo: Cairo University, African Regional Integration Support Project.

Hoff, B. (2016, January 6). Hillary e-mails reveal true motives for Libya intervention. *Foreign Policy Journal.* Retrieved July 25, 2016, from http://www.foreignpolicyjournal.com/2016/01/06/new-hillary-emails-reveal-true-motive-for-libya-intervention/

Khafagy, R. (2005). The Arab Maghreb Union: A puzzling case of regional integration. In M. Ashour & A. Salem (Eds.), *African regional integration: Perspectives and prospects.* Cairo: Cairo University, African Regional Integration Support Project.

Kolster, J., Matondo-Fundani, N., & Santi, E. (2012). Regional integration in North Africa. In E. Santi, S. Ben Romdhane, & W. Sahw (Eds.), *Unlocking North Africa's potential through regional integration.* Abidjan: African Development Bank.

Layachi, A. (2016). Region-building in North Africa. In D. Levine & D. Nagar (Eds.), *Region-building in Africa: Political and economic challenges.* New York: Palgrave Macmillan.

Libyan Prospect. (2016). Libyan parties condemn France's intervention, July 22. Retrieved August 25, 2016, from http://libyaprospect.com/index.php/2016/07/22/libyan-parties-condemn-frances-intervention/

Lustick, I. (1997). The absence of Middle Eastern great powers: Political "backwardness" in historical perspectives. *International Organization, 51*(4), 653–683.

Moroccan American Center for Policy. (2012). *Morocco-US strategic dialogue: Consultation, cooperation, progress.* Washington, DC: Moroccan American Centre for Policy.

Muhlberger, W. (2016). A Thorny Dossier: Egypt's Libya policy. Ahram online, January 27. Retrieved August 25, 2016, from http://english.ahram.org.eg/NewsContentP/4/185840/Opinion/A-Thorny-Dossier-Egypt%E2%80%99s-Libya-Policy.aspx

Nickels, B. (2013, June 23). Mali's regional ramifications. *Sada Journal.* Carnegie Endowment for International Peace, from http://carnegieendowment.org/sada/52018.

Nunez, M. (2012). *5 + 5 Mediterranean security: Shared security.* Framework Document No. 7, Instituto Espanol de Estudios Estrategicos.

Organization of African Unity. (1980). Lagos Plan of Action for the economic development of Africa 1980–2000. Heads of State and Government extraordinary Session, Lagos, Nigeria.

Organization of African Unity. (1991). Treaty establishing the African Economic Community. Assembly of Heads of State and Government, Abuja, Nigeria.

Organization of African Unity. (1993). Cairo declaration on the occasion of the 30th anniversary of the Organisation of African Unity. Assembly of Heads of State and Government, 29th Ordinary session.

Organization of African Unity. (2000). Conference on Security, Stability, Development and Cooperation in Africa (CSSDCA) Solemn Declaration. Heads of State and Government 36th ordinary session, Lome, Togo.
Ould Abdallah, A. (2013). Revitalising the CEN-SAD. Centre for Security. Retrieved August 25, 2016, from http://www.centre4s.org/en/index.php?option=com_content&view=article&id=96%3Arevitalizing-the-cen-sad&Itemid=57
Pinfari, M. (2012). A changing Mediterranean: Regional organizations and North Africa during the Arab spring. *The International Spectator, 47*(1), 134–150.
Reeve, R. (2015). Libya: Between the Sahel-Sahara and the Islamic States crisis. *Briefing Paper*. Oxford: Oxford Research Group.
Reuters. (2016, August 1). US warplanes launch bombing campaign on Islamic State in Libya. Retrieved August 25, 2016, from http://www.reuters.com/article/us-libya-security-idUSKCN10C2NF
Stitou, I. (2015, April 23). Morocco, Algeria compete over Libya. *Al-Monitor*.
Tripoli Post. (2009). Conclusions of the 11th ordinary session of the conference of leaders and heads of state of the Community of Sahel and Saharan State. Retrieved August 24, 2016, from http://www.tripolipost.com/articledetail.asp?c=1&i=3298
United Nations Economic Commission for Africa. (n.d.). CEN-SAD: The Community of Sahel and Saharan State. Retrieved August, 24, 2016, from http://www.uneca.org/oria/pages/cen-sad-community-sahel-saharan-states
Zoubir, Y. (2012). Tipping the balance towards intra-Maghreb unity in light of the Arab spring. *The International Spectator, 47*(3), 83–99.
Zoubir, Y., & Benabdallah-Gambier, K. (2005). The United States and the North Africa Imbroglio: Balancing interests in Algeria, Morocco and the Western Sahara. *Mediterranean Politics, 10*(2), 181–200.

Rawia Tawfik is an assistant professor at the Faculty of Economics and Political Science at Cairo University. She holds a Doctor of Philosophy degree in Politics from the University of Oxford and a Master of Science degree in Politics from Cairo University. Her research interests include issues surrounding African development and regional integration. She was a visiting researcher at the South African Institute for International Affairs, the Africa Institute of South Africa and the German Development Institute. She has published a number of research papers, book chapters and journal articles on Nile hydro-politics, good governance, the New Partnership for Africa's Development and regional integration in Africa, and on Egypt's foreign policy towards Africa.

14

The Ethnography of the Sex Trade among Bini Women in Southern Nigeria

Clementina O. Osezua

Introduction

Several research projects have been conducted by both scholars and non-governmental organisations in the past few years on Nigerian women's involvement in the cross-border sex trade (WHARC 2002; UNICRI/UNDOC 2003). This research has affirmed the degree of damage that the international sex trade has inflicted on the Nigerian government and its people. International development partners and the Nigerian government have taken a combative stance against the phenomenon in a bid to eliminate the trade, which is considered reprehensible by the Nigerian state, while partner development agencies maintain that sex trafficking not only damages the reputation of the Nigerian people, but is inherently exploitative (UNICRI/UNODC 2003). Furthermore, these stakeholders have continued to depict the women involved in transactional sex

C.O. Osezua (✉)
Department of Sociology and Anthropology, Obafemi Awolowo University, Ile-Ife, Osun State, Nigeria

A. Adeniran, L. Ikuteyijo (eds.), *Africa Now!*,
https://doi.org/10.1007/978-3-319-62443-3_14

overseas as victims, thereby lacking agency and vulnerable to a cartel of traffickers who must be brought to book.

Admittedly, an important clarification is imperative about the author's use of the terms 'sex trade' as against 'sex trafficking', which actually constitutes the reference point of this article. This is because the dominant literature in the region has largely ignored the debate prompted by prominent activists in the pro-sex work discourse, where the use of words like 'coercion', 'deception' and 'exploitation' have been seriously contested with regard to their utility in describing the actual experiences of women who have been trafficked from one country to the other for the purpose of sexual exploitation (Alpes 2010)

In December 2000, the United Nations convention against transnational crime adopted a protocol to combat trafficking in persons, especially for women and children. The protocol adopted the definition of 'trafficking' as:

> Recruitment, transportation, harboring, receipt of person by threat or use or force, or other forms of coercion, of abduction, of fraud or deception of the abuse of power of position of vulnerability or of giving or receiving payments or benefits to achieve the consent of a person having control over another person for the purpose of exploitation. Exploitation shall include as a minimum, the exploitation of others or other forms of sexual exploitation of the prostitution of others or other forms of sexual exploitation, forced abduction or services, slavery practices similar to slavery servitude or the removal of organs. (UNICRI/UNDOC 2003) (p. 11)

The Palermo Protocol has become a subject of debate in that the traffic of women excludes women's agency, but emphasises exploitation, coercion or deception, as the case may be.

Noteworthy in this adoption are two basic facts. First, this definition was arrived at from a standpoint of eradicating the phenomenon of sex trafficking and not with the intention to understand the rationale for the emergence of the phenomenon itself. Second, applying certain concepts, such as 'force', 'deception', 'coercion', 'control' and 'lack of consent', to the individuals involved, who are usually women 'deceived' into sexual servitude, consistently obscures the agency of the female sex workers. There are emerging scenarios where women choose to go into the sex trade overseas and consciously seek out traffickers, whom they perceive will help them with travelling documents in order for them to overcome the international migration barriers and access better socio-economic opportunities. This phenomenon has largely been ignored by researchers in this region due to the government's desire to eliminate sex trafficking.

Adopting these concepts, as spelt out by the Palermo Protocol, which was domesticated in 2001, would obliterate the experiences of women in the sex trade who were not coerced or deceived, and who, in some instances, argue that they are more reflexive of their circumstances and are therefore better placed in to negotiate with their *madams* (the women who serve as their agents), in the event of another opportunity to trade sex. Hence the study engaged more inclusive concepts of the sex trade or sexual transactions. The adoption of the concept of sex trade rather than the trafficking of girls and women for the purpose of sexual exploitation makes it possible to include a significant number of Nigerian women who engaged in cross-border sexual transactions without any coercion, but deliberately and consciously.

This chapter is not in denial about the fact that numbers of young women have been lured into the sex trade and are actually trafficked by cartels, but it rather argues that, significantly, the phenomenon of the trafficking of girls and women in Nigeria has undergone a huge change due to its perceived lucrative nature. It supports empirically the fact that as, of now, a significant number of those involved in the international sex trade, who are described as 'victims' when they are apprehended outside the nation's borders, had a good understanding of the sexual component of their prospective jobs prior to their involvement in the sex trade. An average illiterate woman in the study area confirms the lucrative nature of sex trade, and this notion has gained further popularity among a large number of young girls in a developing nation like Nigeria, due to its growing level of impoverishment. These young women are major actors in the international sex trade in Nigeria.

This article also provides an empirical analysis of Nigerian women's involvement in the sex trade by challenging the hegemonic 'victimisation' theory applied to Nigerian women and girls who are involved in the international sex trade. It rather presents data on the huge trajectory of the agency of these women in certain circumstances. This chapter does not intend to give credence to their experiences as legitimate or acceptable; rather, it tries to broaden the context through which women are perceived, and give voice to this social category, whose voices have been largely ignored, as has their structural location within the society. It further shows that, currently, women and girls are actively involved deciding their future in relation to sexual transactions across the Nigerian borders. Many of those women are not trafficked, but rather have agency and are presently responding to an endemic culture of poverty, which necessitates the development of a survival strategy within a hegemonic patriarchal structure that has continued to privilege males.

In addition, this chapter brings to the fore the real manifestation of feminised poverty in a developing nation like Nigeria, the most populous nation in Africa. A high form of gender alienation and 'voicelessness' characterises this type of poverty. The poor lack a voice to change the social order. This is doubly so for women who are poor, and live under a hegemonic patriarchal control, as in the case for the Bini women of southern Nigeria. The poor are seen but not heard. The 'visibility' of this social group and yet their 'inaudibility' is replicated by much research which has focused on 'protectionism' rather than seek a genuine understanding of this social group's experiences.

The ethnographic and an empirical analysis of the Bini women of Nigeria provides a meaningful insight into the experiences and social realities of this social group. It highlights the resilience of this socially disempowered category of women, who seek socio-economic empowerment through an avenue which many consider inappropriate, indecent and immodest—the sex trade. The Bini women of Nigeria were purposely selected as the Bini region has been dubbed the 'sex capital' of Nigeria as many of those involved in international transactional sex are from this ethic region (WHARC 2002 and Osezua 2010).

Theoretical Framework

Oscar Lewis provided a good understanding of the 'culture of poverty' in the 1960s in his publication *La Vida: A Puerto Rican Family in the Culture of Poverty—San Juan and New York*. This provides a basis for examining the present experiences of women who are currently engaged in the international sex trade. Bradbury (1957) intricately explains the close relationship between politics and the economy. Politics determines who gets what, when and how, which also includes economic resources. We often hear terms like 'political economy' or 'politics and economic resource management'. In the process of creating resources and managing the distribution of these, only a few people become rich. This therefore leaves the great majority deprived—the result is that poverty and the culture of deprivation becomes institutionalised. Lewis (1966b) maintains that the poor of a given country have more in common with their compatriots in that country than with the poor in other societies. This underscores the cultural dimension of poverty. In addition, this is true in terms of the poor's cultural views of the world, ideal family and kinship patterns, aspirations, values, and even body movements and language habits.

Another important reference to Lewis' work is the author's stance that poverty breeds family disruption, violence, brutality, cheapness of life, lack of love, lack of education, lack of medical facilities, mental meanness and, in short, a picture of incredible deprivation, the effects of which cannot be wiped out in a single generation. The implication of this is that poverty is a phenomenon that can develop into a cultural system (Lewis 1966b). The concept of the culture of poverty, as proposed by Lewis, suggests that if a culture of poverty is lived among the poor, it is essentially a strategy to cope with their poverty and cannot be, by itself, the cause of that poverty. Consequently, there are possible variations in the strategies engaged in by different people to cope with poverty across various regions of the world.

Lewis identified over 70 traits that characterise the culture of poverty, categorising them into four groups, namely:

- *The relation between the subculture and the larger society*: This involves disengagement and a lack of integration of the poor in the major institutions of empowerment of the larger society.
- *The nature of residential community*: Areas of a residential nature, most of which are characteristically depicted as slums and ghettos, such as poor housing conditions and over-crowding.

- *The nature of the family*: Poor family commitment and solidarity.
- *The attitudes, values, and character structure of the individual in society*: Here also the author pointed out that individuals who grow up in the culture of poverty have strong feelings of fatalism, helplessness, dependence and an inferiority complex.

The utility of this theory lies in the core thesis set out by Lewis (1966a) with respect to the cultural dimension of poverty and its effect on people's perceptions, values, social aspirations, family life and so on. One of the prestige structures that has been established among the Benin people is the value placed on overseas travel. In an earlier work, I argued that that valorising of international travel as a critical index of success in the region has also made the less economically empowered people extremely desirous of overseas travel to escape poverty (Osezua 2011). Similarly, as noted by Lewis in identifying family disruption as characteristic of poverty, evidence continues to suggest that many young girls who have been deported are from poor, dysfunctional, polygynous families, and are economically disempowered in terms of competition and family (Osezua 2011). Consequently, most of those involved in the cross-border sexual trade perceive themselves as economic agents through which generational poverty can be eradicated. The cross-border sex trade remains a remedy for poverty for the poor people of Benin, who sometimes offer their daughters and sisters or even wives to trafficker syndicate squads, whose primary role is to facilitate the migration (Osezua 2010, 2011).

Ethnography of the Bini People of Southern Nigeria

The Binis live in Edo State, which was carved out of the former Mid-Western State, part of the defunct Western region in Nigeria. The total area of Bini-speaking people covers over 4000 km^2 (Bradbury 1957). Benin City is located approximately 250 km east of Lagos, the former capital city of Nigeria. Benin City is also called 'Edo' by its inhabitants and individuals who are from Benin City refer to themselves as 'Oviedo', children of Bini, or 'Ovioba', subjects of the monarch, or Oba (Bradbury

1957). Hence, both Bini and Edo serve as ethnic and linguistic labels for the indigenous inhabitants of the Bini territory (Okpoko and Agbontaen 1993).

Benin City is located in the tropical forest zone of Nigeria. The Bini had a strong pre-colonial presence among the viable kingdoms in West Africa for a long time with an enviable history of dominance and conquest of neighbouring territories, leading to expansion of the kingdom and economic prosperity. The arrival of the Portuguese was a very significant phase of Bini history as the Europeans sought to Christianise them. In contrast, the Binis were more interested in commercial intercourse with the Portuguese (Igbafe 1979). Igbafe further observed that the great demand for slaves by the Portuguese became a valuable means of wealth for the Binis, with the Oba of Bini being the royal monopolist in the sale of slaves to the Portuguese. This was ended by the forceful annexation of the Bini kingdom by the British in 1897. The ruling Oba was then deposed to Calabar, a town in the southern part of Nigeria, where he died. This led to the decline of the empire (Igbafe 1979). Currently, the present Bini Monarch is Oba Erediauwa II. The palace is still regarded by indigenous Bini peoples as the custodian and citadel of tradition in contemporary Benin City.

Traditionally, the economy was agrarian. Several specialised guilds were also predominant. Bini are also known for brass casting, carving, beading and blacksmithing, which were organised into various guilds. Women assisted their husbands on the farms (Bradbury 1957). They also played important roles in the house of Iwebo, the guild that was directly responsible for making special clothes and beads for the Oba. Another significant aspect of pre-conquest Bini economy was the sale of slaves, ivory and pepper. For the pre-colonial Bini, the extended family was the coalescence of the economic, social, judicial, religious and political functions, making it the foundation of Bini social organisation. Polygyny is still predominant (Egharevba 1946). The concept of the first male child—the gate or '*uhro*'—has a played significant role in terms of inheritance and succession in the Bini family. Descent is patrilineal and residences are virilocal. The eldest son, the '*omordion*', in Bini custom plays a significant role in the organisation of the family. Hence, Igbafe (1979) maintained that primogeniture is the general custom in

Bini. In this system, the transmission of rights and ritual offices privileges the eldest son. He is strategic in terms of property inheritance. Older sons share other movable properties. Consequently, the family estate, known as '*igiogbe*', is usually an all-male affair. Women did not inherit their husbands' property. In the same vein, widows could be inherited in the same way properties (Bradbury 1957). The reason why a woman could not inherit property under Bini custom was to ensure that she did not confiscate it for the use of her family members (Lerner 1986). Women could not own estates or landed property prior to colonial contact (Lewis 1966a). However, women can now acquire landed property due to the colonial impact on Bini tradition (Osarumwense 1999).

Marriage holds a central place among the Binis, yet several taboos and restrictions regulate it. Before now, marriage to a 'white man' was seen as abominable (Igbafe 1979). Divorce among Bini women is viewed as the result of colonialism; however, it was commonly granted to women of noble parentage (Usuanlele 1999).

A dominant patriarchal notion about females among the Benin people is the perception that they are an inversion of the ideal; hence, they were made to undergo several rituals and placed under some cultural restrictions (Curnow 1996). Widowhood practices subject women to a mandatory mourning period of 7–21 days after the interment of the widows' late husbands. A widow lies on a bare floor wearing rags with charcoal on her face to mourn her departed husband. Although this practice is eroding fast as a result of Christianity, it has not been totally obliterated as some widows are still compelled to eat with a broken plate and cups, and use their left hands in eating. They were also expected to mourn their dead husbands by their graves. Bini widows declare their innocence from any involvement over their husbands' death before an ancestral shrine. Also of notable social significance is oath-taking among Bini women. It was also revealed that married women in Bini are made to swear oaths of marital faithfulness in the family shrines (Ebohon 1996). Also, on the event of the death of a man, his wife or wives were also made to swear at the village shrine, the '*ogwedion*', to establish their innocence over their husband's death.

Patriarchy and the Feminisation of Poverty among the Bini

The term 'patriarchy' is used to refer to the systematic organisation of male supremacy and female subordination (Kramarae 1992; Stacey 1993; Aina 1998). Patriarchy is therefore a system of male authority which oppresses women through its social, political and economic institutions. Patriarchal structure has been described as a major feature of traditional society. It is conceived as a set of social relations with a material base which enables men to dominate women (Stacey 1993; Lerner 1986; Odejide, 1998). Patriarchy is a system of social stratification and differentiation based on sex, which provides material advantages to males while simultaneously placing severe limits on the roles and activities of females. Consequently, a positive relationship has been established in the literature between patriarchal value entrenchment in a society and the feminisation of poverty. Gender norms for males are built around power and control, independence, violence, risk-taking, early sexual activity and having multiple sexual partners, while gender norms for women include submissiveness, dependence and virginity until marriage and then faithfulness during marriage (Odejide 1998). Boender *et al.* (2004) also maintained that the discriminatory socialisation of women and men into an unequal power relationship is a major cause of the feminisation of poverty: there is a greater tendency for women to be in disadvantaged situations that inhibit their living standards and contributions to development, especially in patriarchal societies like that of the Bini of Nigeria. This trend has resulted in what this paper proposes as a subtle rebellion against patriarchy and poverty by the Bini women through their active engagement in the international sex trade.

Methods

This analysis is based on the qualitative research traditions of ethnography, using life histories of respondents who were selected through snowball sampling technique. Women who had previous experiences of deportation because of their involvement in the sex trade constituted the subjects of this study. The main emphasis of subject recruitment was to examine the family profiles of these deportees and to document their reasons for becoming involved in transactional sex abroad in order to ascertain their agency. For ethical reasons, participation was voluntary. Three of those earlier contacted refused to participate. While all other women who eventually agreed to participate gave a verbal consent, they insisted that their actual names must not be mentioned. Their insistence on confidentiality and anonymity was respected. The research spanned a period of nine months in all—March 2009 to December 2009.

Findings of the Study

Family Profile and Structure

There was a pervasive notion of disrespect exhibited by the women and girls as they blamed their fathers for their present predicament. Many of the young women interviewed disclosed that they were from poor families and that their fathers were the main culprits for their impoverishment. They derided their father's attitude towards polygamous marriage, which they maintained had led to family disorganisation and unhealthy rivalries among siblings of different mothers, thereby making children born into such home very vulnerable to social vices. Osato, a fair-skinned hairdresser in her early twenties, observed:

> Imagine the kind of house in which I grew up in, with three wives with so many brothers and sisters, with no food and protection of any kind. It was terrible. I went to primary school, but could not complete my JSS (Junior Secondary School). (In-depth Interview (IDI), 24 years old, hairdresser, with incomplete junior secondary school education)

Corroborating the earlier assertion that dysfunctional families as a result of competition, rivalry and suspicion characterised the polygamous Bini families, another deportee, a dark-skinned young women in her early twenties, Omoyewen, recollected her days growing up as follows:

> I was born into a family where peace was totally absent. It was a family where a father joins his junior wife to fight the senior wife all the time. I was there until I was 12. It was great pain for me. My mother had to send me away from the house to stay with an aunt of mine and unfortunately for my mum, I was her only child, while the other wife had two boys and three girls.

Omoyewen's decision to live with the father's aunt in order to avoid the pain and disappointment she felt as a young girl ended up one becoming, as she described it, from 'frying pan to fire'. Consequently, she opted for a better option to save her mum from what she considered unfair treatment meted out by her father to her mother. She states:

> I could not just tell my mother since she too was suffering in her husband's house because of the other wife. In fact, going to school was very tough for me. Even though my aunty contributed to my training, mother also did her best. When I saw the level of pain and suffering my mother was exposed to and since I could not also complete my education, I decided to travel abroad.
>
> (IDI, 26 years old, petty trader, with incomplete secondary education)

Omoyewen concluded by blaming the father for her present predicament (at the time, she had just been deported). She enthused:

> I would have been married now and have had children. In this life, it is not even good to have a bad father. If my father had not been troubling my mum, I would not have gone to Italy at all. Look at me now, no mother, no abroad, no children and no marriage.
>
> (IDI, 26 years old petty trader, with incomplete secondary education)

The account of Oghowen, a dark-skinned lady in her late teens, presently a phone call girl, was not different from the first two. Oghowen gave a retrospective account of her family looking very sober as she recounted her family background:

> My father married three wives and he had more children than he could cater for. We were so 'plenty' that one day my father came and drove some

> wives away. My father also drove my mother away and we have to follow our mamma. So that was when our suffering worsened. She said she felt very unhappy. It was hell on earth for my mother and all of us. At that time, I was already in JSS III. My mother had to rent a house out of the meagre earning she got from a sympathiser and rented a room. Every market day, she would sell ice blocks. The funniest part was that we were three children, so you can imagine how life was for us.
> (IDI, 27 years old, phone call girl, with a senior secondary school certificate)

When asked who she had plans for if she had succeeded in her sex trade abroad, she mentioned her mother and her siblings without hesitation. Concerning her father, she added very despitefully:

> I don't have any plan concerning that idiot, apart from my mum and my younger ones.

Coerced or Voluntary?

Contrary to the stance in the literature that women traffickers deceive girls and women into the international sex trade, the findings of this work shows that there has been a shift in this earlier posture as many girls from poor homes are presently negotiating their ways into the global sex trade. Out of the 15 women who were subjects of the life history, 12 of them admitted that they were aware of the sexual component of the trade. Ten (Kramarae 1992) of them admitted that they actually sought out traffickers themselves, with three of them admitting that it was with the co-operation of their mothers. One of them disclosed that her father was the one who initiated the connection to the traffickers. All the others who had mentioned that they sought out traffickers themselves added that the joint efforts of friends and relatives made their search fruitful. For instance, the case of Adesuwa, a 16-year-old who dropped out in primary four and is currently a sales girl in a busy restaurant located in the heart of the city, reveals that they had a prior understanding about the sexual component of their trade:

> For the first time was the good experiences, it was exciting to travel out of Nigeria, the exposure to wearing new clothes before travelling. I like the new clothes. I know say that I will sleep with men. The bad side of the experience one, sleep with too many different men. Sometimes 15 man a day and the last one is the proceeds on a given to the traffickers, even at time I don't even have anything for myself. The bad thing again that the money that I think will make for a long time, those madams used to cheat us too much.
>
> (IDI, 16 years old, sales girl (with incomplete primary education))

Adesuwa was apparently thrilled by the new experience of travelling overseas but had underestimated the demand that the sexual services would place on her. Hence, she complained that there were too many male clients and that she was not even offered all the money that was due to her at the end of each sexual transaction.

Charity, another young woman in her early twenties, described her experience as hard. She, however, admitted that neither of her parents was aware of her decision to get into the sex trade. Charity disclosed that she willingly contacted a friend who knew a viable sponsor. She added that she had seen the life of a whole family in her neighbourhood change with the remittances sent to family members in Nigeria by those who had their daughters abroad. The sex trade became an attraction for her owing to its profitability. Hence, she insisted that she was aware that she was going to become '*ashawo*' abroad.[1]

> My good experience was that at least I was able to see my immediate younger one through his studies in the polytechnics. My bad experience, I went through many insults and pain right there overseas. I knew that I will do '*ashawo*' there. That is why they took me to go and swear an oath, which I did. So the sponsor asked me to pay twice my money. And I did not know. So when I got familiar with the European currency, I now discover that the money was times two. I could not say anything. And anytime, I go out to look for work, if I did not bring back the amount of money my madam asked me to bring, she would beat me, so it was not easy at all.
>
> (IDI, 23 years old apprentice (with senior secondary school certificate))

In the case of Grace, her mother was the main facilitator in the process of providing the necessary links with the traffickers and providing part of the funds needed for securing travelling documents. They were both fed up with the maltreatment meted out to them by the patriarch. Hence, Grace disclosed that she was well aware of the fact that she would be involved in offering sexual services to men, who would in turn pay her. Her mother was also aware:

> My mother tried very much. She sold her trinkets and her gold. She sold most things she had and borrowed money from *osusu*[2] and that was how I eventually made my trip abroad. This thing is common now. Many of us know the kind work we want to do. That is why they want us to swear so that we won't run away. But many times, they overuse us and give us very little money. When you prove stubborn to them, they will set you up by reporting you to the police.
>
> (IDI, 21 years old unemployed woman (with incomplete senior secondary school education))

There were, however, three cases of those who were ignorant that they were going to be involved in the sex trade. A classic case was the story of Paulina, a married woman:

> It is a long story, and anytime I think about it, I feel like killing myself. [There was a pause for some while she cried.] My parents did not know about my travelling. They are innocent. It was my aunty, my father's elder sister that called me into my husband's house and she gave me bad advice that led me astray today.
>
> (IDI, 27 years old unemployed divorcee (with a primary school education))

Although Paulina and her husband were not rich, her aunty was totally dissatisfied with her husband's status, so she decided to take Paulina to Italy to improve the economic condition of her family. She stated that she did not know her aunt was a trafficker, but deceived her that there were things she could do there to improve her economic condition, like plaiting hair. At the time she discovered the reality, she felt it was too late to go back to her husband and family and then decided to cooperate with her aunt:

It was too late for me since I had disrespected my husband and parents without informing them before I left. I was ashamed and so I wanted to make it at all costs. After 2 years, I was deported without a building, a car or something really valuable to my name. Before my deportation, my husband had married a new wife to take care of him and the children I left behind. This husband that I left now owns a beautiful house of his own and a car. I have had to go and beg so that I could come back, but he drove me away and did not even allow me to see my own children. He said I am never to enter his compound again. To make matters worse, the aunt who deceived me and led me astray is the one laughing at me today because no man has come for me again.

(IDI, 27 years old unemployed divorcee (with a primary school education))

Deportees Perspectives on the Sex Trade

A second objective of this paper was to ascertain the willingness of these women to give up the international sex trade despite the series of campaigns against women trafficking in Nigeria and the government's fight against the phenomenon. The findings revealed that more than half of these deported women had painful memories of being violently handled by men, short-changed by the sponsors and traffickers, especially during the process of paying back the money that was used to transport them across the national borders, and harsh working conditions. In spite of these adversities they experienced engaging in the cross-border sex trade, 12 out these women agreed that they would go back if there was another opportunity to travel abroad for the purpose of the sex trade. Many of them are of the opinion that there were very young and ignorant of migration procedures, which the traffickers took advantage of. Hence, they said that they had learnt the requisite experience that would put them above the traffickers. All the women interviewed agreed that sex trade was hazardous yet a lucrative business.

There was a consensus among the women about the greed of the traffickers, who made them pay twice as much as they owed them due to their inability to convert the local currencies, and that this adversely affected the profitability of the business. This, coupled this with stringent migration procedures into destination countries in Europe, reinforced the indispensability of syndicate groups who subjected them to extreme extortion. Some of the women recounted their achievements within a short while and insisted that life abroad was still better than life in Nigeria. They bemoaned their present condition, where they consider themselves unemployed and therefore had to fend for themselves. For one of them, what she earned in a day in the cross-border sex trade, she does not earn in two months in Nigeria. One of the women in her late teens, a hairdresser, stated: 'no business without risk, when you shine your eyes well you will know the "coco" (the core of the matter).'

Discussion and Analysis of Emergent Issues

Life histories conducted among deported Nigerian women who were involved in the international sex trade showed that they were on a rescue mission to save their families from utter impoverishment. There is evidence in the literature that the female child in Bini culture is undervalued and that her male counterpart is preferred (Aghatise 2002). The girl child is further entrapped in a culture with obnoxious traditions and several restrictions that continue to reinforce discrimination meted out against her gender. For instance, the poor status of the girl child can be ascribed to the perception that many Bini men and patriarchs in the lower economic stratum have about female education in the face of economic hardships in the region. Males who are seen as the natural bearers of family names are given preference over and above their female counterparts, who are described as transitory beings.

Most of those who have opted for the sex trade and sought the help of traffickers have limited education, which largely affects their ability to secure a good job. This no doubt influenced their choice to seek a better life overseas, due to the growing rates of unemployment in Nigeria. The majority of them agree that the international sex trade has the propensity for providing them with the desired economic prosperity, which will ultimately enhance their social position within Nigerian society. Moreover, beyond the effects of the structural adjustment policy on the Nigeria economy in the early 1980s, which some scholars opined gave impetus to the phenomenon of sex trafficking in the region, other factors, such as globalisation and the subsequent feminisation of poverty, are important elements within the world system political economy, of which Benin is an integral part. The culture of poverty necessitated a survival strategy. Hence, Sassens (2004) espoused that the concept of feminisation of survival was indicative of the specific reaction of some categories of women to the adverse political economy in Nigeria.

In addition, the study findings confirmed that the Bini people have historically had contact with the outside world, which has culminated in extensive networks in European countries. Hence it was easier for people in this region to draw on the networks that they had built for subsequent illegal migration activities. Again, the evidence that it has been women at the lower socio-economic rungs becoming involved in the international sex trade supports the theory of the culture of poverty, as postulated by Lewis (1966b). Cultural factors like polygyny, male child preference, pervasive traditional beliefs, ancestral worship, excessive cultural taboos and classical patriarchy have contributed significantly to reinforcing the trajectory witnessed in international sex trade among Bini women. However, findings revealed that these women no longer lack agency as the literature indicates; rather, they have devised coping strategies to survive the degree of impoverishment and patriarchal control in the region.

Conclusion

Nigeria is a signatory to many international conversions, including the Beijing Platform for Action New Partnership for African Development Convention on the Elimination of Discrimination against Women, the African Protocol on People's Rights and the Rights of Women. This is coupled with their fact that many local and national non-governmental organisations have channelled their efforts into eradicating trafficking and international prostitution in the region. The establishment of the National Agency for Prohibiting Trafficking in Persons, which is specifically located in Benin City, is meant to stem the incidence of cross-border sexual transactions. Despite these efforts by the Nigerian government and the Edo State government, in particular, to combat the phenomena of trafficking and international prostitution, there are still indications that incidences are under-reported.

The reason for this inability to stem the tide of sex trafficking and the cross-border sex trade simply lies in the decision by the government and relevant stakeholders to engage in protectionism, which always targets eradication rather than giving a voice to this social class. Many of these women perceive the government as unresponsive to their plight and therefore are unwilling to cooperate with government agencies in the campaign against sex trafficking and the cross-border sex trade. This has created a situation where many of the deportees of this trade still maintain that the sex trade is not risk-free, like any other business; this is indicative that a change of behaviour in relation to the sex trade by many of these women is not imminent. It is high time research began to focus on the agency rather than victimising the women involved in sex trade. This way, underlining factors of gender inequality, which are manifested in various forms of cultural discrimination against such women, in relation to access towards critical socio-economic resources, will be truly targeted.

Notes

1. *Ashawo* is an indigenous colloquial word for prostitute.
2. Thrift collection.

References

Aghatise, E. (2002). *Trafficking for prostitution in Italy: Concept paper*. Presented in Expert Group Meeting on "Trafficking in Women and Girls", Glen Cove, New York.

Aina, I. O. (1998). Women, culture and society. In A. Sesay & A. Odebiyi (Eds.), *Nigerian women in society and development*. Ibadan: Dokun Publishing House.

Alpes, M. (2010). Escaping statism: From the paradigm of trafficking to the migration trajectories of West Africa sex workers in Paris. In T. Zheng (Ed.), *Sex trafficking, human rights and social justice* (pp. 117–131). Paris: Routledge.
Boender, C., Santana, D., Santillán, D., Hardee, K., Greene, M. E., & Schuler, S. (2004). *The so what report: A look at whether integrating a gender focus international programme makes a difference to outcomes.* Washington, DC: Intelligence Agency Work Groups.
Bradbury, R. (1957). *The Benin kingdom and the Edo speaking people of South-Western Nigeria*. London: International African Institute.
Curnow, K. (1996). Prestige and gentlemen: Benin's ideal man. *Art Journal, 56*(2), 75–81.
Ebohon, O. (1996). The life of an African woman: The Benin woman in focus. Lecture Delivered on Black Solidarity Day, New York.
Egharevba, J. (1946). *Benin laws and customs.* Port Harcourt, Nigeria: C.M.S. Bookshops, Niger Press.
Igbafe, P. (1979). Slavery and emancipation in Benin (1897–1945). *Journal of African History, 15*(3), 409–429.
Kramarae, C. (1992). The condition of patriarchy. In C. Kramarae & D. Spender (Eds.), *The knowledge explosion: Generation of feminist scholarship*. London: Athen Series, Teachers College Press.
Lerner, G. (1986). *The creation of patriarchy*. Oxford: Oxford University Press.
Lewis, O. (1966a). *La Vida: A Puerto Rican family in the culture of poverty—San Juan and New York*. New York: Random House.
Lewis, O. (1966b, November). The culture of poverty. *Scientific American, 215*(4), 19–25.
Odejide, A. (1998). Women in trade and industry. In A. Sesay & A. Odebiyi (Eds.), *Nigerian women in society and development* (pp. 122–134). Ibadan: Dokun Publishing Co.
Okpoko, A., & Agbontaen, K. (1993). Edo traditions of origin. In A. Okpoko & C. Folorunso (Eds.), *Some Nigerian peoples* (pp. 213–224). Ibadan: Rex Charles Publication.
Osarumwense, N. (1999). The Bini customary law of inheritance of a woman's property(ies) in retrospect in Benin Studies. *Newsletter, 4*(2), 2–3.
Osezua, O. (2011). Cross-border sex trade, transnational remittances and changing family structures among Benin women of Southern Nigeria. *Gender and Behavaviour, 9*(1), 4276–4296.
Osezua, O. C. (2010). *An anthropological analysis of changing family structures, women trafficking and international prostitution among Bini people of Southern Nigeria*. Ile-Ife: Obafemi Awolowo University.

Sassens, S. (2004). Women's burden: Counter-geographies of globalization and the feminization of survival. *Nordic Journal of International Law, 71*, 255–274.

Stacey, J. (1993). Untangling feminist theory. In D. Richardson & V. RoBeninson (Eds.), *Introducing women's studies: Feminist theory and practice*. London: Macmillan.

United Nations Interregional Crimes and Justice (UNICRI)/United Nations Office for Drugs and Crimes (UNODC). (2003). *Trafficking of young women and minors from Nigeria into Italy for the purpose of sexual exploitation*. Report of field survey in Edo State.

Usuanlele, U. (1999). Colonialism and the origin of prostitution among the Benin of Edo State in Benin Studies. *Newsletter, 4*(3), 3–6.

Women's Health and Action Research Centre (WHARC). (2002). Sexual trafficking, a challenge to reproductive health in Nigeria. *Women's Heal Forum, 17*(1), 12–22.

Clementina O. Osezua received a PhD in Social Anthropology from Obafemi Awolowo University, Ile-Ife, Nigeria. She is currently a Senior Lecturer at the same university. A Fellow of both the Carnegie Institute and the Brown University of Institute of Population Research and Development Institute, USA, she has published extensively in reputable local and international outlets. Her research focuses on human trafficking and gendered migration, gender and culture, and gender and religion, and tends towards qualitative ethnographic research.

15

Crime and Human Security in Post-Apartheid South Africa

Shanta Balgobind Singh

Introduction

In South Africa, the threat to human safety and security has taken diverse forms, for instance, crime, political violence, insurgency, terrorism, electoral violence, ethnic clashes, among other threats. Human security focuses on the protection and development of the individual and the community. During the last few decades, more people have been affected by civil wars, the arbitrary exercise of state power, displacement, poverty and hunger than by wars between states. Human security is a widely used concept that refers to seven types of threats that affect people and their lives (Hanns Seidel Foundation 2014:1):

- Economic security
- Food security
- Health security

S.B. Singh (✉)
College of Humanities, University of KwaZulu-Natal, Durban, South Africa

A. Adeniran, L. Ikuteyijo (eds.), *Africa Now!*,
https://doi.org/10.1007/978-3-319-62443-3_15

- Environmental security
- Personal security
- Community security
- Political security

The United Nations Commission on Human Security qualifies the definition of human security as follows (Commission on Human Security 2003: 230):

> Human security in its broadest sense embraces far more than the absence of violent conflict. It encompasses human rights, good governance, access to education and health care and ensuring that each individual has opportunities and choices to fulfil his or her own potential.

Human security ensures the creation of 'political, social, environmental, economic, military and cultural systems that together give people the building blocks of survival, livelihood and dignity'. These systems are often threatened by factors that may emanate from outside the particular state or nation, but which in some cases may have 'internal' or intra-state causes, or may be caused by the state itself. For South Africa, some of these threats to human security include the following (Mlambo 2005: 231):

- Violent crime
- Rape
- Terrorism
- Racially motivated violence
- Politically motivated violence
- Farm murders/attacks and land conflicts
- Illicit arms and drug deals
- Private armies (mercenaries)
- Illegal migration
- Xenophobia-related violence
- Illegal firearms
- Poverty and inequality
- Regional instability

This chapter will address the personal and community security affected by the increase of crime in South Africa. Crime prevention and strategies will also be discussed.

Crime is a phenomenon that has existed throughout time and is prevalent all over the world. It is a violation of human security, both at a micro and a macro level. Although South Africa is 21 years into the transition from apartheid to democracy, this change has not brought with it a system of criminal justice capable of responding to these challenges, especially the increased prevalence of criminal activities in the country. The daily occurrence of crime monopolises newspaper headlines and the country's reputation as a crime capital stretches across national and international borders. The institutions of criminal justice, i.e. the police, courts and correctional centres, remain weighed down by the public belief that they are tools for enforcing the rule of a minority over the majority, rather than for protecting all. Thus, if one takes cognisance of what appears in the media then there is a perception that crime is out of control.

However, this perception is created by the copiousness of publicity given to crime and criminality in the media. While the publicity given to the successful prevention of criminal activity or criminal convictions is not as prominent, overall it seems that criminal violence is itself a manifestation of South Africa's historical traumas as well as its contemporary social ills. Many of the contemporary problems deserve to be addressed in their own right. But South African society will also benefit if measures targeted specifically at the problem of violence are made priorities by the public and the criminal justice system agenda.

Although crime statistics are often better indicators of the prevalence of law enforcement and a willingness to report crime than of actual prevalence, it is clearly evident that there has been an increase in these crimes in South Africa. Since 1994, the national crime statistics for South Africa have been provided in line with the extensive objectives of the National Crime Prevention Strategy, which was adopted in 1996. This involves the collection of reliable crime data together with the expansion of an effective communications approach based on information which would appropriately inform public opinion in the fight against crime. Prior to 2000, crime statistics were released for the period January to September each year. Presently crime statistics are released each September and cover crimes reported to the South African Police Services (SAPS) in the previous financial year (i.e. 1 April of the previous year to 31 March of the current year). This means that, at the time of release, official crime statistics are already approximately six months out of date and do not reflect current crime patterns and trends. These official crime statistics are important to establish historical trends but are not beneficial for the objective of planning or evaluating crime prevention initiatives by either the government or communities.

Human Security and Policy Frameworks

According to Africa (2015: 181), in 1993, with the support of a Transitional Executive Council agreements were reached about the basic principles that should govern a new national security approach. Principles associated with democratic civil security relations, and the broadening of the concept of security, were important shifts in the idea of security. The primacy of human security—the freedom of the individual from threats to physical safety, to a decent quality of life and to human dignity—infused the negotiations discourse and filtered into debates on the role of the security structures, both during the transition and after (Masuku 2004). Approved by the first democratically elected Parliament, the 1994 *White Paper on Intelligence* maintained that while the 'traditional and narrower approach to security has emphasized military threats', the international security agenda was shifting to the 'full range of political, economic, military, social, religious, technological, ethnic and ethical factors that shape security around the world' (RSA 1994). The 1994 *Green Paper on Safety and Security* emphasised the need for democratic control, police accountability and community participation on issues of safety and security, ideas that were carried through to the *White Paper on Safety and Security* adopted in 1998. The *White Paper on Defence*, adopted in 1996, argued that 'security is an all-encompassing condition in which individual citizens live in freedom, peace and safety; participate fully in the process of governance; enjoy the protection of fundamental rights; have access to resources and the basic necessities of life; and inhabit an environment which is not detrimental to their health and well-being' (RSA 1996b).

These policy frameworks, along with the Constitution, provided the context for the post-apartheid laws regulating the security sector. The South African Constitution, adopted in 1996 and replacing the Interim Constitution that had been adopted in 1993, also invoked the language of human security when it urged that 'national security must reflect the resolve of South Africans, as individuals and as a nation, to live as equals, to live in peace and harmony, to be free from fear and want and to seek a better life' (RSA 1996a: s 198a). Rothschild (1995: 55) is of the opinion that human security extends downwards 'to the security of groups and individuals', upward 'to the security of international systems', horizontally from military security 'to political, economic, social, environmental, or 'human security', and in all directions 'upwards to international institutions, downwards to regional or local government, sideways to nongovernmental organizations, to public opinion and the press, and to the abstract forces of nature or of the market'.

Contextualising the Criminal Justice System in South Africa

Prior to 1994 (the end of the apartheid era), the three arms of the criminal justice system comprised the Department of Law and Order, the Department of Justice and the Department of Prisons. As a result of the promulgation of the Interim Constitution Act on 27 April 1994, a number of changes were introduced within this system. The changes included the adoption of the Bill of Rights, which impacted on the admissibility of confessions and admissions by the court. Furthermore, there was a move from reactive policing to proactive policing, which impacted on the functioning of the criminal justice system. The introduction of the Bill of Rights and the abolition of the death penalty created a perception among South African citizens that the government had become 'too soft' on criminals. After the end of apartheid, the SAP was renamed the South African Police Service (SAPS), and the Ministry of Law and Order was renamed the Ministry of Safety and Security. The Department of Prisons was renamed the Department of Correctional Services.

The Constitution of South Africa asserts that SAPS has a responsibility to prevent, combat and investigate crime, maintain public order, protect and secure the inhabitants of the country and their property, uphold and enforce the law, create a safe and secure environment for all people in South Africa, prevent anything that may threaten the safety or security of any community, investigate any crimes that threaten the safety or security of any community, ensure that criminals are brought to justice and participate in efforts to address the causes of crime.

Policing in South Africa

The Minister of Police, Mr Nathi Nhleko, stated in the SAPS Annual Report (2014/2015: 13) that:

> We are enjoined by the National Development Plan…to create a South Africa where all our citizens and those who visit our shores 'are and feel safe'. Given our history and the institutionalised violence that was its main feature, this is a mammoth task. But it is not an impossible task. What makes it possible is that crafted into our democratic dispensation was the understanding and acceptance that policing in our country needs to take a different trajectory. This genetic code started in earnest when the African National Congress was formed in 1912 and finds expression in seminal documents like the Freedom Charter with its rich 60-year history as our pathfinder, the 1992 Ready to Govern and the eventual foundation of our nation, the Constitution as enacted in 1996. Twenty-one years later, we are still positively grappling with this trajectory where human rights reign supreme; where the dignity of each and every person is guaranteed; where policing is centred on the community participation; where the police are held to account for infractions; and where the rule of law reigns supreme.

Elements of the National Development Plan and its progress are encapsulated in this annual report and address the following four pillars:

- Strengthening the criminal justice system

- Professionalising the police
- Demilitarising the police
- Building safety using an integrated approach

Mr Nhleko further stated that 'as we commemorate the 60th year of the Freedom Charter, it becomes relevant to reflect on police performance against some of the critical policing priorities of the African National Congress' Ready to Govern Policy':

- Policing in South Africa is rooted among the communities that the police serve and is accountable to the statutory body, the Community Police Forums that have been democratically elected by their communities, whose legislated objects of existence and functioning include: establishing and maintaining a partnership between the community and the service; improving the rendering of police services to the community at national, provincial, area and local levels; improving transparency and accountability of the service to the community; and promoting joint problem identification and problem-solving by the service and the community. More work still needs to be done to advance the gains this government has demonstrated in this regard.
- The Department of Police, in heeding the call by the President to take government to the people and in ensuring that policing is subject to public scrutiny and open debate, has created a platform of Izimbizo and outreach programmes with key stakeholders and interest groups. This office alone has held scores of sessions with communities, business and other critical agents in order to create awareness, share information and obtain feedback from the recipients of our services.
- The Independent Police Investigative Directorate is dealing with allegations of police misconduct, and SAPS in keeping with the ethos of human rights has implemented 978 out of 1000 recommendations. Furthermore, SAPS also arrested 686 police officers for various types of crime in 2014/2015 to show it is committed to professionalism and is intolerant of criminal conduct by its employees.
- The police service shall continue to strive for high performance standards. This is evidenced by the improved performance of SAPS as contained in the annual report being presented to the public.

Crime Statistics in South Africa

The SAPS crime statistics report covers only crimes that have been reported to the police (metropolitan police agencies do not record or release crime statistics but open criminal cases with the SAPS, which forms part of the statistics). For SAPS reporting/documentation purposes, crimes are defined by specific codes and sub-codes. This information is entered onto police dockets when a case is opened, capturing the nature or type of crime reported. In order to create the national crime statistics report, SAPS collects and collates such data from approximately 1130 police stations (the exact number of stations fluctuates slightly each year) from the nine provinces in South Africa. The annual SAPS crime statistics report provides total national reported crime data, together with the same information further broken down by province and by police station. No information is provided about the exact location of reported crimes within each precinct and it should be noted that many police stations cover large and socio-economically diverse areas.

In South Africa crimes are recorded and tabulated as statistics by the Department of Statistics, based on data provided by the police, the courts and the correctional services. The data is then processed—a report is published annually in which all the details of the number of crime increases/decreases and other information on crime in South Africa are given.

There are specific crimes that continue to rise. The escalating level of public unrest also draws local members to conduct basic policing to support public policing and ease the strain on resources. Although the Government has invested enormously in the policing environment, there is currently a total personnel strength of 197,946, with a comparative police to population ratio of 1:336.

Police Statistics

Police statistics describes the crimes that have occurred. Although these records represent the most reliable and comprehensive sources of information on crime and are useful for crime studies, not all crimes are reported to police. Police statistics provides details of how and when the crime was committed, who committed it, who the victim was and what injuries or losses were suffered.

Court Statistics

Although court statistics deal with the persons involved in crime, they include only material brought forward in court. Therefore, the numbers of crimes recorded by these statistics are lower than in police statistics.

Correctional Statistics

Correctional statistics refers to crimes where imprisonment was imposed as a punishment. The number of offenders who are sentenced to imprisonment by the courts and the release policy of the Department of Correctional Services influence the size of the South African prison population. Statistics are usually collected on offences, offenders and victims.

Crime and Human Security

A major cause of insecurity in South Africa is the high rate of crime, especially violent crime such as murder, rape and hijacking. A National Victims of Crime Survey carried out by the Institute for Security Studies found that between September 2002 and August 2003 nearly one quarter of all South Africans had been victims of crime. The survey also found that although police reports indicate that reported crime decreased between 1998 and 2003, that is, feeling of insecurity increasing from 25% in 1998 to 58% in 2003. However, official reports continue to emphasise government crime prevention strategies, which, they say, are having a positive effect on crime. These include the National Crime Prevention Strategy (NCPS), the National Crime Combating Strategy and the Integrated Justice System (IJS). These measures have had the effect of stabilising and reducing some crime trends. One official report shows that, with the exception of common assault and theft, the 20 most serious crimes in South Africa have been decreasing.

Perceptions of crime do not always coincide with the reality on the ground. SAPS are of the opinion that crime has been declining in absolute terms since 1994. With the exception of robbery with aggravating circumstances, illegal possession of firearms and drug-related crime, all crime figures are reportedly declining (Mlambo 2005: 247).

Some selected figures of actual police crime statistics are given in the table below.

Reasons for the Increase in Crime

Some of the reasons for the increase in crime are:

- Unemployment
- Poverty
- The post-apartheid influx of black African migrants
- Culture of violence borne out of the nature of apartheid and the anti-apartheid struggle
- Faulty and ill-equipped criminal justice system
- Rising inequality
- Availability of drugs and firearms

Violent crimes have important implications for the perception of security or insecurity among South Africans—this category constitutes approximately a third of South Africa's recorded crime.

Categories of Crime

A growing trend worldwide is that people are setting their security above all else, including above the protection of human rights. This has seen international public tolerance for suspending civil rights in the face of threats to public security—both criminal threats to personal security and terrorist threats to national security—creating new challenges for human rights activists. It has also been pointed out that in their preoccupation with the rights of perpetrators, human rights activists have ignored the rights of victims or potential future victims, and are thus failing to advance social justice (Mlambo 2005: 230).

The annual SAPS crime statistics report tends to focus on what it terms 'serious crime'. How this category is broken down by SAPS tends to differ slightly with each annual report, which has added to confusion and uncertainty about crime trends. However, these crimes can broadly be divided into:

- *Crimes against a person*—generally referred to as 'contact crime' where a person or people are injured/harmed or threatened with injury/harm. A further sub-category of 'contact-related crime' is used for violent crimes committed against property with the intention of causing damage to a person, for example arson or malicious damage to property.
- *Crimes against property*—crime that occurs in the absence of a victim or where the victim is unaware of the crime at the time (i.e. where no person is directly or immediately harmed or threatened), for example theft of or from an unattended vehicle.
- *Other serious crimes*—includes commercial and financial crimes, ranging from large-scale fraud and corruption to small-scale incidents such as shoplifting.
- *Crime detected by police action*—these are crimes that are not reported by the public but are detected through direct police action such as roadblocks and SAPS intelligence operations. Crimes under this category include the illegal possession of firearms and driving under the influence (of drugs or alcohol); and the use, possession or trade of illegal drugs. Increases in the latter two categories (which have been notable over preceding reporting periods) can be directly attributed to intensified police activity rather than an indication of increased activity of this nature.

These broad categories are not arbitrary—they are important when it comes to understanding, or attempting to understand, both motive and mode of the crimes, both of which are necessary to understand if effective crime prevention strategies are to be deployed. However, within these general categories there is often further confusion about or between specific crimes, which can lead to misinformation or the misinterpretation of data. During 2014 and 2015 approximately 2.2 million crimes were reported in South Africa (SAPS 2014/2015).

There is a common perception in South Africa that crime only began to increase from 1990 onwards, in conjunction with the political transition. In fact, most serious crimes, notably murder, robbery and housebreaking, began to increase from the mid-1980s. South Africa's crime problem is not recent. Crime increased significantly during the 1980s, when the apartheid state was being most strongly challenged. The increase in levels of crime peaked in 1990, the year in which the political transition began. Reported crime levels in South Africa began to in the mid-1980s, peaking in 2002/2003. During that time, and particularly in the period between 1994 and 2002, South Africa developed a reputation as a country with one the highest levels of violent crime in the world. Between 2002/2003 and 2014/2015, reported crime levels (for certain crimes) steadily declined but public perceptions about crime and safety have not reflected this change. This is partly due to inadequate communication from government and police, selective reporting in the media and the increase in the sort of social media commentary which contributes to the dissemination of myth and misinformation around crime, rather than accurate and contextualised information.

This matters for a number of reasons. The real or perceived risks of crime may alter decisions that are made about personal safety: how one lives day to day, which political party to vote for, and how one (individually or as an electorate) chooses to spend resources on fighting crime. Many in government see an interest in down-playing the problem that crime represents. Both the flaws and the accurate elements need to be understood. Understanding what crime statistics can—and cannot—tell us is a vital starting point for anyone wanting to contribute to tackling crime.

One of the primary obstacles facing political stability and human rights in South Africa is crime. High incidents of violent crime and media coverage results in people living in a state of perpetual fear and insecurity, despite official reports indicating a decrease in certain crimes. Human rights can be divided into six categories (Nickel 2003):

Security rights: protecting people against murder, torture etc.
Due process rights: protecting people against abuses of the legal system.
Liberty rights: protecting freedoms such as religion, speech etc.
Political rights: protecting political participation.
Equality rights: protecting equally with regard to citizenship, the law and non-discrimination.
Social rights: protecting access to education, healthcare etc.

Crime compromises the defence of human rights within a society. It poses a threat to public security and thus provides potential for political instability. Furthermore, crime creates an environment which is not conducive to the maintenance of human rights standards. Crime impacts on human rights in three fundamental ways. First, any crime directly violates the rights of the victim. This is to suggest that an act of crime compromises the liberties of the victim where the criminal has no such right to do so. Second, crime creates an environment of civil insecurity and builds negative public sentiment. This makes the maintenance of human rights standards exceptionally difficult in that the line between the rights of the victim and the rights of the offender becomes permanently blurred. Third, crime creates policing problems—in order to maintain public support, government often needs to resort to measures which are authoritarian in nature and thus undermine a democratic transition. The table below represents the number of crimes committed in the Republic of South Africa from 2005 to 2015 (Table 15.1).

From the table above, which depicts the increase/decrease of crime in South Africa for the periods 2005–2015, it is noted that certain crimes have decreased whereas other crimes have increased. 'Trio crimes' are incidences of violent organised crime, with a particular focus on house robberies, business robberies and vehicle hijackings. These crimes are regarded as high priority crimes. Given the deplorable and violent nature of these crimes, these are the types of crime that threaten the feeling of safety and security that citizens have in their homes, in their businesses and when commuting between the two. A reduction in these types of

Table 15.1 Crime in South Africa 2005–2015

Republic of South Africa												
Financial years (2005–2006 to 2014–2015)												
											Comparison 2013–2014 with 2014–2015	
Crime category	April 2005 to March 2006	April 2006 to March 2007	April 2007 to March 2008	April 2008 to March 2009	April 2009 to March 2010	April 2010 to March 2011	April 2011 to March 2012	April 2012 to March 2013	April 2013 to March 2014	April 2014 to March 2015	Case difference	% Change
Contact crimes (Crimes against the person)												
Murder	18,455	19,106	18,400	18,084	16,767	15,893	15,554	16,213	17,023	17,805	782	4.6
Total sexual offences	67,064	64,071	62,484	69,197	66,992	64,921	60,539	60,888	56,680	53,617	−3063	−5.4
Attempted murder	20,369	19,957	18,643	18,140	17,247	15,360	14,730	16,236	16,989	17,537	548	3.2
Assault with the intent to inflict grievous bodily harm	225,655	216,754	208,705	202,328	203,807	197,470	191,612	185,050	182,333	182,556	223	0.1
Common assault	225,434	207,869	195,885	190,709	194,922	184,103	180,165	171,653	166,081	161,486	−4595	−2.8
Common robbery	74,221	70,598	64,417	58,764	56,993	54,442	52,566	53,196	53,505	54,927	1422	2.7
Robbery with aggravating circumstances	119,242	126,038	117,760	120,920	113,200	101,039	100,769	105,488	118,963	129,045	10,082	8.5

(continued)

Table 15.1 (continued)

Republic of South Africa												
Financial years (2005–2006 to 2014–2015)												
											Comparison 2013–2014 with 2014–2015	
Crime category	April 2005 to March 2006	April 2006 to March 2007	April 2007 to March 2008	April 2008 to March 2009	April 2009 to March 2010	April 2010 to March 2011	April 2011 to March 2012	April 2012 to March 2013	April 2013 to March 2014	April 2014 to March 2015	Case difference	% Change
Total contact crimes (Crimes against the person)	**750,440**	**724,393**	**686,294**	**678,142**	**669,928**	**633,228**	**615,935**	**608,724**	**611,574**	**616,973**	**5399**	**0.9**
Contact-related crimes												
Arson	7247	7438	6948	6445	6304	6157	5996	5665	5458	5127	−331	−6.1
Malicious damage to property	141,776	140,713	134,251	131,469	129,343	122,814	119,907	119,026	117,983	120,662	2679	2.3
Total contact-related crimes	**149,023**	**148,151**	**141,199**	**137,914**	**135,647**	**128,971**	**125,903**	**124,691**	**123,441**	**125,789**	**2348**	**1.9**
Property-related crimes												
Burglary at non-residential premises	54,217	58,240	62,756	69,829	71,544	68,907	69,902	73,492	73,464	74,358	894	1.2

(continued)

Table 15.1 (continued)

Republic of South Africa												
Financial years (2005–2006 to 2014–2015)												
											Comparison 2013–2014 with 2014–2015	
Crime category	April 2005 to March 2006	April 2006 to March 2007	April 2007 to March 2008	April 2008 to March 2009	April 2009 to March 2010	April 2010 to March 2011	April 2011 to March 2012	April 2012 to March 2013	April 2013 to March 2014	April 2014 to March 2015	Case difference	% Change
Burglary at residential premises	261,402	248,462	236,638	245,465	255,278	246,612	244,667	261,319	259,784	253,716	−6068	−2.3
Theft of motor vehicle and motorcycle	85,595	85,979	79,970	75,630	71,449	64,162	58,800	58,102	56,645	55,090	−1555	−2.7
Theft out of or from motor vehicle	138,582	123,361	110,988	108,909	120,054	122,334	129,644	138,956	143,801	145,358	1557	1.1
Stock-theft	26,526	26,155	26,053	27,255	29,428	26,942	27,611	26,465	24,534	24,965	431	1.8
Total property-related crimes	**566,322**	**542,197**	**516,405**	**527,088**	**547,753**	**528,957**	**530,624**	**558,334**	**558,228**	**553,487**	**−4741**	**−0.8**
Crime detected as a result of police action												
Illegal possession of firearms and ammunition	13,239	14,160	13,335	13,918	14,430	14,385	14,372	14,813	15,362	15,116	−246	−1.6

(*continued*)

Table 15.1 (continued)

Republic of South Africa												
Financial years (2005–2006 to 2014–2015)												
											Comparison 2013–2014 with 2014–2015	
Crime category	April 2005 to March 2006	April 2006 to March 2007	April 2007 to March 2008	April 2008 to March 2009	April 2009 to March 2010	April 2010 to March 2011	April 2011 to March 2012	April 2012 to March 2013	April 2013 to March 2014	April 2014 to March 2015	Case difference	% Change
Drug-related crime	94,792	104,369	108,902	116,949	134,687	150,561	176,218	206,721	260,596	266,902	6306	2.4
Driving under the influence of alcohol or drugs	33,076	38,210	48,338	56,121	62,904	66,645	69,410	71,025	69,725	68,561	−1164	−1.7
Sexual offences detected as result of police action	–	–	–	–	–	–	2726	4175	4720	6340	1620	34.3
Total crime detected as a result of police action	**141,107**	**156,739**	**170,575**	**186,988**	**212,021**	**231,591**	**262,726**	**296,734**	**350,403**	**356,919**	**6516**	**1.9**
Other serious crimes												
All theft not mentioned elsewhere	424,677	407,714	387,554	386,858	360,120	361,222	370,916	356,847	363,517	360,541	−2976	−0.8

(continued)

Table 15.1 (continued)

Republic of South Africa												
Financial years (2005–2006 to 2014–2015)												
											Comparison 2013–2014 with 2014–2015	
Crime category	April 2005 to March 2006	April 2006 to March 2007	April 2007 to March 2008	April 2008 to March 2009	April 2009 to March 2010	April 2010 to March 2011	April 2011 to March 2012	April 2012 to March 2013	April 2013 to March 2014	April 2014 to March 2015	Case difference	% Change
Commercial crime	51,911	59,637	63,233	75,166	82,414	85,646	85,570	89,138	76,744	67,830	−8914	−11.6
Shoplifting	64,433	65,428	66,934	80,713	88,568	78,326	71,810	71,267	70,487	71,327	840	1.2
Total other serious crimes	**541,021**	**532,779**	**517,721**	**542,737**	**531,102**	**525,194**	**528,296**	**517,252**	**510,748**	**499,698**	**−11,050**	**−2.2**
Subcategories of aggravated robbery												
Carjacking	12,783	13,534	14,152	14,855	13,852	10,541	9417	9931	11,180	12,773	1593	14.2
Truck hijacking	829	892	1245	1437	1412	999	821	943	991	1279	288	29.1
Robbery at residential premises	10,173	12,761	14,481	18,438	18,786	16,889	16,766	17,950	19,284	20,281	997	5.2
Robbery at non-residential premises	4384	6675	9836	13,885	14,504	14,637	15,912	16,343	18,573	19,170	597	3.2

crime throughout the country will do a great deal to restore the feeling of safety and security, as well as restoring public confidence in the effectiveness and efficiency of South African security agencies.

Smaller, single-owner and emerging black businesses in the old central business district areas, townships and suburbs remain vulnerable to this kind of crime, which can have a devastating impact on one of the primary job-creating sectors in South Africa. Business robberies increased from 4384 in 2005 to 19,170 in 2015. Many of these robberies happened at small businesses, including plaza shops. In the same period, robberies at residential property increased from 10,173 to 20,281, an increase of almost 100 %. Drug-related crimes increased from 94,792 in 2005 to 266,902 in 2015.

Research conducted by the author in 2012 on the public perceptions of crime revealed that most people regard crime and violence as an issue of concern, and generally regard themselves as greatly at risk of becoming crime victims. Over the past decade one million whites emigrated from South Africa to other countries and the main reasons cited for this was the high crime rate and affirmation action. When respondents were asked what their perceptions of crime were for the area that they lived in, many respondents reported that the fear of crime greatly impacted on the quality of their lives, and the majority of them held negative views about the effectiveness of police and government action; for example, 70 % of females believed that crime was out of control. A response was:

> South Africa has some of the most violent crime in the world. Instead of stealing and robbing a person, South Africans are shot, raped, and murdered. Crime in South Africa highlights the South African criminal's lack of respect of the value of human life and fellow citizens.

Initiatives to Combat Crime and the Promotion of Human Security

The government of South Africa has initiated a number of initiatives to combat crime, including the NCPS of 1996, the National Crime Combating Strategy of 1998, the White Paper on Safety and Security of

1998–2003, the IJS, the adoption of the Prevention of Organized Crime Act 121 of 1998, the adoption of the National Prosecuting Authority Act 32 of 1998, and the National Prosecuting Authority Amendment Act 61 of 2000. Furthermore, there were initiatives that resulted in changes in the investigation of crime.

The NCPS was promulgated in 1996. It was the first comprehensive strategy for the prevention of crime. The NCPS lists seven crimes which pose the greatest threat to the citizens and the welfare of the South Africa:

- Crime with firearms, which has allowed violence to increase to a great extent
- Organised crime, such as drug smuggling and illegal immigration, which has risen sharply since South Africa's re-admission to the international community
- White collar crime
- Gender violence and crimes against children
- Violence added to conflict between groups such as that in KwaZulu-Natal
- Vehicle theft and hijacking
- Corruption in the criminal justice system

The NCPS is based on the premise that there should be a wider responsibility for crime prevention and that there should be a shift in emphasis from reactive crime control, which deploys most resources towards responding to crimes have already been committed, to proactive crime prevention, aimed at preventing crime from occurring at all.

Conclusion

As South Africa moves into its third decade of democracy, over the past 20 years, following periods of colonialism and apartheid, the twenty-first century has commenced with increased levels of serious and violent crimes in the country. General analysis of crime patterns depicts a discernible correlation between crime and challenging socio-economic conditions. Of serious concern in the policing environment is the observed violent social behaviour, which continues to prevail in the majority of serious crime categories, some of which persists in the face of action by law enforcement agencies and requires a collective effort from all stakeholders affected, including South African own communities. Prevention in policing should be a partnership engagement to ensure that the domino effect of crime, where small crimes that should have been addressed at points of origin are allowed to escalate into more serious crime, is prevented (SAPS 2014/2015: 6).

The law enforcement components, i.e. the police as well as other functionaries of the criminal justice system, should be sensitised towards crime against human security by way of well-structured training programmes. Such training programmes should include violations of human rights. The reality is that the underlying causes of crime are deep-rooted in prevailing socio-economic conditions often beyond the control of the police. Building a safer South Africa requires partnerships and open communication with the community. Personal and community safety necessitates the sharing of responsibility among the police and all stakeholders within government, municipalities, businesses, civil society and other interest groups. This should include the reduction of contact crimes, an efficient and effective criminal justice system, addressing corruption in the public and private sector and ensuring that South Africa's borders are controlled and well managed.

References

Africa, S. (2015). Human security in South Africa. *Strategic Review for Southern Africa, 37*(1). Department of Political Sciences, University of Pretoria.

Commission on Human Security. (2003). *Human security now: Protecting and empowering people*. New York: Commission on Human Security.

Masuku, S. A. (2004, September). World of crime: Youth views on crime in the Nelson Mandela Metro. *SA Crime Quarterly*, no. 9, Institute for Security Studies, Pretoria.

Mlambo, N. (2005). Perceptions of human security in democratic South Africa. Chap. 6., pp. 225–262. Institute for Security Studies.

Nickel, J. W. (2003). Making sense of human rights revised edition, September 11. Retrieved January 3, 2016, from www.ddp.org.za

Rothschild, E. (1995). What is security. *Daedalus, 124*(3), 53–98.
RSA. (1994). White Paper on Intelligence. Retrieved November 15, 2015, from http://www.ssa.gov.za
RSA. (1996a). The Constitution of the Republic of South Africa (Act 108 of 1996), as adopted on 8 May 1996 and amended on 11 October 1996 by the Constituent Assembly.
RSA. (1996b). Defense in a democracy. White Paper on National Defense for the Republic of South Africa, May 8. Retrieved November 15, 2015, from http://www.dod.mil.za
RSA. (1998). White Paper on Safety and Security. "In service of safety", September. Retrieved November 15, 2015, from http://www.policesecretariat.gov.za

Shanta Balgobind Singh is a Professor in the programme of Criminology and Forensic Studies at the University of KwaZulu-Natal. Her research areas are diverse and include: the criminal justice system; correctional services internationally and nationally; the consequences of overcrowding of correctional centres; women and health issues; maritime challenges; youth and offending; drug use/abuse; the Indian diaspora in the twenty-first century; and indigenous knowledge systems.

16

Environmental Conflicts and Forced Migration in the Nigerian Niger Delta

Iwebunor Okwechime

Introduction

This chapter focuses on environmental conflicts and forced migration in Nigeria, with particular emphasis on the oil-producing communities of the Niger Delta region. Although discourses on the Niger Delta have dealt substantially with the various aspects of the crises bedevilling the region, inadequate attention has been paid to life-threatening concerns such as environmental conflicts and forced migration in the region. This chapter takes the position that the dynamics of environmental conflicts and forced migration in the Niger Delta are deeply rooted in the contradictions thrown up by the oil-dependent global capitalist system and the processes of liberal globalisation, which have ensured that the people of the Niger Delta are denied environmental justice. At the intersection of environmental conflicts and forced migration in the Niger Delta lie issues

I. Okwechime (✉)
Department of International Relations, Obafemi Awolowo University, Ile-Ife, Osun State, Nigeria

A. Adeniran, L. Ikuteyijo (eds.), *Africa Now!*,
https://doi.org/10.1007/978-3-319-62443-3_16

of development (modernisation), the capitalist mode of production, environmental exploitation and the political superstructures at local and global levels, all of which define and are themselves defined by the values and needs of developed societies. Thus, underpinning the phenomenon of environmental conflicts and forced migration in the oil-producing communities is its 'situatedness' at the local–global intersection, where the local is subordinated to the global. As Kakonen (1993) has so convincingly argued:

> The exploitation of nature as well as the changes in its utilization always depends on whose values define the use of nature in order to satisfy human needs and also whose needs will be satisfied...The needs of the European or the Christian world have defined how to develop the relation between human beings and nature. This means that Euro-American values and needs have defined the utilization of nature.

Although environmental conflicts and forced migration have gone on since the arrival of the oil and gas industry 61 years ago in the Delta region, such conflicts and migrations are not entirely unique to the oil-producing communities. Events over the past decade have underscored the universal nature of the problem across the Nigerian nation state. The repeated cases of violent and deadly conflicts between pastoralists (or herdsmen), on the one hand, and farmers, on the other, across the country attest to this. Indeed, farmer–grazer conflict has become a defining feature of relations between Fulani pastoralists and indigenous peasant communities across the six geopolitical zones of the federation. Meanwhile, farming communities in states such as Benue, Plateau, Nasarawa, in the north-central area; Anambra and Enugu, in the south-east; Ekiti, Ogun and Oyo in the south-west; and Delta, in the south-south, have been engaged in bloody clashes with the Fulani herdsmen, resulting in the destruction of whole communities by the more heavily armed pastoralists.

The sources/causes of the resource use or environmental conflicts are varied and interwoven, as they 'articulate with other ethnic, political, and religious conflicts' (Moritz 2010). Over the years, population growth, the expansion of agricultural activities, as well as erosion and social and industrial pressures on land have limited greatly the area the nomad population can utilise. Similarly, the past two decades have witnessed a substantial rise in the population of Fulani pastoralists migrating southwards to escape the deadly effects of climate change and desertification on their herds (Adisa and Adekunle 2010; Ademola 2016; Omotuyi 2016). At the height of farmer–herder conflicts in the north-central state of Jos, former President Olusegun Obasanjo was compelled to declare a state of emergency there, in 2004, after reprisal attacks by both Christians and Muslims had left thousands dead, with 'more than 20,000 refugees fleeing to neighbouring Cameroon' (Moritz 2010: 138–148). His successor, President Goodluck Jonathan, took similar steps when the conflicts recurred during his administration.

In the specific case of the Niger Delta region, it is fair to observe that, prior to the 1980s, environmental conflicts among the oil-bearing communities were relatively unknown. This is not to suggest, however, that the very factors that have spawned environmental conflicts and forced migration in the Niger Delta were non-existent before the 1980s. On the contrary, an explanation of the upsurge in environmental conflicts and the resultant forced migration in the region should be located in the contradictions embedded in the globalisation of capital in a predominantly agrarian environment. These contradictions, and the environmental and social costs of resource extraction, have materialised in ways that have blighted the environment and livelihoods of the peasant communities of the Delta region. For instance, the environmental degradation and pollution emanating from oil spills and gas flaring are major sources of environmental conflicts between local communities and the oil companies/Nigerian government. The contradictions are also defined in terms of the politics and management of oil revenues, which are skimmed or appropriated by the political elites, both at the local and national levels. Lastly, these communities that bear the social and environmental costs of resource extraction in their communities are, according to Lubeck *et al.* (2007) 'underrepresented or even excluded from power while the resource flows are captured by a political oligarchy skilled at spreading the petro-rents among a network of cronies'.

The Nigerian State: An Overview

The analysis of environmental conflicts and forced migration in Nigeria will be incomplete without a look at the nature of the post-colonial state in Nigeria, and how this determines its relationship to the various social classes, on the one hand, and, on the other, its relationship to external forces. It also explains how the state in Nigeria plays its role as 'the authoritative allocator of scares resources, as a mediator, and also, paradoxically, as a contestant in conflict' (Obi 1999a, b). By nature, the state in Nigeria is as undemocratic, alien, exploitative and violent as its predecessor, the colonial state. As Ake (2000) has so convincingly stated: 'the post-colonial sate in Africa is very much like its colonial predecessor, its power over economy and society is enormous, arbitrary and it is largely privatized. For all but few of its citizens, it is alien and remote, uncaring and oppressive.' For Alavi (1979), the essential problem about the state in in post-colonial societies stems from the fact that it was established not by an ascendant native bourgeoisie, but by a foreign imperialist bourgeoisie. What this implies, therefore, is that, generally, the state in Africa is a creation and function of capital, with its dominant classes joined in an unholy alliance with global capital for the local accumulation of money (Obi 1997a). Elaborating, Obi observed that:

> First of all, we need to understand the nature of the state in Africa as a captive of a coalition of social forces whose interactions broadly reflect the social relations of conflict. The state and the social forces that define its hegemony intervene directly in the African ecosystem at the behest of global, national and local capital (1997).

Obi's position is buttressed by Beckman (1982) when he asserts that 'The primary role of the Nigerian state is to establish, maintain, protect and expand the conditions of capitalist accumulation in general, without which neither foreign nor local capitalists can prosper.'

Since the Nigerian state is, as Ake has noted, instituted in such way that it cannot but be immersed in the struggle of the classes, it is reasonable to argue that this characteristic lies at the heart of the crises that have plagued it since independence. This also explains why the Nigerian state has proved incapable of mediating the crisis in the oil-rich Delta region.

The Niger Delta: The Context

According to the United Nations Development Programme (UNDP) (2006: 19), the Niger Delta comprises the area covered by the natural delta of the Niger River to the east and west. The Niger Delta Environmental Survey (NDES) defines the Niger Delta as the low lying arcuate deltaic plain with a northern apex a little to the south of Onitsha at Aboh (503° 44″N, 603° 38″E), a western apex by the estuary of the Benin River (5033″E), and an eastern apex at Imo River estuary (4027° 16″N, 7035° 27″E) (NDES 2000). Put another way, the western and eastern boundaries of the delta can be found around the Benin River and the Imo River, while its northern boundaries are situated close to the bifurcation of the Niger River at Aboh.

Through the Niger Delta, Africa's largest wetland and the world's third largest mangrove forest, eight major rivers, 21 estuaries and a dense network of distributaries empty their waters and sediments into the Atlantic Ocean. Apart from being one of the most important biological areas in the world, the Niger Delta is also rich in mineral resources, notably oil and gas. While there is no one-line definition of the Niger Delta, extant definitions have been either 'cartographic' or 'political' (Roberts 2005: 95; International IDEA 2003; Azaiki 2003). According to Roberts, the cartographic definition highlights the 'natural Niger Delta', while the political definition is based, essentially, on the component states in the region—this is sometimes extended to cover all the all oil-producing areas; there are also content-specific 'operational definitions'. Estimates regarding the size of the Niger Delta vary from 25,000 (UNDP Report 2006: 19) to 30,000 square kilometres (Osuntokun 1999, cited in Roberts, 2005: 97).

For our purposes in this chapter, we shall adopt the UNDP (2006: 19) definition of the Niger Delta as embracing 'all the oil-producing areas and others considered relevant for reasons of administrative convenience, political expedience and developmental objectives'. Included in this definition are all the nine oil-producing states—Abia, Akwa Ibom, Baylsa, Cross River, Delta, Edo, Imo, Ondo and Rivers. Based on this 'political' definition, the Niger Delta land area extends to 75,000 square kilometres (UNDP 2006: 19; Roberts 2005: 95). Four distinct ecological zones can be identified, namely coastal ridge barriers, mangrove forests, fresh water swamp forests and lowland rainforests (Onosode 2003: 63; UNDP 2006: 19–20). It has been shown (Roberts 2005: 96–97) that beyond this formation, the natural environment of the Niger Delta is divided into two ecological zones—'the riverine' and the 'upland'. According to Roberts, while the lower vegetation in the lower delta (which is riverine) consists of mangroves, that of the upper delta consists of fresh water swamp rainforest. As should be expected, the distribution of occupations has been largely determined by the division of the natural environment. Roberts has noted that 'the drier lands towards the head of the delta mainly supports farming, the collection and processing of palm fruits, hunting and some fish, while much further south, the riverine economy revolves mainly around fishing and trading'. The UNDP Report cited states that agriculture, fishing and forestry account for about 44 % of employment in the region.

As a multi-ethnic, multilingual and multi-cultural society, the Niger Delta comprises over 20 different ethnic and sub-ethnic groups, many of whom trace their origins to the Yoruba, Igbo, Edo and Ijaw. It is these groups, together with the Delta Cross, that constitute the five major linguistic and cultural groups in the region (UNDP 2006: 21). Based on the 2005 census, the population of the people in the Niger Delta is over 30 million. Among the ethnic groups in the region, the Ijaw, who are distributed among six states of the Nigerian federation, constitute the largest single ethnic group in the region and the fourth largest in the country.

Environmental Conflicts in the Niger Delta: A Multi-Level Analysis

Given its natural location, multi-ethnic, multilingual and multi-cultural character, the Niger Delta is no stranger to environmental conflict. As Ifeka (2001: 101) has noted, conflict over the ownership and distribution of resources has been a defining feature of the Niger Delta's economic history. For example, in *The Great Ponds*, a novel set in pre-crude oil Niger Delta society, Elechi Amadi (1968) captures the violent dispute between two coastal communities over the ownership of fish ponds. Similarly, trade disputes over the control of trading posts and routes were not uncommon among the pre-colonial communities of the Niger Delta (Alagoa 2004; Ikime 2005). However, the presence of oil companies in the Niger Delta over the last 50 years has not only altered the nature and dynamics of environmental conflicts in the region, but also exacerbated them in ways that have defied a lasting solution (Human Rights Watch 2002; Amnesty International 2005a, b). It has been argued (Onosode 2003; Adekunbi 2008; Ikelegbe 2010; Adeyemi 2008; Okwechime 2011; Obayiuwana 2012; Ebie 2013) that oil and gas development has led to the intensification of conflict in the Niger Delta and the growth of environmental concerns, both of which are inextricably linked to the politics of resource ownership, its mode of appropriation and use. In other words, conflicts over the oil-bearing environment are directly linked to the social relations that underpin the exploration and production of oil and gas in this volatile region. As the renowned British environmentalist, Andrew Rowell (1994), elaborates:

> Oil and environmental conflicts are rooted in the inadequate social relations of production and distribution of profits from oil, and its adverse impact on the fragile ecosystem of the Niger Delta. It involves the Nigerian state and oil companies on one side and six million people of the estimated eight hundred oil-producing communities concentrated in the seventy thousand square kilometres, on the other.

The first level of conflict over the environment is that between oil companies and the oil-producing communities and centres on corporate neglect, environmental degradation and the disruption of social and economic life, lack of developmental projects and violation of human rights, compensation, employment opportunities and contracts, among others. According to a Shell-commissioned study, land acquisition, oil spill compensation, hiring and contracting are aspects of Shell's and, by extension, other oil companies' policies that can 'feed into, or even create conflict' (WAC Global Services 2003). For instance, the state-owned petroleum company, the Nigerian National Petroleum Company (NNPC), puts the amount of oil spilled into the environment yearly at 2300 cubic metres, with an average of 300 individual spills annually (Ero 2008: 56). The UNDP's Niger Delta Development Report of 2006 has revealed that the region recorded over 6800 oil spills between 1976 and 2001, with a loss of approximately three million barrels of oil (UNDP 2006). These statistics are buttressed by Amnesty International's report that, between 1960 and 2010, a period of 50 years, the oil-producing communities of the Niger Delta have experienced oil spills equal to the Exxon Valdez every year. According to the report, between 9 and 13 million barrels were spilled during the period (Amnesty International 2009). A more recent report emanating from the Ministry of Environment underscores not only the scale of oil spillage in the region, but also the accuracy of data in this respect. According to a former Minister of Environment, Mr John Odey, Nigeria recorded 3203 cases of oil spillages in the country between 2006 and 2010. The minister revealed that out of the above-reported cases 23 % were caused by equipment failure, operational/maintenance error and corrosion, while 45 % of the oil spillages that occurred were attributed to sabotage/vandalism (Lawrence *et al.* 2011: 1). Oil spills are widespread in the Niger Delta—and a good number of the spills that occur are not reported at all.

From the perspective of oil companies, spills are a function of sabotage perpetrated by oil thieves and other criminal gangs, rather than equipment failure, as the inhabitants of the region are wont to argue. With many of the pipes rusty with age, it would seem to make good business sense for oil companies like Shell operating in the region to claim that a majority of their oil spills are caused by sabotage, since such claims legally absolve them from the obligation of compensating affected communities or injured parties. This is not to deny the existence of oil thieves in the creeks, who commit acts of vandalism as part of their criminal enterprise. Critics have argued that some of these acts of vandalism have been known to have been perpetrated with the connivance of oil company workers (Omeje 2005; Human Rights Watch 2005). Furthermore, critics have cited lax enforcement of laws in the oil sector and the fact that the Nigerian state regards the companies as the producer of the national wealth as factors that have made it very difficult for the government to control or even sanction these companies.

Moreover, it has been observed that state regulatory agencies, such as the Department of Petroleum Resources (DPR), charged with the responsibility of monitoring the oil companies have themselves had to rely on oil companies' equipment and facilities in their monitoring efforts. Shedding further light on this, the Executive Director of Environmental Rights Action, Nnimmo Bassey, notes:

> I was in conversation with an official of the Directorate of Petroleum Resources and learned that sometimes when they call for pipelines to be replaced, the companies reach out to political officers in Abuja. They inform the officers that if they embark on replacing the pipelines it translates to halting some level of production. Of course the politicians who are more or less drunk on crude oil cash would not want any reduction on the avalanche of naira and dollars and would the regulator's suggestion aside (2009).

Apart from the issue of oil spills, gas flaring constitutes a major source of conflict between local communities and oil companies operating in their areas. It is estimated that as many as 2.5 billion cubic feet of gas per year is flared by the Nigerian state and oil multinational corporations in the Niger Delta. In fiscal terms, this translates to $2.5 billion in lost revenues to the government of Nigeria (Amanze 2009: 29). Citing the World Bank report of 2004, Adeyemi (2008) maintains that Nigeria flares 75 % of the gas it produces. Indeed, it is mind-boggling to note that the volume of gas flared in Nigeria, according to analysts, approximates the same quantity Trinidad and Tobago utilises for both domestic use and export (Amanze 2009: 29). Although gas flaring was outlawed by the Nigerian government in 1984, the oil companies have nevertheless persisted in this unsavoury practice.

Despite the introduction of sanctions, the companies have preferred the cheaper option of flaring gas to adopting the gas re-injection techniques which are technologically available. While the DPR estimates that there are 117 gas flare sites across the Niger Delta, independent research conducted by the Department of Microbiology, Abia State University, Nigeria, indicates that that there are over 200 gas flaring sites across the length and breadth of the oil-producing region (Adeyemi 2008). In 2005,

Shell appealed against a High Court ruling that oil companies should end gas flaring, insisting that it would end the practice in 2009, a year after the federal government's flare-out terminal date (*THISDAY* 29 May 2007: 60). However, reports emanating from official circles indicate that the government is no longer interested in the 2009 fare-out terminal date. The head of the DPR, Mr Osten Olorunshola, justifies his government's vacillation on the issue of gas flaring in these terms:

> Flaring has dropped. We used to have roughly about 25 per cent of pure gas flared as at the end of last year [2011], but it has now dropped to 18 per cent. Gas, like you know, is becoming very exciting both in terms of development and revenue generation. We are actively working towards what is called a Gas Network Code which is purely a protocol that is going to regulate how the players as well as the molecules interplay within the pipelines. We are also going to regulate that, we are also trying to conclude the codes. (Kalejaye 2012: 29)

One of the leading local newspapers, in what could be described as a reaction to the minister's comment, wrote a scathing editorial entitled 'The high cost of gas flaring in Nigeria' in which it lamented that 'Nigeria's tolerant disposition towards the perennial gas flaring in the Niger Delta region has become a constant reminder of the profligacy that attends the management of the country's resources' (*The PUNCH* [editorial] 22 October 2012: 18). Citing a former military Head of State, the editorial noted that, '…last year [2011], this country flared over 460 billion standard cubic feet of gas that, if processed and exported, would have fetched the country over $2 billion and minimized the health and environmental impact of gas flares'. According to the editorial:

> It will be missing the point to believe that the multinational oil companies are solely guilty of gas flaring in Nigeria. Since they are all statutorily in joint venture partnerships with the Nigerian National Petroleum Corporation, a government company, it means the Nigerian government is equally culpable in the missed target for the ending of gas flaring.

On this reading, the conflict between oil communities and the Nigerian government is a product of the government's habitual equivocation on matters that border on the very survival of the inhabitants of the oil-producing region. What it further demonstrates is the centrality of the environment as a major source of conflict between the oil companies and the host communities.

Flowing from this is the second level of conflict which takes place between the Nigerian state and oil-producing communities. Essentially, the dynamics of this conflict revolve around the contradictions between the Nigerian state and the embattled oil minorities in terms of access to centres of political and economic power, control of resources (oil rents), worsening levels of poverty and economic underdevelopment, widespread unemployment among the youth population, social dislocation and the absence of revenue transparency and accountability. As the UNDP Niger Delta Human Development Report notes, 'The critical issue is not only the increasing incidence of poverty, but also the intense feeling among the people of the region that they ought to do better… This to a large extent explains why there is so much frustration and indignation in the region' (UNDP 2006: 36).

It was the sense of marginalisation and deprivation felt by the mass of the people that ultimately led to an upsurge in militant activities marked, as it were, by vandalism of oil industry facilities, especially pipelines, fatal and near-fatal attacks on oil workers as well as hostage-taking by criminal youth groups. Arguably, the most formidable among the numerous militant groups is the Movement for the Emancipation of the Niger Delta (MEND), which emerged from the ashes of the Dokubo-Asari-led Niger Delta Peoples Volunteer Force. In 2006, the year that recorded the worst incidence of militancy in the region, Nigeria was losing over 872,000 barrels of oil per day as a result of attacks by MEND on the country's oil pipelines and facilities (*Nigerian Tribune* 2006). The hallmarks of MEND's campaign of violence and destruction were the kidnapping of oil company workers, especially expatriates, and the blowing up of oil installations.

According to Ibiyemi (2009: 1), the Nigerian government and its Joint Venture (JV) partners of the NNPC lost a total of $50 billion to the insurgency in the Niger Delta between 2006 and 2009. Similarly, between February 2006 and August 2009, 500 expatriates were kidnapped, while about 89 cases of oil fields located onshore and offshore, including Shell's Bonga oil field, were closed down to prevent attacks by the militants (Ibiyemi 2009: 2). A report by the Niger Delta Development Monitoring

Corporate Watch showed that 155 persons were kidnapped by militant groups in 2006 alone. This was three years before the late President Yaradua granted an amnesty to the militants in 2009. In its 2006 Niger Delta Human Development Report, the UNDP noted that '[h]ostage-taking is not only a stress on foreign captives, their families and the company they work for, but also a challenge to international diplomacy' (UNDP 2006). With regard to pipeline vandalism, the former head of the NNPC, Sanusi Barkindo, disclosed that Nigeria lost over 174 billion naira to pipeline vandals in over 16,083 pipeline breaks between 2000 and 2010 (Yusuf 2010: 21).

Relatedly, the third level of conflict in the Niger Delta takes place at intra- and inter-community levels. As noted, the struggle for benefits derivable from oil exploitation has been a fundamental factor in the outbreak and exacerbation of inter-and intra-community conflicts throughout the Delta region. Such benefits usually range from compensations to contracts, to scholarships, infrastructure and to the social status which they confer (Gbemudu 2006). According to Shell's important baseline study (2003):

> Access to benefits derived from SCIN's presence is the prominent trigger for company-communities conflict, conflict between communities and conflict within communities. In the resource scarce environment of the Niger Delta, individuals and groups attempt to position themselves to access cash, contracts and legitimacy to mention some. This leads to leadership tussles, conflicts over boundaries between villages, over who are 'genuine' inhabitants of villages and over what kind of development projects a community receives.

Across the oil-producing communities, there is a consensus among local people that land acquisition and land ownership are easily the greatest potential sources of conflict. Several reasons account for this. Agriculture remains the major community-based economic activity in a region where the physical terrain makes arable land a very scarce resource. Moreover, agricultural land degradation, occasioned largely by the activities of oil companies, is severe, and this is further worsened by the fact that population growth across the region is putting intense pressure on the available land and other resources (Omeje 2005). With the advent of oil, the premium on land has become even greater, largely because land ownership now determines if a community is designated as 'host' community, a status linked to benefits such as employment opportunities, community development projects and contracting opportunities.

Yet another dimension of environmental conflict that has gained currency among communities across the region, but which is largely glossed over in the discourse, is the non-violent strain, typified largely by advocacy, petition-writing and endless legal tussles, of which land ownership remains a critical aspect. With respect to advocacy, communities involved usually attempt to convince the general public within and beyond the oil communities about the justification of their cause relative to that of an opposing or rival community. As Ikporukpo explains, 'it could be verbal, although, in most cases it is published in newspapers' (Ikporukpo 2003: 12) In his view, legal tussles have, over the years, become a very popular expression of conflict in the Niger Delta region. These conflicts, usually accompanied by wanton destruction of life and property, invariably result in forced migration by victims from affected communities.

Forced Migration

Migration is one of the oldest coping strategies for dealing with environmentally induced change. As a concept, migration is usually conceived of in generic terms. Considered as such, it embraces all forms of movement, voluntary and involuntary, and across or within national boundaries. Forced migration can be triggered by a variety of causal factors, including persecution, natural and industrial disasters, development projects, environmental degradation, war and conflict, and ethnic discrimination, to mention but a few. As Akokpari (2000) has observed, migration in this context is sometimes used interchangeably with displacement because the conditions that precipitate displacement are always those that engender migration. Forced migrants belong to the category designated as internally displaced people. The United Nations, cited by Mason (2000), defines internally displaced people as persons or groups of persons who have been forced or obliged to flee or leave their homes or places of habitual residence, particularly as a result of or in order to avoid the effects of armed conflict, situations of generalised violence, violations of human rights or natural or man-made disasters, and who have not crossed an internationally recognised state border.

Thus, forced migration becomes, in the context of the Niger Delta, a process in the general scheme of environmental conflicts attendant on the crisis of environmental governance spawned in the region. The worst cases of forced migration arising from environmental conflicts in the oil-producing communities were recorded in the 1990s under the military regimes of General Ibrahim Babangida and General Sani Abacha, and during the civilian administration (1999 and 2007) of President Olusegun Obasanjo, who was himself a former military Head of State between 1976 and 1979. Cases of environmental conflicts under during the administrations of Umaru Yara'Adua and Goodluck Jonathan were overshadowed by the amnesty programme which was introduced in 2009.

Some of worst cases of forced migration associated with environmental conflicts in the region include the Umuechem-clash, 1990; the Egi-Shell conflict, 1990, the Ogoni-Shell crisis, the Ijaw-IIlaje conflict, 1998; the Emadike-Epebu conflict, 1998; the Kalabari/Bille conflict, 2000/2001; the Liama-Shell clash, 2002; the Odi invasion, 1999; the Odioma raid, 2005, to mention a few (Human Rights Watch 2002; *ERAction* 2005; Rowell *et al.* 2005). In a lecture entitled 'Energy, Environment and Disasters', the late Governor of Bayelsa State, D. S. P. Alamieyeseigha (2005), listed 73 incidents of environmental conflicts between 1971 and 2005 in the oil-producing communities and their outcomes. It needs to be stressed that Alamieyeseigha's list is by no means exhaustive.

Owing to fears about access to resources, forced migrations in the oil-producing communities invariably trigger secondary conflicts between and within communities, thereby generating further environmental stress and insecurity in the region. The major beneficiaries of this kind of situation are usually oil companies and the government since it enables them to divide and rule the communities, while oil production continues apace. As should be expected, the communities, especially the most vulnerable members, such as women and children, are often the worst victims of forced migration in the Niger Delta. For instance, the long-drawn-out conflict between the Nigerian state and Ogoni resulted in the destruction of 30 villages, over 2000 deaths, more than 3000 injured and about 100,000 internal refugees, many of whom were forced to migrate to other cities like Port Harcourt, Lagos, Benin, Ibadan, Calabar, Owerri, Aba, Abuja and other places within the country (Wiwa 1997; Saro-Wiwa 1995; Ekine 2001; Kpalap, int. 2005). Similarly, the Ijaw-Ilaje conflict of 1998 generated a huge population of internally displaced persons as thousands of Ijaw citizens were forced to migrate to Bayelsa State, and similar numbers of Ilaje people were also forced to seek refuge in the relative safety of mainland Ondo and its environs (General Three Lions, interview 2015) The aftermath of the military Joint Task Force operations launched against the militants in the deep creeks of Gbaramatu, Warri South West Council of Delta State in 2009 left no fewer than 5000 internally displaced (Olimogbe and Ebiri 2009). It is important to note that government's response to the humanitarian crisis generated by the hostilities was, irrespective of the government in power, that of total indifference. Civil society groups have always more quickly to humanitarian crisis arising from environmental conflicts (Three Lions, interview 2015).

Niger Delta vs. the Environmental Conflict Debate: The Limits of Neoliberal Rhetoric

The causal factors behind environmental conflicts are often complex processes that hinge, as Obi puts it, on issues of access, control, equity and sustainability (Obi 1997a: 16, b). These issues, he argues, revolve around ecology and politics, and are linked to the environmental scarcities conflict or the environmental scarcities nexus. However, studies by scholars like Kaplan, Klare, Homer-Dixon, Brown and Jacobson have tried to explain environmental conflicts in the Third World in terms of the population–conflict nexus. In Homer-Dixon's view, for example, environmental scarcities are 'scarcities of vital renewable resources such as soil, water, forests, and fish' and, as he further explains, the human induced resources scarcities may combine with 'bulging populations' and 'land stress' to cause violent conflicts (Homer-Dixon 1998). As he elaborates:

> Group-identity conflicts are likely to rise from large scale movements of populations caused by environmental security. As different ethnic and cultural groups are pushed together by these migrations, we should expect people in these groups to see themselves and their neighbours in terms of 'we and 'they', in other words, they will use identity of their own group to judge the worth of other groups, often negatively. Such attitudes can lead to violence.

It would appear Homer-Dixon is ignorant of the critical issues involved in the debate on environmental conflict, especially in resource-rich, rural, pre-capitalist societies like the Niger Delta. If it is granted that a rapidly growing population relative to the available resources can create scarcities and consequently environmental conflict, it should equally be reckoned that this would depend on several factors, including the nature of social forces/actors who are engaged in the struggle to control access to environmental resources, the social distribution of power over environmental resources and the dominant mode of production in that particular environment. These factors are crucial for an objective analysis of the phenomenon of environmental conflict in Niger Delta-type societies. As Obi again observes:

> Apart from the limitation of Homer-Dixon's model which fails to address the structural basis of the environmental scarcities on which the model is constructed, it offers no deep insight into the structural basis of environmental conflict in the developing world. By concentrating on the external manifestations of eco-refugees 'spilling across borders and disrupting relations among ethnic groups, documenting tensions arising between countries over access and control of water resources, while ignoring the economic system which fuels the scarcities in the midst plenty, the very root of conflict is glossed over, and the victim blamed for engaging in violent conflicts which threaten global security (Obi 1997a: 16).

The Niger Delta epitomises the picture portrayed above, because conflicts in the oil communities are a function of the contradictions generated by the expansion of global oil capital as it plunders the oil-rich environment of the Niger Delta. Despite the benefits claimed for globalisation and the market by their proponents, Third World countries have yet to reap the much-vaunted benefits; rather, indigenous peoples like the oil-producing communities of Nigeria have known nothing but mindless exploitation, environmental destruction, worsening levels of poverty and underdevelopment, and recurrent cycles of boom and bust as their economies become more and more integrated into the global capitalist market economy. Ifeka (2006) sheds light on the issue when she notes that:

> Thousands of kilometres of global capitalism's oil technology—pipe lines, flow stations, huge oil terminals and company compounds—sit literally cheek by jowl with fishing communities organized for the most part within non-capitalist relations of production for subsistence and seasonal sale of surplus, eking out a living in the oil polluted swamps and soil of planet earth's second largest wetlands. Here, capital-intensive systems of resource extraction and processing that rely on highly skilled labour imported from overseas and elsewhere in Nigeria exist alongside, and in general do not mesh directly with, non-capitalist relations of production between households. (pp. 721–722)

The foregoing clearly explains why the issue of global security looms large in the analysis of Kaplan and his like-minded colleagues. In their thinking, environmental security is synonymous with national security. In this light, Kaplan's description of the environment as 'a hostile power' should come as no surprise because the environment is a site for all kinds of struggle that represent the unequal power relations that exist between peasant communities, on the one hand and, on the other, the industrialised capitalist nations of the world. Kaplan's conception of the environment as an entity of 'hostile power' reveals his realist predispositions. According to him:

> It is time to understand the 'environment' for what is: *the* national-security issue of the early twenty-first century. The political and strategic impact of surging populations, spreading disease, deforestation and soil erosion, water pollution, air pollution, and, possibly, rising sea levels in critical, overcrowded regions like the Niger Delta and Bangladesh—developments that will prompt mass migrations and, in turn, incite group conflicts—will be the core foreign policy challenge from which most others will ultimately emanate, arousing the public and uniting assorted interests left over from the Cold War (Kaplan 2001).

The significance of these sorts of assertion lies largely in the fact that they give morale to the status quo forces that are blind to the environmental challenges generated in non-capitalist communities like Niger Delta by policies that are motivated by greed and materialism (see Okoye 2014). Consequently, environmental justice, which ought to have provided the basis for any advocacy on environmental security, is sacrificed at the altar of capitalist ideology.

Concluding Comments

The thrust of this chapter has been an investigation of environmental conflicts and forced migration in Nigeria, with special reference to the Niger Delta. It was established that the rising incidence of environmental conflicts and the concomitant forced migration that Nigeria has been experiencing over the last two-odd decades are multi-causal in nature, ranging from natural to man-made causes. The frequent clashes between Fulani pastoralists and the farmers in various parts of the country have largely been dictated by environmental change, increases in population and economic development, all of which have led to conflicts between the different social groups. It was noted that the contradictions of the capitalist economic system are nowhere greater than in the oil communities, where capital-intensive systems of resource extraction have laid waste to the once-pristine environment of the Niger River Delta.

The chapter also laid bare the ideological basis underlying the arguments of the population –conflict theorists. It was demonstrated that the population–conflict thesis which privileges global environmental security at the expense of global environmental justice has proved grossly inadequate in terms of helping us to understand the sources and causes of environmental conflict in Niger Delta-type societies. Rather than heap the blame on the economic system, which ensures the depletion of resources with grave consequences for humankind, as well as flora and fauna, the advocates this destructive economic system have laid the blame on the doorsteps of the victims of the system. This is unacceptable because their postulations usually wield enormous influence over the policies of their home countries' decision-makers. It becomes even more objectionable because these policies are normally packaged by dyed-in-the-wool neoliberal economists as the panacea to our development crisis, without pausing to think whether these policies are suitable to the distinctive needs and problems of the people upon whom they are being imposed.

The Nigerian state and its political elites, as protectors of this system and its capital, have been complicit by joining forces with their Western paymasters to dispossess the Niger Delta inhabitants of their environment and their resources, thereby laying the foundation for environmental conflicts and forced migration in the region. The Nigerian state has a responsibility to ensure that the rights of local communities are not violated, either by government officials or by corporate personnel. It behoves the government of Nigeria to ensure that companies operate according to internationally acceptable standards, and according to Nigerian environmental laws, as enshrined in the Constitution. Finally, comprehensive but genuine remedial actions must be undertaken by the government and oil companies to restore the Niger Delta environment. In this way, environmental conflicts and forced migration will be drastically reduced.

References

Ademola, A. (2016). Killing for pasture or water? In R. I. Ako-Nai, M. O. Omotosho, & S. Omotuyi (Eds.), *One hundred years of Nigerian nationhood: Governance, foreign policy and development*. Ibadan: College Press.

Adeyemi, A. (2008, August 16). Black gold as a curse. *Tell*, Lagos.

Adisa, S., & Adekunle, O. A. (2010). Farmer-Herdsmen conflicts: A factor analysis of socio-economic conflict variables among Arable crop farmers in North Central Nigeria. *Journal of Human Ecology, 30*(1), 1–9.

Ake, C. (2000). *The feasibility of democracy in Africa*. Dakar: CODESRIA.

Akokpari, J. K. (2000). The political economy of migration. *Journal of Identity, Culture and Politics, 1*(1), 57–77.

Alagoa, E. J. (2004). *The uses of hindsight and as foresights: Reflections on Niger Delta and and Nigerian history*. Port Harcourt: Onyoma Publications.

Alamieyeseigha, D. S. P. (2005). Energy, Environment and Disaster: The Niger Delta Experience. A Lecture Delivered at the *International Conference on Energy, "Environment and Disaters"*, Chalotte, North Carolina, United States of America, July 24.

Alavi, H. (1979). The state in post-colonial societies: Pakistan and Bangladesh. In H. Goulboourn (Ed.), *Politics and State in the Third World*. London: The Macmillan Press Limited.
Amadi, E. (1968). *The great ponds*. Ibadan: Heinemann.
Amanze, C. (2009, December 21). Gas flaring: FG may extend deadline. *THISDAY*, Lagos.
Amnesty International. (2005a). Nigeria: Claiming rights and resources: Injustice, oil and violence in Nigeria. Amnesty International AFR 44/025/2005 (Public) News Service.
Amnesty International. (2005b). Nigeria: Ten years on: Injustice and violence haunt the oil delta. Amnesty International AFR 44/025/2005 (Public) News Service.
Amnesty International. (2009). *Nigeria: Petroleum, poverty and pollution in the Niger Delta*. London: Amnesty International.
Azaiki, S. (2003). *Inequities in Nigerian politics: The Niger Delta, resource control, underdevelopment and youth restiveness*. Yenagoa: Treasure Books.
Bassey, N. (2009). *Knee deep in crude: Environmental field reports* (Vol. 2). Benin: Environmental Rights Action/Friends of the Earth Nigeria.
Ebie, M. A. (2013, August 4). Environmental degradation and economic impact of oil exploration in Isokoland. *The Guardian*, Lagos.
Ekine, S. (2001). *Blood and oil*. London: Centre for Democracy and Development.
Ero, A. (1997, February 18). 50 years of oil is shit. *Tell*, Lagos.
Ero, A. (2010, August 16). Echoes from the Wasteland. *Tell*, Lagos.
Gbemudu, E. (2006, March). New dawn for resettled communities. *Daily Independent*, Lagos.
General Three Lions was leader, Western Zone Division, The Niger Delta Peoples Volunteer Force, Headquartered in Arogbo Ijaw, Ondo State, Nigeria. He was interviewed in Arogbo ijaw on September 28, 2015.
Homer-Dixon, T. (1998). Environmental security and inter-group. In M. T. Klare et al. (Eds.), *World security challenges for a new century* (3rd ed.). New York: St. Martins Press.
Human Rights Watch. (2002). *The Niger Delta: No democratic dividend* (Vol. 14, No. 7). Washington: Human Rights Watch.
Human Rights Watch. (2005). Rivers and blood: Guns, oil and power in Nigeria's rivers state. *A Human Rights Watch Briefing Paper*, February.
Ibiyemi, S. (2009, August 3). FG, JV Partners Lose N8 to N/Delta Unrest. *Financial Standard*, Lagos.

Ifeka, C. (2001). Oil, NGOs and youth struggles for resource control in the Niger. *Review of African Political Economy, 28*(87), 99–105.
Ifeka, C. (2006). Youth cultures and the fetishization of violence in Nigeria. *Review of African Political Economy, 110*, 721–736.
Ikelegbe, A. O. (2010). *Oil, resource conflicts and the post-conflict transition in the Niger Delta region: Beyond the Amnesty*. Benin City: Centre for Population and Environmental Development (CEPED).
Ikime, O. (2005). *History, the historian and the nation: The voice of a Nigerian historian*. Ibadan: Heinemann Educational Books Nigeria Publishers PLC.
Ikporukpo, C. (2003). When victims become destroyers: The oil industry and communal self-destruction in the Niger Delta. *ERAction*, July–September.
Kakonen, J. (1993). “Environmental Conflicts and Third World System” in Social Science and Conflict Analysis. Proceedings of the *International of the Conflict Resolutions*, held in Helsinki.
Kalejaye, K. (2012, November 6). Nigeria's oil reserve stands at 36.2 bn barrels: Gas flare drops 18 Per cent. *Vanguard*, Lagos.
Kaplan, R. D. (2001). *The coming arnachy: Shattering the dreams of the post-cold war*. New York: Vintage Books.
Kpalap, B. Ara, Ogoni Activist, and formerly Information Officer of MOSOP. He was interviewed in Port Harcourt in April 2005.
Lawrence, F., Musari, A., & Okara, R. (2011, August 2010). Nigeria records 3, 203 oil spillage cases in four years. *The Guardian*, Lagos.
Lubeck, P. M., Watts, M. J., & Lipchutz, R. (2007). Convergent interests: U.S. energy security and the “securing of Nigerian democracy”. *International Policy Report.*
Mason, E. (2000). Forced migration studies: Surveying the reference landscape. *Libri, 50*, 241–251.
Moritz, M. (2010). Understanding Herder-farmer conflicts in West Africa: Outline of a processional approach. *Human Organization, 69*(2), 138–148.
Niger Delta Development Survey (NDES). (2000). *Niger Delta Development Priorities and Action Plan, phase II report, volume 2.*
Obayiuwana, O. (2012, January). The rape of paradise. *New African.*
Obi, C. (1997a). Economic adjustment and deeping of environmental conflict in Africa. *Lesotho Social Science Review, 3*(1), 13–29.
Obi, C. (1997b). Globalization and local resistance: The case of Ogoni versus Shell. *New Political Economy, 2*(1), 137–148.
Obi, C. (1999a). Resources, population and conflicts: Two African case studies. *Africa Development, XXIV*(3 & 4), 47–70.

Obi, C. (1999b). *The crisis of environmental governance in The Niger Delta 1985–1996.* Occasional Paper Series, Vol. 3, No. 3.
Okoye, C. A. (2014, May). An evaluation of the evolved African conception of the environment. *African Identities, 12*(2), 139–151.
Okwechime, I. (2011). Oil multinationals and the politics of corporate social responsibility in the Niger Delta. In A. Alao & R. Taiwo (Eds.), *Perspectives on African studies: Essays in honour of Toyin Falola*. Muenchen: LINCOM Publishers.
Olimogbe, H., & Ebiri, K. (2009, May 22). The battle of the creeks renders residents homeless. *The Guardian*, Lagos.
Omeje, K. (2005, September). Oil conflict in Nigeria: Contending issues and perspectives of the local Niger Delta people. *New Political Economy, 10*(3), 321–334.
Omotuyi, S. (2016). Killing for pasture or water? In R. I. Ako-Nai, M. O. Omotosho, & S. Omotuyi (Eds.), *One hundred years of Nigerian nationhood: Governance, foreign policy and development*. Ibadan: College Press.
Onosode, G. (2003). *Environmental issues and challenges of the Niger Delta: Perspective from the Niger Delta*. Lagos: Lily Bank Property and Trust Limited.
Osuntokun, A. (1999). *Environmental problems of Nigeria*. Ibadan: Davidson Press; In Association with Fredrick Ebert Foundation, University of Lagos.
Roberts, N. F. O. (2005). Environment and livelihood in the Niger Delta. In K. Fayemi, S. Amodi, & O. Bamidele (Eds.), *Towards an integrated developmement of the Niger Delta*. Lagos: Centre for Democracy and Development.
Rowell, A. (1994). *Shell—Shocked: The environmental and social costs of living with shell in Nigeria.* Greenpeace Report, July.
Rowell, A., Marriott, J., & Stockman, L. (2005). *The next gulf: London, Washington and oil conflict in Nigeria*. London: Constable and Robinson.
Saro-Wiwa, K. (1995). *A month and a day: A detention dairy*. Ibadan: Spectrum Books Limited.
The PUNCH. (2012). [Editorial] The high cost cost of flaring gas in Nigeria, October 22, 2012.
THISDAY, Lagos, May 29, 2007.
United Nations Development Programme (UNDP). (2006). *Niger Delta Human Development Report.* Abuja: United Nations Development Progamme.
WAC Global Services. (2003). *Peace and security in the Niger Delta: Conflict Expert Group Report.* Working Paper for SPDC, December.
Wiwa, D. (1997, October). The role of women in the struggle for enviromntal justice in Ogoni. *Delta*.
Yusuf, A. (2010, March 3). Nigeria loses N174b to pipeline Vandals in 10 years. *Daily Independent*, Lagos.

Iwebunor Okwechime is a Senior Lecturer in International Relations at Obafemi Awolowo University, Ile-Ife, where he teaches courses on global environmental politics, international economic relations and the politics of energy resources. He studied at the University of Lagos, Lagos, and Obafemi Awolowo University, where he received his PhD. He is a former A.G. Leventis Foundation Postdoctoral Fellow at the School of Oriental and African Studies, University of London, United Kingdom, and a former African Humanities Programme Postdoctoral Fellow at the College of Humanities, University of Dar es Salaam, Tanzania. His articles have been published in various journals focusing on international relations, oil and environmental politics in Nigeria, and the press and democratic struggles in Nigeria. His recent publications include 'Africa and the global economic and financial crisis' (2016), and 'Oil and human security challenges in Nigeria' (2016). He has recently co-edited *State, Governance and Security in Africa* (2016).

17

Informal Immigrant Traders in Johannesburg: The Scorned Cornerstone in the Southern African Development Community Integration Project

Christopher Changwe Nshimbi and Inocent Moyo

Introduction

Southern African countries have crafted several regional instruments that demonstrate Southern African Development Community (SADC) member states' aspirations for an integrated region. Based on a case study of informal immigrant cross-border traders (hereafter, 'immigrant traders') from the SADC region operating in the City of Johannesburg (hereafter, 'Johannesburg'), South Africa, the extent to which the legislative apparatus at national and city level of a SADC member state promotes or inhibits regional integration through human mobility is explored. This is

C.C. Nshimbi (✉)
Department of Political Sciences, Centre for the Study of Governance Innovation (GovInn), University of Pretoria, Pretoria, South Africa

I. Moyo
Department of Geography & Environmental Science, University of Zululand, KwaDlangezwa, South Africa

A. Adeniran, L. Ikuteyijo (eds.), *Africa Now!*,
https://doi.org/10.1007/978-3-319-62443-3_17

relevant because the free cross-border movement of persons is among four factors, including capital, labour, goods and services, that drive regional integration. Attention is given to those immigrant traders driven out of their home countries and compelled to engage in informal trade in South Africa by, among other things, political and economic crises. It is thus appropriate to raise questions concerning the activities of these traders in Johannesburg. Do the socio-economic legislative frameworks of a local authority, such as Johannesburg, in a SADC member state promote the activities of immigrant traders? Does such legislation take regional instruments into consideration in their formulation? Are immigrant traders in Johannesburg subjected to or faced with attempts by local authorities to curtail the traders' business activities? If so, how does this affect regional efforts to integrate Southern Africa, given that South Africa, as the regional economic giant, is the major business hub for the traders? Are immigrant traders ignored or unwelcome in South Africa?

These questions arose from the need to explore the legislative and institutional regulatory regime for trading in Johannesburg, South Africa, vis-à-vis immigrant traders from the SADC region pushed by political and socio-economic crises in their home countries into informal cross-border trade. Further, the questions are raised in an attempt to show whether the regime permits SADC immigrant traders and activities and, by extension, if this approach enhances regional integration.

Following this introduction, the section 'Methodology' steps down to the regional level to examine the status of integration in Southern Africa, the governing legislation and the place of human mobility and immigrant traders in SADC's integration agenda. The section 'Regional Human Mobility and Regional Integration: Conceptual Clarifications' zeroes in on South Africa and Johannesburg to assess national and local legislation, policies and realities on the ground and how these affect regional integration. Empirical evidence concerning immigrant traders' experiences with national and local legislations and enforcers of the legislation is also presented here and in the section 'Continental Integration and Human Mobility: The AU's Integration Agenda', which continues the reflections on immigrant traders' experiences in Johannesburg. The section 'AU Policy Frameworks on Migration' concludes and provides some recommendations.

Methodology

In pursuing the research goals, existing regional, national and local legislation relating to human mobility in the SADC and legislations governing the conduct of immigrant traders from Southern Africa in Johannesburg was reviewed. In-depth qualitative interviews with the traders operating in Johannesburg were supplemented with the information from secondary sources. The interviews were conducted between August 2012 and July 2013. The interview method was triangulated with field observations. Deployment of the interviews was based on the understanding that such an approach uses inductive logic, in terms of people's understanding and interpretation of their lives and experiences (Creswell 2009; Teddlie and Tashakkori 2009). Additionally, the approach helped in understanding the human experiences of 'attachment, dislocation, alienation and exile, as the human experiences constitute experiential reality and are integral parts of migration' (Entrekin and Tepple 2006: 32). Johannesburg is a major destination for immigrants (Landau and Gindrey 2008), which made it a suitable site for exploring Southern African immigrant traders' activities vis-à-vis regional migration and integration.

Table 17.1 lists the number of immigrant traders interviewed by nationality and gender. A total of 30, including 21 males and nine females, were interviewed. They originated from the Democratic Republic of Congo (DRC), Malawi, Mozambique, Tanzania, Zambia and Zimbabwe.

Table 17.1 Nationalities of informal migrant traders in the Johannesburg Metropolitan District

	Sex		
Nationality	Male	Female	Total
Congolese (DRC)	1	2	3
Malawian	6	0	6
Mozambican	1	0	1
Tanzanian	8	0	8
Zambian	1	3	4
Zimbabwean	4	4	8
Total	**21**	**9**	**30**

Source: Authors, fieldwork interviews

Interview questions focused on the traders' immigration status in South Africa; why they engaged in informal cross-border trade and whether they were self-employed or employed by someone else; why they had migrated to Johannesburg; their experiences of discrimination, hostility and xenophobia during the conduct of their business; whether they received any support from Johannesburg; the nature of the business they engaged in; and the extent of their trans-national travel and networks.

Of the 30 traders, all three Congolese (i.e. from the DRC) were in South Africa on asylum permits. For the Zimbabweans, except for one who possessed a temporary residence permit, the other seven were all on visitors' permits (Table 17.2). All those from, respectively, Malawi (6), Tanzania (8) and Zambia (4) were on visitors' visas. The only Mozambican respondent in the sample had a South African permanent residence permit. Respondents disclosed that political and economic difficulties in their home countries motivated them to migrate to Johannesburg. Specifically, the DRC and Zimbabwean respondents emphasised that political and economic instability had forced them to migrate to South Africa. Although countries like Malawi, Mozambique, Tanzania and Zambia enjoy peace and relative economic stability, respondents from these countries said the desire for a better life and to advance themselves and their families at home had driven them to migrate.

All the 30 respondents disclosed that they were self-employed. Among them, some ordered goods from wholesalers in and around

Table 17.2 Types of permits held by respondents

	Type of permit					Total
Nationality	Permanent resident	Temporary resident	Visitors	Asylum/ Refugee	None	
Congolese (DRC)				3		3
Malawian			6			6
Mozambican	1					1
Tanzanian			8			8
Zambian			4			4
Zimbabwean		1	7			8
Total	1	1	25	3	0	30

Source: Authors, fieldwork interviews

Johannesburg for retail within Johannesburg and in other SADC countries. Others made and sold beadwork, embroidery, sandals, various artefacts and crafts while others dealt in vegetables, cosmetics, clothes, blankets, shoes, mobile phones and mobile phone accessories, computers and information technology accessories and appliances. The traders exported goods to other SADC and their home countries, making their trade trans-national. A Zimbabwean trader, for example, said he owned a shop in Johannesburg, another in Bulawayo, Zimbabwe, and also travelled to Malawi and Zambia to sell goods there. The immigrant traders also imported goods such as traditional African wear, artefacts and crafts into South Africa from their home countries. They indicated that some wholesale businesses in Johannesburg had survived in recent years because of the business the immigrant traders had brought from the SADC region. The context within which these informal immigrant traders operate in Johannesburg is, therefore, worth exploring.

Regional Human Mobility and Regional Integration: Conceptual Clarifications

The term 'immigrant informal cross-border traders' here refers to people of non-South African origin, but who are nevertheless from Southern Africa and are citizens of SADC member states, and who live in South Africa and regularly travel between South Africa and their home and other SADC countries for business. Their form of business is primarily informal trade. The traders regularly travel around the SADC region in search of business, economic opportunities and prospects for personal advancement through trade.

Regional economic integration principally progresses through four stages (Balassa 1961). First, integrating countries mutually remove trade restrictions and establish a free trade area (FTA) between them. These countries, however, maintain trade tariffs with third-party countries. The integrating countries then, second, adopt a common external tariff, which they apply against third-party countries and thus form a customs union (CU). Third, the CU progresses to a 'common market' when member states remove all restrictions on the movement of capital, labour, goods and services within the group. Fourth, the countries establish an economic and monetary union through supranational authorities, which coordinate their economic policies. As part of this union, participating countries use a single currency, a central bank and unified fiscal system. A common economic policy also defines an economic and monetary union. Complete integration may come as the final stage, establishing a supranational authority that oversees all policies affecting all member states of the integration scheme. These stages of integration emphasise the synthesis of national markets. Moreover, regional integration is a process and a means, not an end in itself (Nshimbi 2005).[1]

The mobility of economically active persons across borders is significant to regional integration. As demonstrated in the stages of integration, the movement of capital, labour, goods and services is vital. SADC is committed to formulating regional instruments to remove obstacles to the free movement of these productive factors and of people in general among member states (Declaration and Treaty of SADC 1992: 6). Currently, however, SADC has not formalised a mechanism to guide the regional movement of labour (Nshimbi and Fioramonti 2013). Immigrant traders move goods, capital and services across borders in the SADC and play an important role in its integration project. The economic activities of immigrant traders in SADC states such as South Africa are significant. SADC's informal cross-border traders, for instance, traded goods amounting to US$2500 per capita per month in 2008 (Southern African Research and Documentation Centre 2008).

The experience of the European Union (EU) exemplifies the importance of the free movement of people, including immigrant traders, to regional integration. The free movement of people in the EU is absolute and no longer restricted to conventional labour. The evolution of the European integration project sheds light on the notion and significance of cross-border human mobility and how this relates to regional integration. Citing the European Commission (2012a, b), Nshimbi and Fioramonti (2013) indicate that the free movement (of persons, capital goods and services) in the EU is enshrined in the EC Treaty, and that as a fundamental principal of the EU, free movement is a cornerstone of the EU's Single Market. 'When considered from a regional integration perspective and the implied goal of Common or Single Markets, discussion of freedom of movement connotes one of four freedoms: free movement of goods, services, capital and persons' (Nshimbi and Fioramonti 2013: 30). The term 'free movement' vis-à-vis human mobility in the European integration project and the Single Market context initially applied to the free movement of workers only, and gradually evolved to connote freedom of movement of people in general (Nita forthcoming, as cited in Nshimbi and Fioramonti 2013). Today, the mobility of EU citizens across that region has evolved to a stage where freedom of movement is a right enshrined in the Maastricht Treaty and confers on citizens of member states of the EU the rights to move across national borders and to live, work and establish themselves in another member state.

Applying the EU experience to the SADC region poses fundamental challenges, however. First, the factors behind the EU's formation, including, among other things, the desire to end longstanding hostilities between France and Germany, differ from SADC's initial aim to reduce the founding member states' economic dependence on apartheid South Africa. Second, the evolutionary path that the EU has followed, based on economic cooperation in the European Coal and Steel Company and leading to functional spillovers into other areas of cooperation, and the eventual establishment of European citizenship, differs from the approach and political rationale that founding members employed in establishing SADC. SADC's predecessor, the Southern African Development Coordinating Committee (SADCC), operated on a project basis, with each member state tasked to coordinate a specific sector, such as energy, mining, tourism, etc.

Considering the universal applicability of economic laws and the principles of supply and demand, however, and that most SADC members liberalised their economies in the late 1980s and early 1990s, the challenges SADC presently faces can be addressed through the lenses of SADC labour and migration frameworks and, specifically, migrant traders' experiences and how a city in a SADC member state handles immigrant entrepreneurs operating in the area under local authority jurisdiction. Clearly, this does not imply that the conclusions of this approach are applicable to other member states of the SADC, let alone to other cities in the country under focus. However, the example highlights the existing challenges and the gaps between regional legislation, informed by regional objectives, with the local legislation and practices.

Continental Integration and Human Mobility: The AU's Integration Agenda

The AU is seeking to establish an Africa Economic Community (AEC) by 2028 through the 1991 Abuja Treaty for the Establishment of the African Economic Community. The strategy for attaining the AEC is through regional economic communities (RECs), which are used as building blocks. Eight such RECs constitute the pillars for moving Africa towards the AEC.[2] The progress SADC makes in its integration project impacts the AU's continental integration agenda. Human mobility and, in particular, the free movement of economically active persons across a region's borders constitutes one of four productive factors essential to enhancing regional integration. Moreover, the AU's New Partnership for Development cites the migration–development nexus as a sectoral priority in the New Partnership's Human Resources Development Initiative (African Union Executive Council 2006a, b). In this regard, the role of immigrant traders in regional integration, poverty reduction and regional development is relevant.

AU Policy Frameworks on Migration

The AU realises that migration contributes to development and defines its approach to migration through two key policy frameworks: the Migration Policy Framework for Africa and the African Common Position on Migration and Development. The Migration Policy Framework for Africa addresses nine key migration issues, including labour migration, border management, irregular migration, forced displacement, the human rights of migrants, internal migration, migration data, migration and development, and inter-state cooperation and partnerships (AU 2006a). The African Common Position on Migration and Development, on the other hand, raises 11 priority policy issues and recommendations for national, continental and international action. These include migration and development; human resources and the brain drain; remittances, trade, migration and peace; security and stability; migration and human rights; gender; regional initiatives; and access to social services. The AU encourages its members to adopt employment policies that allow the free movement of people within the envisaged AEC. The Abuja Treaty provides for member states to gradually remove obstacles to the free movement of persons, capital, goods and services and to give the member states' citizens the right of residence and establishment in second AU states. Accomplishing the free movement of persons in the SADC region, therefore, contributes to the realisation of the AU's goals.

Regional Integration and Human Mobility in Southern Africa

SADC currently has a membership of 15.[3] It started out in April 1980 with nine countries as SADCC. The Front Line States movement established SADCC as self-reliant and reduced members' dependence on apartheid South Africa. The SADC Treaty, signed in August 1992, transformed SADCC into SADC. The SADC Treaty seeks regional development through formal regional institutions and to establish an economic community through the successive stages of regional integration discussed in the section 'Regional Human Mobility and Regional Integration: Conceptual Clarifications' above. This has consequences for the movement of factors of production in Southern Africa, particularly labour. A regionally harmonised and well-managed regional migration governance system can enhance cooperation, foster integration and contribute to regional development.

SADC Legislations Regarding Human Mobility

Overall, SADC aims to enhance economic and social integration in Southern Africa. It seeks, among other things, 'to strengthen and consolidate the long standing historical, social and cultural affinities and links among the people of the region' (Declaration and Treaty of SADC 1992: Article 5.1.h.). In working towards its objectives, SADC commits to developing 'policies aimed at the progressive elimination of obstacles to the free movement of capital and labor, goods and services, and of the people of the Region generally, among Member States' (ibid., 1992: Article 5.2.d.). The 2005 Draft Protocol on the Facilitation of Movement of Persons in the SADC is the legal instrument that implements the provisions of the SADC Treaty (Declaration and Treaty of SADC, 1992: Article 10.3) on the movement of persons in the region. The protocol commits to support, assist and promote the AU's efforts to eventually establish the AEC through, among other things, free movement of persons (Draft Protocol on the Facilitation of Movement of Persons in the SADC 2005: 1).

Accordingly, the overall objective of this protocol is to develop policies that will progressively eliminate obstacles to human mobility into and within SADC members' territories (ibid., 2005: 3). As of 2014, only six SADC states—Botswana, Lesotho, Mozambique, South Africa, Swaziland and Zambia—had ratified the Facilitation of Movement Protocol (Nshimbi and Fioramonti 2014). The protocol is not yet in force, therefore, because SADC has not reached the required two thirds majority threshold of member state ratifications. In the absence of a protocol to govern regional labour migration, national laws of the respective SADC states regulate human mobility in the region.

Table 17.3 outlines and summarises the aims and ratification statuses of several other SADC instruments that affect human mobility and govern regional migration in SADC, albeit to limited extents. Only those instruments that member states have ratified are binding and enforceable, however, with possibilities for benefiting immigrant traders. For example, the Code on Social Security (Table 17.3, viii) is one of very few SADC instruments that accommodate so-called illegal and undocumented migrants. Article 17.3 of the code provides for

Table 17.3 Migration-related legislation and instruments in the SADC region

Instrument	Aim	Status
1. SADC Declaration and Treaty	Remove obstacles to the free movement of capital, labour, goods and services, and of people among member states	In force
2. Protocol on the Facilitation of Movement of Persons	Progressively eliminate obstacles to human mobility into and within SADC States	Not ratified
3. Charter of Fundamental Social Rights	Facilitate the promotion of labour policies, practices and measures which facilitate labour mobility, remove distortions in labour markets and enhance industrial harmony and increase productivity in SADC member states	In force
4. Protocol on Education and Training	Facilitate freer movement of students and staff in the SADC region for studies, teaching, research, etc. Cooperation in higher education and training among SADC States to encourage the establishment of measures to facilitate student credit transfers among SADC universities	In force
5. SADC Labour Migration Action Plan 2013–2015	Facilitate labour migration in the SADC	Adopted 2013
6. SADC Labour Migration Policy Framework	Promote regular migration in the region Development and implementation of national labour migration policies Development and conclusion of multilateral and bilateral labour agreements Protection of migrant workers' rights in the workplace and in the community Harmonisation and standardisation of national labour migration policies Integration of the intra-regional labour migration data into the SADC Labour Migration Information System	Approved by SADC Employment and Labour Sector: 2014

(*continued*)

Table 17.3 (continued)

Instrument	Aim	Status
7. SADC Decent Work Programme 2013–2019	Prioritise the strengthening of capacities in data collection and reporting on progress in employment and labour-related issues in member states	Signed and agreed
8. SADC Protocol on Employment and Labour	Basis for harmonising employment and labour standards in the SADC region. Equal treatment and social protection of workers and members of their families and the protection of migrant workers	Signed 2014
9. Code on Social Security	Urges member states to work towards free movement of persons and to progressively reduce immigration controls in the SADC Protection of all lawfully employed immigrants through six core principles to be indigenised into the respective SADC states' national laws and included in bi- or multilateral agreements between SADC states. Provide illegal and undocumented migrants with basic minimum protection and extend social protection to refugees in line with regional and international instruments	Adopted 2007
10. Protocol on Trade	Permit and facilitate cross-border movement and establishment of traders in services to establish cargo, clearing and forwarding offices in second member states	In force
11. Protocol on Immunities and Privileges	Visa-free entry privileges for SADC officials into all SADC states	In force
12. Protocol on the Development of Tourism	Facilitate intra-regional travel by easing or removing travel and visa restrictions, and harmonising immigration procedures	In force
13. Protocol on Transport, Communication and Meteorology	Strategically eliminate or reduce hindrances and impediments to the movement of persons, goods, equipment and services in the region	In force

Source: Compiled by authors from various sources

illegal residents and undocumented migrants to receive basic protection and enjoy coverage according to the laws of the host country. The code also provides for member states to ensure the protection of immigrant workers—regarding social security—through a set of core principles, among which member states should ensure coverage of self-employed migrant workers on the same basis as employed migrants (Article 17.2(f)).

The SADC Labour Migration Action Plan 2013–2015 and the SADC Labour Migration Policy Framework, on the other hand, are the latest and most elaborate instruments relating to labour migration in the SADC. Though not yet in force, the Labour Migration Policy Framework, whose objectives are similar to those of the Action Plan, was approved and signed by the Employment and Labour Sector of the SADC in July 2014. Like the Action Plan, it aims to promote regular migration in the SADC region; develop and implement national labour migration policies; harmonise and standardise national labour migration policies and integrate intra-regional labour migration data into the SADC Labour Migration Information System; and promote the conclusion of multilateral and bilateral labour agreements. Unlike the Action Plan, the policy framework further seeks the development and conclusion of multilateral and bilateral labour agreements and the protection of migrant workers' rights in the workplace and in the community.

The instruments in Table 17.3 also give an insight into the thinking of SADC governments on human mobility, and the role and relevance to Southern African integration. Four brief observations can be made about the instruments. First, the instruments aim at removing obstacles or hindrances to movement of, among other things, labour in the SADC. Second, the instruments seek to facilitate movement. Third, despite seeking to facilitate movement, the instruments do not allow free movement but only seem to allow for the managed entry of regional citizens into other member states. Fourth, neither the instruments nor SADC specifically addresses facilitated, managed or free movement of immigrant traders across member states' borders.

The Status of Integration in SADC

SADC has significantly progressed in the initial stages of integration especially. Most members liberalised their economies in the 1980s and 1990s under the auspices of international financial institutions. Since 2000, SADC states have also reduced trade tariffs, as prescribed by the SADC Protocol on Trade. Apart from bottlenecks stemming from non-tariff barriers, SADC has attained the minimum conditions for an FTA as most members have completed the tariff phase-down schedule. This primarily involves respective members ratifying and enforcing the Trade Protocol (Table 17.3).

SADC, however, defaulted on establishing a CU by 2010. The 2001 SADC Regional Indicative Strategic Development Plan, adopted in 2003, outlines a comprehensive 15-year regional integration roadmap for SADC—running from 2005 to 2020—that envisaged an FTA by 2008, a CU by 2010, a common market by 2015 and an economic union by 2018, preceded by a monetary union from 2016. Going by the provisions of the Protocol on Trade, more remains for SADC to do to deepen integration beyond FTA accomplishments. The protocol touches on such issues as services, trade facilitation, investments, competition policy, product standards and intellectual property rights, all of which indicate the need to regulate SADC's business environment. To this can be added the need to address human mobility to foster deeper integration.

In aiming to liberalise trade in goods and services and contribute to cross-border and foreign investment, the SADC Protocol on Trade bears on and relates to other regional instruments. Cross-borders traders in services whom the Trade Protocol permits to establish cargo, clearing and forwarding offices in other member state territories (Table 17.3, x) should benefit from the facilitation of labour migration planned in the SADC Regional Labour Migration Action Plan 2013–2015, for example (Nshimbi 2013; see also Table 17.3, v). Despite the reality of cross-border human mobility in the SADC region and the activities of immigrant traders in second member states, regional instruments such as those outlined in Table 17.3 generally do not directly address, let alone appreciate, the contribution of these economically active actors to the regional economy, the economies of immigrant traders' host countries and the overall SADC integration project.

The SADC Integration Project and Informal Cross-Border Traders

Informal cross-border trade is an important and integral part of migration in the SADC region. It contributes to poverty reduction, as it is a source of livelihood for many people (Southern Africa Trust (SAT) 2013). Despite near omission in SADC and most member states' policies, informal cross-border traders and activities are a practical approach to regional economic integration from below. The value that informal cross-border traders place on their trading activities is evident in the formation of the Southern African Cross Border Traders Association (SACBTA), an association of informal cross-border trade associations from SADC member states, in 2009 (SAT 2009). SACBTA campaigns for the promotion of informal cross-border trader activities in the SADC region. In a meeting held in Mozambique in April 2013, SACBTA, in collaboration with SAT and Oxfam, discussed its place in the Tripartite FTA negotiations.[4] Informal cross-border traders in SACBTA further seek opportunities and to overcome challenges they encounter in relation to the SADC FTA, the COMESA Simplified Trade Regime and the EAC (SAT 2013). Despite SACBTA forums and recommendations, legislation and policies that hinder cross-border trader activities in individual SADC member states remain, however. SACBTA currently cannot determine individual SADC countries' agendas regarding the activities of informal cross-border traders. Cross-border traders, for example, have difficulties obtaining trading permits and these permits are expensive where available. In Zambia, for instance, such permits cost approximately US $200 apiece (SAT 2013: 27), which is high for an informal trader.

South Africa: National and Local Legislations and Human Mobility

In discussing Southern African immigrant traders operating in Johannesburg, their business activities and related issues, and impact on regional integration, it is necessary to briefly review South Africa's and Johannesburg's legislation relating to trade and human mobility.

South African immigration laws and policies after 1994 did not fundamentally depart from apartheid legislation until after 2002. The belief immediately after 1994 was that a deliberate policy that encouraged immigration into South Africa would threaten the interests of the new dispensation (Crush *et al.* 2006). This thinking influenced immigration policies at that time. The 1995 Amendment Act altered the Aliens Control Act of 1991 by just removing the openly racist and undemocratic provisions (Peberdy and Crush 1998). Essentially, the Aliens Control Act of 1991 remained in force in the 1990s, prompting assertions that South Africa was 'preparing tomorrow's transformation with yesterday's tools' (Segatti 2006: 42). Fundamental problems of discrimination against foreign nationals remained (Nyamnjoh 2006) because of embedded impediments to the legal recognition and the acceptance of foreign immigrants in South Africa.

Immigration Act 13 of 2002 was the major legislation that replaced the Aliens Control Act 1991. The former replaced the word 'alien' with 'foreigner' and introduced many provisions that encouraged immigration. Section 26 and 27 of Immigration Act 13 also simplified the granting of different types of permits to foreigners. Additionally, Immigration Act 13 of 2002 (as amended by Act 19 of 2004) came into effect in July 2005 with a commitment to make the issuing of permits to foreigners more efficient.

Immigration Act 13 has been amended several times since 2002, the latest being the Immigration Amendment Bill 2010, approved March 2011, and enacted as Immigration Amendment Act, 2011 (No. 13 of 2011). The amendments generally encourage legal immigration to South Africa. However, a report in *The Star* (3 May 2011) said of this act that it fundamentally made the entry qualifications for foreigners into South Africa more stringent, and included attempts to clamp down on undocumented immigrants and to protect South Africa from immigration influxes.

Local Legislation, Law Enforcement and Immigrant Traders in Johannesburg

Section 13(a) of the Local Government, Municipal Systems Act (No. 32 of 2000) governs informal trading in Johannesburg. Based on this, Local

Authority Notice 328 published the Informal Trading By-laws 2009 for the Johannesburg municipality on 14 March 2012 to: promote social and economic development; a safe and healthy environment; and municipal planning, trading regulations, licensing and control of undertakings that sell food to the public, markets, public places, municipal roads and street trading (Gauteng Provincial Administration 2012: 1). Johannesburg recognises the need for a developmental approach that, among other things, provides job and entrepreneurial opportunities in the informal trading sector, harmonises the relationship between the informal and formal trading sectors, and facilitates the graduation of informal entrepreneurs to the formal sector.

Johannesburg's policy on informal trade, therefore, seeks to create opportunities for the informal trading sector to share in the benefits of and further contribute to Johannesburg's economic growth. The policy regulates informal trading within Johannesburg's jurisdictional area in a manner that recognises and enhances the city's constitutional and other statutory obligations (ibid., 2012). The Johannesburg municipality, therefore, offers excellent opportunities for informal traders, including immigrant traders. However, do such policy provisions materialise in relation to immigrant traders' operations in Johannesburg?

Attitudes Towards and Treatment of Immigrant Traders in Johannesburg

The Johannesburg municipality, officers from the South African Police Service and, especially, the Johannesburg Metropolitan Police Department (JMPD), and ordinary South African citizens subjected immigrant traders in the Johannesburg inner city to various forms of discrimination, hostility and xenophobia. Evidence from interviews with immigrant traders suggests the traders felt they were 'not wanted' in South Africa because it was almost impossible for them to acquire business permits from the Department of Home Affairs. According to the traders, the South African government, represented this department, was less supportive of them and their economic activities.

At municipal level, respondents claimed that officials at the Johannesburg Metropolitan Trading Company (JMTC) were hostile to them, even though they possessed the required documentation for applying for trading stalls. This alleged resistance from JMTC is diametrically opposed to the stated aims and objectives of Section 13(a) of the Local Government, Municipal Systems Act (No. 32 of 2000). Under the section dealing with freedom to engage in informal trade, Notice 328 allows *any member* of the Community of Johannesburg to engage in informal trade, provided they comply with the provisions of the informal trading By-law, the Act and any other applicable laws (Gauteng Provincial Administration 2012: 9). The Municipality notably states that any member of the Community of Johannesburg can apply and be allotted a trading stall. The immigrant traders' claims to the contrary suggest they face hurdles in their daily operations and indicate the gap between local authorities, national and regional legislations and actual practice at the local level. Only 30 % (including three females and six males) of the immigrant traders interviewed had been allotted trading stalls in Johannesburg. The rest, who were in the majority, representing 70 % (including six females and 15 males), resorted to street and pavement trading. Conditions in street and pavement trading were such that JMPD officers and some South African citizens harassed and looted the immigrant traders' goods. However, the few immigrant traders allotted trading stalls also adamantly asserted that JMPD officers harassed them and looted their goods. The traders claimed they resorted to bribing the JMPD officers in order to survive this challenge, resulting in a difficult working space for them.

The immigrant traders who were not allotted trading stalls suffered xenophobic treatment from some South African citizens, who took the traders' goods without paying. South African citizens who engaged in similar trade were not treated in the same way. JMPD officers also impounded immigrant traders' goods for allegedly operating in non-trading zones; 70 % (or 21) of the immigrant traders who experienced this said they survived by bribing the JMPD officers.

In summary, immigrant traders have experienced and perceive hostility from the South African government, citizens, the Johannesburg municipality and law enforcement officers. The hostility has occurred despite the immigrant traders demonstrating an economic and entrepreneurial contribution to Johannesburg, as illustrated by Tanzanians and

Malawians, who excel in beadwork and embroidery and make monthly incomes ranging from approximately R7000 to R12,000 from the trade, and others who sell traditional African attire. Further, 20 % of the immigrant traders claimed they import their goods (clothes, blankets, shoes and various household goods) from China and in large quantities.

Reflections on Immigrant Traders' Experiences in Johannesburg

The South African government theoretically welcomes and embraces all immigrants from the SADC region and beyond. Immigration Act 13 of 2002, Immigration Act 19 of 2004 and Immigration Amendment Act, 2011 (No. 13 of 2011) principally promote immigration. Also, the South African government through the education sector promotes the movement of academic staff and students from the SADC region into South Africa for teaching, research and training purposes. Interview responses, however, paint a picture on the ground where immigrant traders have difficulties acquiring trading permits. Consequently, they operate and live on the edge, are exposed to harassment and lured into corruption. This violates the Code on Social Security in SADC (Table 17.3, ix), which encourages member states to promote free movement of persons and reduce immigration controls (Article 17.1). It also violates the aspirations of the SADC Labour Migration Action Plan and the labour policy framework discussed in Section 3.2 (Table 17.3, v & vi). South Africa's national Immigration Acts do not expressly recognise the existence of immigrant traders within the immigration scenario. This handicaps the activities and contribution of these regional economic actors to South Africa and the SADC region. As a country, South Africa has in this regard not moved towards indigenising the SADC Declaration and Treaty with respect to the movement of people, particularly immigrant traders.

Despite its current status, the Draft Protocol on the Facilitation of Movement of Persons in the SADC has a great bearing on regional movement especially when it is enforced (if it ever will be). The six SADC states that have ratified the protocol ought to implement its provisions even if the protocol is not yet enforced. The JMTC,

Johannesburg and their officials' actions towards immigrant traders in Johannesburg, along with some aspects of South Africa's Immigration Acts (e.g. attempts to clamp down on undocumented immigrants and to protect South Africa from immigration influxes), suggest a reluctance, on the part of South Africa, to encourage the free movement of persons in the SADC. Though ratified and enforced, the SADC Protocol on Trade ignores immigrant traders while formal businesses and organisations enjoy the benefits of this protocol. When asked, immigrant traders expressed an ignorance of the existence of this instrument.

South Africa seems realist in its foreign policy towards the rest of Africa. South Africa may be gaining, economically, in African countries where it has engaged in peace-building, and South Africa's foreign ministry—the Department of International Relations and Co-operation—stresses that South African companies are more aggressive in securing commercial interests in such countries (*Mail & Guardian* 2013). The government has even produced an economic diplomacy tool for its diplomats, according to the department's minister, Maite Nkoana-Mashabane. 'Our White Paper on international relations says our diplomats must play an important role in advancing the cause of South African business beyond our borders and attracting investments in their respective countries of accreditation,' Nkoana-Mashabane said in an interview (ibid., 2013). Formal companies (e.g. MTN, Game Stores, Shoprite, etc.) constitute some of South Africa's formal cross-border businesses in Africa. Could it be that immigrant cross-border traders and activities are insignificant or their contribution too negligible to South Africa's economy to attract government interest?

At local level, interviews on the actions of the Johannesburg's officials towards immigrant traders revealed a gap between regional instruments and practice. Going by Section 13(a) of the Local Government Municipal Systems Act (No. 32 of 2000), Johannesburg offers excellent opportunities for immigrant traders. In practice, this is different, however, because the law apparently does not protect the immigrant traders. The people supposed to enforce the law, such as JMPD officers, reportedly violate it by harassing and accepting bribes from the traders. Therefore, a gap exists in the SADC regional integration project between regional instruments and South Africa's national and local legislations and policies. Other realities on the ground suggest gaps, especially to the extent that immigrant traders face difficulties in acquiring trading permits and stalls. The dynamics on the ground, where immigrant traders actively participate in SADC's integration process, point to the need for more effort to ensure that regional, national and local legislations and polices harmonise.

Immigrant Traders in a SADC Member State: Implications for Regional Integration

Human mobility, cross-border trade involving immigrants and related issues are currently absent from SADC's regional economic integration and trade agenda. SADC has only recently formalised its recognition of the importance of a formal and managed regional labour migration system (Nshimbi 2013; see also Table 17.1, v & vi). Though welcome, this does not accommodate immigrant traders, however. Neglecting immigrant traders and their role in regional economic integration does not benefit the SADC integration process. SADC and member states should treat immigrant traders as an essential part of the integration project because the traders actually move finance, goods and services across SADC.

At national level, and as a SADC member, South Africa demonstrates that national legislations such as Immigration Act 13 of 2002, Immigration Act 19 of 2004 and Immigration Amendment Act, 2011 (No. 13 of 2011) theoretically support immigrants, including those from the SADC. However, based on immigrant cross-border traders' claims that South African immigration officials find the traders irritating at border entry points en route to Johannesburg on the traders' many business trips, a discrepancy exists between South Africa's immigration laws and the enforcement of the laws.

Johannesburg by-laws in principle support immigrant trading activities. A mismatch between the by-laws and their implementation exists, however. City officials, who should facilitate the issuance of trading permits, are unwilling, xenophobic, or both, thus frustrating the immigrant traders, who resort to unauthorised street trading. A Zimbabwean street trader's views amplify this:

> the reality on the ground is that it is not easy to be allocated a trading stall. Even when you have the required documentation, when you arrive at the offices, the officials ignore you or tell you that you don't qualify. The fact is that, as an immigrant, one cannot challenge the officials. The only option is to sell from the streets, this is what is happening, not some baseless theories about how well we are treated. (M. Shumba, Zimbabwean, personal interview, Johannesburg inner city, September 2012)

Law enforcement officers indiscriminately harass and loot immigrant traders' goods or accept bribes from them. A Congolese immigrant trader commented during interview that:

> if the Johannesburg Metro Police were fair and professional, without regarding immigrant traders as moving ATMs [Automated Teller Machines], our [immigrant traders'] trading activities would yield greater economic benefits to Johannesburg specifically, and South Africa and the SADC region in general. (D. Kasongo, Congolese, personal interview, Johannesburg inner city, April 2013)

Conclusion and Recommendations

The SADC regional economic integration project is an appropriate strategy for enhancing Southern Africa's development. Theoretical and practical examples of benefits that come with regional economic integration abound, as demonstrated in the presentation of economic theory on integration, and the EU case where regional human mobility has evolved to the level of regional citizenship. SADC, however, overlooks the vital element of human mobility involving immigrant cross-border traders and corresponding activities. At a regional level, SADC has crafted various and ratified instruments to enhance regional economic integration. Regional legislation specifically designed to address, manage and govern regional human mobility, however, remains unenforced. Existing trade and legislations that touch on migration are almost silent on issues relating to informal and immigrant traders. SADC's categorical pronouncement of the need to enhance the movement of labour in the region ignores immigrant cross-border traders.

The traders are economically active participants in Southern Africa's integration project, just as the formal labour whose movement SADC seeks to facilitate in the Regional Labour Action Plan 2013–2015 and the SADC Labour Migration Policy Framework. Immigrant cross-border traders are on record as reducing poverty and are a source of livelihood in many SADC countries (SAT 2013: 1). Given SADC's development objective through regional economic integration, Southern Africa cannot continue to ignore immigrant traders. The traders open up markets for and participate in moving capital, goods and services in their transnational trading activities, in accordance with the synthesis-of-markets aim of regional integration.

The following recommendations build on the research findings and preceding discussion.

(a) SADC and regional integration policy instruments should categorically capture and fully accommodate immigrant traders, their role and activities in Southern Africa's integration process.
(b) South Africa, and by extension all SADC states, should allow for accessible business permits for immigrant traders. In addition to enhancing regional integration and regional and national development, business permits for such traders will reduce incidences of harassment and corruption and ease the traders' movement and trading activities. Immigrant traders in the SADC region are the grassroots implementers of regional economic integration, because they actually move money, goods and services across the region's borders. SADC should institute immigration policies that recognise, support and make immigrant traders' movements easy.
(c) Members of the SADC should operationalise and strengthen the capacity of localised, domestic SADC institutions to oversee, coordinate and ensure the domestication, implementation and monitoring

of regional instruments (or even establish new instruments). Under the existing SADC Structure, SADC National Committees and SADC National Contact Points are responsible for technical matters, including, among other things, regional policy and strategy initiation and formulation, in respective member states. The case of Johannesburg suggests that such committees are yet to fully fulfil their mandates. Local authorities in member states, such as Johannesburg, should, in conjunction with the domestic institutions proposed to oversee and ensure domestication of regional legislation and policies, design robust frameworks for the use of trading stalls and follow up on immigrant traders' operations at such stalls in order to regulate their activities. Formally established reporting, follow-up and feedback mechanisms will ensure immigrant traders are not victimised by unscrupulous municipal and law enforcement officers, provide traders with recourse to law and legal protection, while feeding back to the region for purposes of evaluating and monitoring the implementation of regional policies.

Notes

1. As a means, regional integration provides opportunities for member states of a regional bloc, for example, to meet the objective of expanding markets for domestically produced goods and thus contribute to development.
2. The eight RECs include the Community of Sahel-Saharan States (CEN-SAD), the Common Market for Eastern and Southern Africa (COMESA), the East African Community (EAC), Economic Community of Central African States (ECCAS), the Economic Community of West African States (ECOWAS), the Intergovernmental Authority for Development (IGAD), SADC, the Union du Maghreb Arabe (UMA). See http://www.au.int/en/recs/. [Accessed 26 November 2012].
3. Angola, Botswana, the DRC, Lesotho, Madagascar, Malawi, Mauritius, Mozambique, Namibia, Seychelles, South Africa, Swaziland, Tanzania, Zambia and Zimbabwe.

4. The Tripartite Free Trade Area, dubbed the COMESA-EAC-SADC FTA, is an initiative of the Common Market for Eastern and Southern Africa (COMESA), East African Community (EAC) and SADC that seeks to integrate and establish an FTA of the three regional organisations and 26 member states of the three RECs on 10 June 2015.

References

African Union (AU). (2006a). *The migration policy framework for Africa.* Executive Council, Ninth Ordinary Session, Banjul, The Gambia, June, 25–29.

African Union (AU). (2006b). *African common position on migration and development.* Executive Council, Ninth Ordinary Session, Banjul, The Gambia, June, 25–29.

Balassa, B. (1961). *The theory of economic integration.* Homewood, IL: Richard D. Irwin.

Creswell, J. H. (2009). *Research design: Qualitative, quantitative and mixed method approaches.* Singapore: Sage.

Crush, J., Campbell, E., Green, T., Nangulah, S., & Simelane, H. (2006). States of vulnerability: The future brain drain to South Africa. *Migration Policy Series, No 42.* Cape Town: IDASA.

Entrekin, J. N., & Tepple, J. H. (2006). Humanism and democratic place making. In S. Aitken & G. Valentine (Eds.), *Approaches to human geography* (pp. 30–41). London: Sage.

European Commission. (2012a). *The EU single market: General policy framework.* Retrieved December 24, 2012, from http://ec.europa.eu/internal_market/top_layer/index_en.htm

European Commission. (2012b). *Employment, social affairs and inclusion: Enlargement…transitional provisions.* Retrieved December 10, 2012, from http://ec.europa.eu/social/main.jsp?catId=466&langId=en

Gauteng Provincial Administration. (2012). Informal trading by-laws, 2009. *Provincial Gazette*, 14 March 2012. Retrieved September 2, 2013, from http://www.joburg.org.za/images/stories/2012/Sept/local_government_-_municipal_systems_act_32-2000_-_city_of_johannesburg_-_informal_trading_by-laws_final%20%20by%20laws%203.pdf

Landau, L., & Gindrey, V. (2008). Migration and population trends in Gauteng Province 1996–2055. *Migration Studies Working Paper Series # 42*. Forced Migration Studies Programme, University of Witwatersrand. Retrieved March 4, 2011, from http://www.migration.org.za/research-outputs/working papers

Mail & Guardian. (2013, November 15–21). We are not looting the continent, we're cultivating peace and trade, pp. 12–13.

Nita, S. (forthcoming). Regional free movement of people: The case of African Regional Economic Communities. In *Regions and cohesion: Special edition*. UNESCO-UNU Chair on Regional Integration, Migration and Free Movement of People.

Nshimbi, C. C. (2005). Is regionalization in Southern Africa the appropriate strategic response to globalization? In A. Ahmed (Ed.), *World sustainable development outlook 2005, global competitiveness: A common goal in a digital society*. Bradford: Emerald Group Publishing Limited.

Nshimbi, C. C. (2013). State of denial. *OpenDemocracy*. Retrieved December 2, 2013, from http://www.opendemocracy.net/chris-nshimbi/state-of-denial

Nshimbi, C. C., & Fioramonti, L. (2013). *A region without borders? Policy frameworks for regional labour migration towards South Africa*. Johannesburg: African Centre for Migration and Society, University of the Witwatersrand.

Nshimbi, C. C., & Fioramonti, L. (2014). The will to integrate: South Africa's responses to regional migration from the SADC region. *African Development Review, 26*(S1), 52–63.

Nyamnjoh, F. B. (2006). *Insiders and outsiders: Citizenship and xenophobia in contemporary South Africa*. Dakar; London; New York: CODESIRA and Zed Books.

Segatti, A. (2006). Reforming South African immigration policy in post-apartheid period (1990–2006): What it means and what it takes. In *Migration in post-apartheid South Africa: Challenges and questions to policy makers* (pp. 33–79). Paris: FASOPO.

Southern Africa Trust (SAT). (2009). Retrieved August 31, 2013, from http://www.southernafricatrust.org/docs/Informal_Cross_Border_Traders_Association_launched_20090716.pdf

Southern Africa Trust (SAT). (2013). Regional cross border Trade Stakeholder Dialogue, Maputo, Mozambique, April 23–24, 2013. Retrieved August 31, 2013, from http://www.southernafricatrust.org/docs/regional_cross_border_trade_stakeholder_dialogue_20130716.pdf:27

Southern African Development Community. (1992). *Treaty of the Southern African Development Community*. Gaborone, Botswana: SADC.
Southern African Development Community. (2005). *Protocols on facilitation and movement of persons*. Gaborone, Botswana: SADC.
Southern African Research and Documentation Centre (SARDC). (2008). *Optimising regional integration in Southern Africa: Assessing informal cross border trade in SADC*. Final Report.
Teddlie, C., & Tashakkori, A. (2009). *Foundations of mixed methods research: Integrating quantitative and qualitative approaches in the social and behavioural sciences*. Los Angeles; London; New Delhi; Singapore; Washington, DC: Sage.
The Star. (2011, May 3). *Law will strangle growth in South Africa*, p. 10.

Christopher Changwe Nshimbi is the Deputy Director of the Centre for the Study of Governance Innovation, Department of Political Sciences, at the University of Pretoria. He researches the informal economy, cross-border trade, social cohesion, migration, water and regional integration. Dr Nshimbi teaches graduate courses on African regions and politics, regionalism in Africa and regionalism in Southeast Asia, and on the Jean Monnet Comparative Regionalism Masters' Programme. He also teaches undergraduate courses on African states and governance, and political dynamics. He sits on several regional and international technical working groups and committees on labour, migration, the water sector and social protection.

Inocent Moyo is a Lecturer in the Department of Geography and Environmental Studies, University of Zululand, South Africa. He is a seasoned researcher on migration and development, immigration politics, cross-border traders, regional integration, transnationalism, borders and borderlands, and urban informality and governance. He has published and presented internationally on these topics. He is passionate about the political economy of reform and the grassroots forms of economic activity; in his view, these could provide an alternative paradigm and/or trajectory of progressive politics for improving the human condition. Dr Moyo has a deep attachment to critical emancipatory research, which he believes should be used to challenge customary modes of understanding and responding to the problems affecting humanity today and in the future.

18

Emergent Dynamics in Female Migration in West Africa

Timothy Abayomi Atoyebi

Introduction

Migration, which is an important demographic phenomenon, is not restricted to Africa; rather, it is a common phenomenon that has socio-economic and political implications. As a global phenomenon, however, it applies to both developed and less developed countries. Globally, the number of female migrants has been rapidly increasing, rising from 41 million in 1975 to 1995 million in 2005 (Bourguignon 2006). Studies have also shown that women have been actively engaging in international commercial migration. For example, it has been estimated that the world's migrant population in 2005 was more than 80 million people, and that about half of these were women (Bourguignon 2006). The records in 2005 showed that 47.4 % of the 17 million immigrants in Africa are women, as against 42.3 % in 1960. However, between 1990 and 2013, the number of international migrants rose by over 77 million, or by 50 %.

T.A. Atoyebi (✉)
Department of Sociology, Bowen University, Iwo, Osun State, Nigeria

A. Adeniran, L. Ikuteyijo (eds.), *Africa Now!*,
https://doi.org/10.1007/978-3-319-62443-3_18

Analysis of this statistics revealed that in 2013, women comprised 48 % of all international migrants worldwide (International Migration Report 2013). Most of these women migrated within the region; only a few moved to North America and Europe.

A World Bank study in 2010 revealed that France tops the list of destination country for emigrants from Africa (14 % of migrants). It is followed by Côte d'Ivoire with 9 %; Saudi Arabia has a total of 8.1 %, while South Africa and USA account for 5 % of emigrants from Africa. Statistics presented reveal that the total share of African emigrants living in Western Europe, Canada, USA and Australia was about 37 %.

Generally, individuals (men or women) tend to move from countries with fewer socio-economic opportunities to those with more opportunities. Adepoju (2010: 9) summarises the characteristics of contemporary African migration in the following words:

> Distinctive forms of migration characterize the different sub-region: labour migration from western and central Africa to other locations within the regions, as well as to other OECD (Organisation for Economic Cooperation and Development) countries and those of the oil rich countries of the Middle East; refugee flows from eastern and southern African countries to south Africa, and cross-border clandestine migration of seasonal workers and nomads in West and East Africa.

Aside from the characteristics explained her, the literature has also shown that an estimated one third of the people of West Africa are living in places other than their places of birth and the majority of them are involved in circular, cross-border and commercial migration (Adepoju 2010; Adepoju and Wiel 2010). Human Development Report (2009) emphasised that most international migrants (more than three quarters) prefer to relocate to a country with a higher level of 'human development' than their country of origin, especially if this country is within their respective sub-region (Adepoju 2005; Adeniran 2012).

The fundamental basis for migration is the fact that no country has everything it requires, or can fully satisfy the interests, needs and basic requirements of all its citizens. It is only natural that, where intense scarcity and unequal resource distribution exists, there will be waves of immigrants seeking better opportunities beyond the borders of their own country or nation (Adesina and Adebayo 2009). Adeniran and Olutayo (2015) indicated that when factors attracting people into another society are more active than those discouraging them, they tend to move.

Migration, whether at internal, regional and international levels, is a demographic and socio-economic phenomenon and is usually affected by both internal and external factors (De Haas 2006). These factors may vary from country to country, depending on whether the country is migrant-sending or migrant-receiving, but would include rapid population growth, political instability, escalation of ethnic conflicts, persistence economic decline, retrenchment of public sector workers, poverty and environmental deterioration (Adepoju 2008).

Other related factors are the distance and porosity of borders between the countries; colonial links and shared culture and languages; and immigration policies and relevant legislation in the host countries. Important as well are the age structure of the population in the country of origin and available socio-political networks. Besides these factors, Adepoju (2008) further emphasises that pull factors are essentially high wages in the receiving countries, quality of social security systems, political stability, as well as a past colonial relationship between the host and country of origin.

For several decades, women's migratory process have not been given adequate attention in related studies. Their socio-economic relevance, contributions and unique experiences were not given sufficient attention in the discussions. Before the 1960s, most migration theories assumed that most migrants were male and that women were merely wives and dependants who traditionally followed their husbands. In other words, migration was a purely male phenomenon. If women played any role at all, it was as spouses, daughters and dependants of male migrants (Kymzer 2009; Ndaye 1999).

However, notable developments during pre-colonial and post-colonial periods have changed this. A significant proportion of women are now seeking socio-economic advancement across borders. As is argued in this study, women are no longer mere 'associates' of men, but independent actors. In fact, the feminisation of migration, which refers to the growing flow of women migrating globally, is gradually becoming an undeniable reality in Africa (Piper 2005). Most women leave the (dis)comfort of their homes to become economic providers or breadwinners for their families (UNDP 2009).

Motivating Factors for Female Migration in West Africa

In West Africa, a number of factors, such as poverty, diseases, natural disasters, land degradation and an increased rate of unemployment, are motivating women towards migration. The women, just like their male counterparts, migrate in the hope of generating income to secure better living conditions for themselves and their families. For example, available records reveal that at the time of independence in 1960, Nigerians, particularly women, were already noted for migration to foreign lands in search of better living conditions. Motivations such as the desire to enhance standards of living, avoid social turmoil/political repression and escape natural disasters have acted as push factors, leading to large-scale movements of Nigerians across national boundaries (Adeniran and Olutayo 2015; Kivisto and Faist 2010; Meagher 2007).

However, studies have shown that poverty alone does not always determine the decisions of women to migrate. These depend, among others, on community settings and traditions, as well as the circumstances surrounding the individual (Adepoju 2010). In fact, it has been suggested that when examining female commercial migration, it is important to fully understand the role preferences in a society where commercial migration occurs. Roles given to males and females in a patriarchal society vary from community to community and so does the status mobility of the genders. Similarly, female migration would differ in traditional societies as against industrial. Marphatia *et al.* (2000) reveal that religion and cultural values have influenced the commercial migration of traditional Malian women, for instance.

Other economic factors such as increasing labour demand on the service market, family obligations, unemployment, low wages, limited social and economic opportunities and the desire to expand their horizons are equally motivating women into migration. A number of income-earning opportunities can empower women and loosen the traditional restrictions on women's mobile capability. In addition, economic and social upheaval can also serve as an impetus for female migration. However, non-economic factors, such as surveillance by communities, patriarchal structure, traditions, forced marriage, domestic violence, and discrimination against certain groups of women—single mothers, unmarried women, widows or divorcees—can also engender female migration (UNDP 2009).

Regardless of the various factors motivating women, migration involves several trends that remain complex in the context of women's cross-border migration in Africa. This article therefore highlights the experiences of migration that are unique to women and equally reveals some trends that remain hidden in the context of movement of women across borders, particularly in the West Africa sub-region. To some extent, this article discusses the evolution, prevalent trends and implications of female migration, with a particular focus on Ejigbo-Yoruba women, who have adopted migration as a coping mechanism for poverty reduction.

Methods of Data Collection

Due to the itinerary of women migrants and illiteracy which characterise cross-border migration, qualitative methods of data collection were therefore considered appropriate for this type of research. The following qualitative research instruments were triangulated in this study: direct observation, in-depth interviews (IDIs) and key informant interviews (KIIs). Equally, secondary data was sourced from the archives, published journals and articles, periodicals and the internet. Observations were carried out alongside interviews done in markets, shops, churches and mosques in Abidjan (Côte d'Ivoire) and Ejigbo (Nigeria). By virtue of such observations, relevant data on dynamics and trends of female migration, interactions and actions among the women migrants in the course of their movement and other socio-economic activities was captured and recorded in the field notes.

In all, 40 IDIs were conducted in Côte d'Ivoire with Ejigbo-Yoruba women immigrants who have stayed in Côte d'Ivoire for a minimum of five years and who have been involved in trans-border trading activities between Ejigbo and Abidjan for a minimum of two years. These migrant traders were purposively selected based on years of settlement, number of apprentices, scale of trading activity, trading experience, availability, convenience and consent. The 40 IDIs were conducted with women migrant traders in the six selected 'communes' in Abidjan, Côte d'Ivoire: Abobo, Adjame, Cocody, Yopougon, Treichville and Marcory.

This method enabled the researcher and his assistants the opportunity to probe in-depth the migratory experiences of Ejigbo-Yoruba women migrants. The in-depth interview guide was designed in English and interpreted in Yoruba for Ejigbo-Yoruba women migrants who could not communicate in English.

The KII, which provided a flexible way of obtaining some detailed information, was used to gather qualitative data from nine respondents. These individuals were purposively selected for the KIIs, based on their knowledge of the subject matter, relevant experience, availability, convenience and consent. Precisely, three relevant Ivorians (two men and one woman) who interact with Ejigbo-Yoruba women migrants were interviewed in Abidjan, Côte d'Ivoire, particularly shop owners and traders who patronised Nigerian women's traders.

Also, three Yoruba traditional leaders were interviewed (three of them were interviewed in Abidjan). As a matter of fact, these traditional leaders are more or less the most prominent individuals within Nigeria's migrant community in the area of study. Study has shown that when immigrants settle in new places or location, associational lives constructed and structured around ethnic, religious and national interests often develop (Cohen 1966). In addition, three KIIs were conducted for some men (husbands of women migrant's traders) on their perceptions and views about the dynamics and trends of women migration among Ejigbo-Yoruba women in Abidjan, Côte d'Ivoire.

Social Action Theory

The adoption of Social Action Theory in this work is to provide an understanding of individual behaviour in the course of trans-border migration. To Max Weber (1964), sociology is a science which attempts the interpretative understanding of social action in order to arrive at causal explanations (Elwell 2005). This is based on the notion that the study of society is 'an attempt to understand the subject matter by identifying oneself with the human actors and imaginatively experience their intentions and motives' (Labinjoh 2002: 18–19). The actor's purpose and judgements are, therefore, causally relevant in explaining the action undertaken.

Weber's Social Action Theory (1964) typically deals with subjective understanding of social phenomena. In other words, human action is directed by meaning and motives. For any behaviour to become an 'action', it must be deliberate; action in turn must not be interpreted objectively but in terms of the subjectively intended meaning of the actor, which includes the effect the actor wants the action to have on 'others'. When this happens, such action becomes a social action (Rex 1961; Menzies 1982). From another perspective (Ritzer 2008), social action involves the intervention of thought process between the occurrence of a stimulus and the ultimate response. Recognising that human action is social in nature, Weber identifies four ideal types of social actions: the goal-oriented rational action, value-oriented rational action, affective or emotional motivations, and traditional action.

According to Labinjoh (2002), traditional action refers to all forms of behaviour which are driven by traditional cultural values. Affective action connotes all behaviour that aligns with feelings and emotions. Purposive rational action describes behaviour that is guided by the consideration of ends, means and secondary consequences. The value-oriented rational action represents behaviour that is driven by the desire to achieve a specific goal. The rational social action, according to Weber (1968), has a clear definition of goal, a deliberate and systematic evaluation or assessment of the possible means of attaining the goal, and a careful selection of the most appropriate means of attaining the goal.

This theory implies that social action is rational when an appropriate means is employed to achieve a given end. The agent may use his expectations of the behaviour of external objects and other human beings as a 'condition' or 'means' to achieve his rationally pursued and calculated purposes (Runciman 1972). A person acts rationally when his action is guided by a consideration of end, means and secondary consequences. When acting, he rationally assesses means in relation to ends, ends in relation to secondary consequences, and finally the various possible ends in relation to each other. The pursuit of a goal, therefore, involves an actor rationally calculating the available means and choosing the most appropriate one.

The argument of this research is anchored in the notion that the migratory activity of women migrants is rooted in Social Action Theory as any individual, before opting for cross- and trans-border migration, will have a clear definition of the goal, a deliberate and systematic evaluation or assessment of the possible means of attaining the goal, and a careful selection of the most appropriate means to attain the goal.

To understand the motives of Ejigbo-Yoruba women in becoming involved in trans-border migration, attention must be paid to the motive and the appropriate means that women choose to attain their goals. From the prism of the ideal types (traditional, affective, goal-oriented and value-oriented rational actions), the choice of trans-border migration must involve a network of social relationships in which a series of actions fall within different categories of social actions at different stages. For instance, their engagement in trans-border migration might be linked to the need to earn an income, address poverty or support the family, among others.

Pre-Colonial Trends in Migration in West Africa

Available records have shown that pre-colonial migration in West Africa occurred largely in search of security, or to discover new land safe for settlement and fertile for farming. The migration occurred without administrative and legal barriers and the interactions went beyond the trade in commodities to include intermarriages between powerful kingdoms, exchange of slaves and military alliances (Adeniran 2012). The relationships were basically oriented towards the exploitation of labour and the institutionalisation of religious pilgrimages. The absence of strict legal restrictions, as well as the lack of clearly defined borders among existing empires and kingdoms, made it possible for nomads, farm workers and traders to travel freely and easily cross borders, even during the colonial era (Adedokun 2003; Adepoju 2005).

Migration studies (Adeniran 2012; Mahadi 1989; Akinjogbin 1980) have argued that the main factor that facilitated population movements across West Africa and led to the establishment of various migrant settlements was the outbreak of wars and the fall of most of the existing kingdoms and empires. These studies further noted that, for some reasons, most of the ethnic groups that left the old Bornu Empire came to occupy parts of some territories belonging to Francophone countries. In the same vein, most of the ethnic groups that left the old Oyo Empire, particularly the Yoruba, also preferred to settle in both Francophone and Anglophone countries neighbouring what is now Nigeria.

A fundamental factor in trans-border migration from parts of Yoruba to other West African countries was the Fulani invasion. Also, the existence of the trade in slaves in many parts of present-day Nigeria, including Yoruba land, which began around 1500, resulted in the forced displacement of populations across the region and led to the trans-border migration of people to locations they considered safe (Adepoju 1996). A number of Yoruba kingdoms were displaced and the citizens were forced to scatter to other parts of West Africa and, indeed, to other parts of the world.

In the West African sub-region, the issue of female migration is economically built around trade and agricultural production. In fact, most agricultural region witnessed the influx of female cross-border labour migrants during the period that is generally referred to as the 'boom period' of the 1930s. During this period, the production of cocoa and coffee became a lucrative business. As such, countries like Ghana and Côte d'Ivoire became attractive to migrant women, who either moved with their husbands or alone. It important to emphasise that migrants, particularly women from other neighbouring countries such as Burkina Faso, Mali, Senegal and Niger Republic moved into Ghana and Côte d'Ivoire, took advantage of the boom to enhance their living conditions (Adeniran 2012).

Post-Colonial Trends in Women Migration in West Africa

The post-colonial migratory process in West Africa reflected a new trend with the emergence of free migration of labour within the West African region. Most adults, including women, migrated to work for wages in any atmosphere characterised by a complex grid of social relations and inter-dependences. Migrants cared less about the artificial borders inherited from the colonial regime, which were in any case considered to limit or inhibit migration. The introduction of cash crops, to increase foreign exchange earnings needed for the various development projects in the various countries, further resulted in the flow of labour into the cocoa belts in Ghana and Côte d'Ivoire, coffee in Côte d'Ivoire, and groundnuts and cotton in Senegambia. These zones attracted labour (men and women) from Burkina Faso, Niger, Chad, Mali, Guinea, Cape Verde and Togo (Adepoju 2010). In the same vein, studies indicated that the Yoruba of south-west Nigeria emigrated to some West African countries, such as Ghana, Dahomey (the Republic of Benin), Togo and Côte d'Ivoire, especially to Abidjan (Adegbola 1972; Afolayan 1991).

Generally, intra-regional mobility in West Africa is dominated by a predominantly north–south movement (from countries such as Mali, Burkina Faso, Niger and Chad), to the more prosperous plantations, mines and cities of coastal West Africa (predominantly Côte d'Ivoire, Liberia, Ghana, Nigeria, Senegal and The Gambia). In other words, the relatively prosperous economies of the Ghana–Côte d'Ivoire migration pole attracted large numbers of internal labour migrants as well as international migrants from countries such as Togo and Nigeria (mainly to Ghana), Guinea (mainly to Côte d'Ivoire) and Burkina Faso, Niger and Mali (to both). In fact, the Presidents of Ghana and Côte d'Ivoire, in the spirit of pan-Africanism, opened their borders to immigrants and allowed them to work and stay (Anarfi and Kwankye 2003).

In Southern Africa, while movement was allowed for male labour migrants and local mine workers, movement of their female counterparts was restricted during apartheid. However, the post-apartheid era witnessed an unprecedented increase in female migration, resulting in important changes in economic, social and cultural dynamics. This phenomenon of labour migration equally affected other countries within the Southern African sub-region—Zimbabwe, Bostwana, Namibia, Mozambique and Zambia (Muzenda 2015).

One important feature that has played a fundamental role in the pattern of emigration in Africa, particularly in the West African sub-region, is shared linguistic and historical roots. This reflects the fact that most

African countries share extensive commonalities, especially common linguistic and historical roots in their border areas which can and, in fact, did facilitate movement across these borders. For example, countries like Burkina Faso, Mali, Guinea and Senegal are located in the same region with Côte d'Ivoire and are linked together by strong ethnic, religious and linguistic ties. In fact, Côte d'Ivoire is regarded as the melting pot for these neighbouring countries (Adepoju 2010).

Prevalent Trends in Female Migration in West Africa and Its Implications

Generally, the economic downturn witnessed by most countries has altered the patterns of migration in almost all the continents of the world. In fact, the late 1960s and early 1970s witnessed a socio-economic downturn that propelled migration imbalances across the regions. When many of these countries introduced structural or macro-economic adjustment measures, this led to a sharp decrease in the number of entrants into the labour market and generated a living situation which caused mass emigration to better areas. For example, in the West African sub-region, a noticeable number of labour migrations were observed in countries like Nigeria, Ghana, Togo and The Gambia.

By the 1970s, a number of skilled migrants, such as doctors, nurses, teachers, lecturers and engineers, migrated from Ghana to Nigeria and later to other African countries, as well as to Europe and North America. They were attracted by relatively higher salaries and better prospective living conditions. Statistics or data generated from the Network of Surveys on Migration and Urbanization in West Africa revealed that between 1988 and 1992, more than 6.4 million migratory movements were recorded between across the network of countries made up by Burkina Faso, Côte d'Ivoire, Guinea, Mali, Mauritania, Niger and Senegal (Bocquier and Traoré 2000).

As most of the countries in Africa witnessed economic and political crises, most of their citizens were compelled to seek greener pastures elsewhere. In the instance of Nigeria, the collapse of the oil price actually acted as a push factor that compelled migration of both skilled and unskilled labour abroad. With globalisation greatly impacting and bringing changes in the orientation of people towards their understanding of migration, the situation expedited the movement of people across the various regions in Africa and to other regions outside the continent (Akanle 2009; Akanmu and Adesina 2009). As people move to other regions, the resultant effects are a growing need for employment or job opportunities, urban poverty and insecurity. While various governments in the various sub-regions have attempted to meet the needs of migration through the development of migratory policies, a missing link that requires urgent attention is how to handle the issue of women who are opting for migration as a means of escaping from growing urban and rural poverty. Migration that is driven by poverty has been identified as the only social or demographic process that has the potential for disrupting and destabilising Africa (Cross and Omoluabi 2006). Hence, this study specifically examines the case of Ejigbo-Yoruba women who have adopted trans-border migration as a means of escaping from poverty.

Case Analysis of Ejigbo-Yoruba in Côte d'Ivoire

Generally, when it comes to migration, traditional research on migration downplays the involvement of women in migratory decisions. Recently, Lutz (2010) alludes to the traditional research on migration, which also indicates the absence of women in 1970s, therefore questioning the assumption of women's 'passive roles' in migration. While some research indicated a dominance of men in migration and disregarded female participation, there were indications that Yoruba migrant women had been historically involved in long-distance migration and trade across West Africa and had reached Ivory Coast (now Côte d'Ivoire) before the outbreak of the Second World War. By the 1940s, records show that migrant traders (men and women) from Oyo, Ijebu, Iwo and Ejigbo in Nigeria were already settled in Ivory Coast. This was noted in the following narratives on Ejigbo people's migration tendencies:

> My father told me that by 1942, they had settled in Ivory Coast and that messages were sent by their people in Ejigbo in Nigeria, advising them to

stay in their country of destination until after the war (KII, Male, 69 years, trader, Abobo-Abidjan, 10 May 2014).

This means that the peoples' migration could have even started before the First and Second World Wars. What is also clear is that by the latter half of the twentieth century, the Ejigbo-Yoruba traders might have moved across West Africa into countries such as Niger, Burkina Faso and Senegal (Eades 1980):

> Ejigbo people migrated to Ivory Coast when the Ivorians were still living in the 'pre-modern' or 'dark era'. During socio-economic interactions, the Yoruba made use of stone to explain the real value of a particular good to an Ivorian. An Ivorian could purchase a pair of shoe twice. (KII, Yoruba, Male, Retiree, 70 years, St Michael (Temidire), Abidjan, 12 May 2014).

The peoples' migration into Côte d'Ivoire started at a time when most Ivorians were not well educated and, apparently, exchanges were marked by the inability of the host community to handle money calculations properly, either because they could not read the labels on goods or they could not understand written French or spoken English. In line with Adeniran's observation (2012), the Ejigbo-Yoruba who presently occupy a prime position in the economic exchange structure developed a preference for Côte d'Ivoire at a time when rudimentary and subsistence patterns of livelihood were prevalent in most parts of the sub-region. Consequently, those migrating from Nigeria were labourers and artisans who, given a favourable economic outlook, later took to trading. Stories told to others about their economic success were enough to motivate the 'others' into participating in migration.

Poverty and Movement to Côte d'Ivoire

There was a general consensus among the migrant traders and community leaders concerning the history of trans-border migration among Ejigbo-Yoruba people. It was generally emphasised that extensive long-distance trade routes through West Africa were well established before the arrival of Europeans on the continent. For example, the caravan trade developed systems of credit, transport, information exchange, the settlement of business disputes and insurance among separate people without modern institutions to perform these services (Cohen 1968), promoting economic integration among ethnic groups. It is worth noting that poverty in the migrants' country of origin and the anticipated surplus gains that they could derive from various forms of economic activities in Côte d'Ivoire facilitated the movement of Yoruba, particularly the Ejigbo-Yoruba, to Côte d'Ivoire. As self-employed, they could address issue of poverty that they experienced:

> My late father told me that it was poverty that brought them to this place and that they came here (Côte d'Ivoire) purposely to trade so that they could sustain themselves and their family (KII, Male, Traditional ruler, 68 years, St Michael (Temidire), Abidjan).

In the light of this, one could possibly infer that the action of the early Ejigbo-Yoruba migrants who migrated to Côte d'Ivoire could be viewed as part of the normal process of human activity aiming at fulfilling individual or group needs and aspirations (Rhodes 1991). According to the oral tradition extant in the community, the movement of Ejigbo-Yoruba people to Côte d'Ivoire prior to the colonial era assisted in entrenching them in important socio-economic niches, which could be of great later advantage. Such entrenchment actually created opportunities for the migrants to access the large plantations and the rich mineral deposits in Côte d'Ivoire (Adesina and Adebayo 2009). This entrenchment is usually considered to be more secured than any in their country of origin. Hence, in the course of an interview, some informants suggested that Côte d'Ivoire has been attracting a lot of migrants, especially women, because of the peaceful nature of the country. Within this setting, they can raise their children to higher social classes by giving them an education and securing for them financially more rewarding jobs. According to an interviewee:

> Our people came to Côte d'Ivoire to better their fortunes as well as hide themselves. To be frank, Côte d'Ivoire is a peaceful and comfortable place

> where one could hide his head, build a house and have sufficient number of children without any interference from the people at home. Culturally, the children are saved from witches and wizards (IDI, Female, Trader, 48 years, Temidire-Abidjan, 25 May 2014).

From the above data, several reasons are given for settling down in Côte d'Ivoire; they include a better context for self-development, the avoidance of unwarranted competition and rivalry, the avoidance of crippling kin demands that stunt the accumulation of capital, and the negating influences of the spiritual forces embodied in characters referred to as 'witches' and 'wizards'. Basically, a contrast is set up between the country of origin and the host country: the country of origin is negative to development while the host or destination is positive to development. The country of origin emphasises a greater sense of communalism than the host country, which allows for some sort of individuality. The culture at home is not conducive to business but that outside is. A further confirmation of the positive value of the host country is this:

> My aunty, who brought me to Abidjan, has eight children, which may not be possible back at home in Nigeria. They may not allow all of them to survive in our place in Nigeria. But here in Abidjan they are doing fine, some are kept in the shop as '*morunsogba*', while some go to collect money from customers (IDI, Female, 48 years, Trans-border trader, Temidire-Abidjan, 28 May 2014).

Many of our informants actually emphasised that most of the early Yoruba migrants, especially women, saw their immigration to Côte d'Ivoire as a way of escaping from communal and family problems. In other words, the desire to be free from domestic and family restraints within their immediate communities encouraged their immigration. The movement of Ejigbo-Yoruba people to Côte d'Ivoire prior to the colonial development and post-colonial era actually assisted the people in taking up socio-economic opportunities as they participated actively in the Ivorian economy. Many of these migrants, especially women, saw immigration as a good channel through which they could secure better economic fortunes in distant places which they considered to be a secure environment.

Migration, Gender and Family Relationships

However, there is a general consensus among the community leaders that most of the Yoruba migrant traders who were attracted to Côte d'Ivoire before the outbreak of the Second World War were men, and that they migrated with their wives and other members of their families. A key informant explains that:

> When the early Yoruba migrants from Ejigbo came to Ivory Coast, they came in groups together with their wives. While the men as the head of the family purchased goods from different sources, their wives were occupied with the task of selling the goods to the final consumers (the Ivorians) (KII, Male, 72 years, trader, Abobo-Abidjan, 14 May 2014).

From the above data, it is clear that the early Ejigbo-Yoruba migrants (men and women) were driven by common factors and as such they moved in groups and were supported by favourable family decisions that encouraged the emigration. The informant also hinted at the occupational role which migrants adopted in their host country. The family structure of Ejigbo migrants in Côte d'Ivoire is not different from the type being operated by the Yoruba in Nigeria. In a patriarchal society, roles are allocated based on gender. While men are involved in agriculture, women engage in the harvesting, processing and marketing of the farm produce. In traditional Nigerian Yoruba society, women do not own land, but process and sell farm produce in the markets. Their role is mostly in the markets as key factors in the distributive economic system. They have done this to support the family and ensure its survival in economically difficult times. When they entered into the trans-border migration they became income-earners, supportive agents in the family affairs and participants in the decision-making process in the family.

There was a general consensus among the various informants and respondents about the fundamental factor for a successful transition to a new life in Côte d'Ivoire. A male informant, as an example, said:

> Our parents who came early to Côte d'Ivoire were determined and united. In fact, it was this unity and strong determination that made them trek to some nearby communities before joining some lorries that travelled for

several days to Ivory Coast (Côte d'Ivoire) (KII, Male, Traditional ruler, 68 years, Temidire-Abidjan, 30 May 2014).

The above assertion shows that sheer determination to succeed in spite of all odds, to break away from a restrictive environment into a liberating one, to take risks in spite of the danger involved, marked the difference between success and failure, a life of stagnancy and one of change. Rather than being helped, they could now help others, as this elderly informant explains:

> My father told me that when he came to Ivory Coast with his wife, they started a small business and the business prospered. He went back home to bring his brothers and siblings to assist him (KII, Male, 68 years, trader, Temidire-Abidjan, 11 May 2014).

They started small, but usually ended up big. Initially, the business involved only the husband and wife, but as it expanded, it included the children and later the kin. This organisation is central to the Ejigbo-Yoruba's success in trans-border migration and trade. For example, an Ivorian buttresses this point, as follows:

> The Yoruba are the most enterprising people in Côte d'Ivoire. They are hardworking and committed to their trade. You see those Yoruba women over there, I mean those four women, only one started the trade in that place; she later brought all those three to join her. The most senior of them usually travel to Nigeria to buy goods, while the rest will stay in the shop. The Yoruba are followed by the Lebanese, Mauritanians, Burkinabe, Senegalese and Ghanaians. Our people only like to do bureau (office) work (KII, Male, trader, 30 years, Ivorian, Adjame-Abidjan, 18 May 2014).

Trans-Border Trading and Changing Patterns in Family Relationships

Another issue of concern is that of family relationships in the host country. It is common knowledge among the Yoruba migrant community that most Yoruba men in Côte d'Ivoire do not stick to one woman only. Of course, the society that they come from practices polygyny. However, they have a pool of women to select from, since many of the migrant women have now left their husbands back at home (in Nigeria). These men 'take care' of them in Côte d'Ivoire (observation/interactions, Yopugon, Abidjan, 26 May 2014). In recent times, a number of women now come to Côte d'Ivoire with their children, and appear to be much more interested in earning a living and taking care of their children than in caring for a husband who could live without them, or who may depend on them for assistance:

> My husband is in Ibadan (Nigeria) and I travel regularly to Lagos to purchase goods and in most cases I don't go to Ibadan to see him. I have to struggle to take care of my children. To be frank, we may not even see for one year and we are surviving here in Côte d'Ivoire (IDI, Female, 48 years, Trans-border trader, Yopugon-Abidjan, 26 May 2014).

For many months, they may not see their husbands, in what appears as a trade-off: the husband uses the wife to get living funds, the wife funds the husband in order to have her freedom:

> My wife stays in Abidjan while I live in Ejigbo. She is really making it in Abidjan because her business is booming. Whenever I am in need or broke, I will ask her to send money to me from Abidjan and if she refuse I will ask her to bring my children to Nigeria within the next two weeks. She will then apologise and quickly send the money to me (KII, Male, 65 years, Retiree, Ejigbo, Nigeria, 20 July 2014).

From what the informant says, the children are a tool of negotiation on both sides of the divide, especially when the going is rough, or there is an unwilling bargainer who seems to be breaking loose completely from the marital connection. This is evident in the train of thought of one Focus Group Discussion (FGD) participant, who emphasised:

> In our house (in Nigeria) we have been taught that a woman must work, even if her husband is very rich, she must not rely on the wealth of the man, but she must work with her own hands because men are not to be trusted. In fact, our men here are unreliable and they are not responsible.

> Any woman who refuses to work and depends on them will suffer (FGD, Female, 45 years, Trans-border trader, Yopugon-Abidjan, 16 May 2014).

A critical factor influencing marital relationship is trust. Women who have come to think that their husbands are 'useless' exploiters want their independence. The independence enables them to control their personal accounts. Men who are the husbands of these migrant traders, in the view of the respondents, are not ready to shoulder family responsibilities. This has necessitated changes in the attitudes of the women to their husband, and by extension changes in the process of decision-making in the family, particularly with women now financially independent of their husbands. There is no doubt that most of them have used the relational crisis in the family to engage in trans-border trade purposely to earn income to support themselves, the siblings or their birth family. A woman migrant trader narrated her experience as follows:

> When I first came to Côte d`Ivoire, I trusted my husband so much that I left my shop for him to manage with a lot of goods while I was engaging only in trans-border trade popularly known as '*Ona ni na*' in Abidjan. Unfortunately, he mismanaged the business and later left me to marry another woman here in Abidjan. I have to take care of my two children alone. The same thing happened to my sister who handed over her business to her husband, now the man has abandoned her after several beatings. He is now with another woman. In fact, our men here are not to be trusted, that is why you see that I am serious with my business. I cannot trust any man again (IDI, Female, Trans-border trader, 45 years, Yopugon-Abidjan, 27 May 2014).

The developments within the family, which occasion trust or distrust, cooperation or conflict, impact on the trans-border trade in two ways. It is either that business is jointly owned or individually owned. But the latter is much preferred by embattled women. It makes them financially and socially independent of husbands' control, enables them to start and control their own business and claim the profits, and also creates the opportunity to avoid the tension of a polygamous marriage. Hence, the comment below:

> Most of our women here join trans-border trade primarily because of their family. Most of our men here have more than one wife and do not have time for us and our children (FGD, Female, Trans-border trader, 40 years, Yopugon-Abidjan, 29 May 2014).

It is not that men are unaware of such need for independence, for in an interview with Oba Yoruba in Treichiville (popularly called Temidire), the seriousness of the issue of the changing pattern in women's role within the family, in Côte d'Ivoire, has brought about a change in the family structure. This is what he has to say:

> Here in Côte d'Ivoire, women are now the breadwinners in most homes. Gone are the days when women sit at home and husbands fend for the family. These days, they keep the home and sustain the family especially through their involvement in different economic activities such as trans-border trade (KII, traditional ruler, 68 years, Treichiville, Abidjan, 30 May 2014).

There are, of course, some informants that have divergent views about the issue, but there is still some gender notion of the 'others'. For instance, an interviewed woman said:

> Men are not the same; some are good while some are bad. For me, my husband does not give me any problem. Whenever I am going on trans-border trade, he will take care of the shop and the children. I trust my husband and he trusts me too (IDI, Female, Trans-border trader, 40 years, Yopugon-Abidjan, 30 May 2014).

The woman interviewed has a joint husband and wife venture, in which individual roles in ensuring success do not conflict. The same goes for the male informant that we interviewed:

> I raised the capital and introduced my wife to trans-border trade, especially the trade between Nigeria and Côte d'Ivoire. The trade is lucrative and most of our people especially women engage in it and they are doing well. For me, I trust my wife and I know she cannot engage in all the 'nonsense' that goes on in trans-border trade. To be sincere, there is money

> in trans-border trade but it is also a risky business. If one is not careful, the business can easily break the family. To keep my family, we now interchange our travelling or business trips. If the road is good and there is no problem, I allow my wife to travel, while I stay in our shop. If the road is not good, I will rather travel and allow my wife to stay in the shop (KII, Male, husband of trans-border trader, 58 years, Adjame-Abidjan, 2 June 2014).

The informant does not specify the kind of 'nonsense' that goes on in trans-border trade, but we can reasonably infer that it touches on sexuality and the implications for economic and political independence and gain. Here is the view of another male informant:

> Women who are involved in trans-border trade are perceived to be promiscuous. Most of our people here believe that majority of these women befriend the drivers and they do anything to get their goods to their destinations. In fact, here in Abidjan, when you see a Nigerian woman who is now changing clothes and looking fine, people will ask such a woman whether she has joined the trans-border trade (KII, Male, husband of trans-border trader, 45 years Adjame-Abidjan, 26 May 2014).

From what the informant says, vehicle drivers could take sexual or economic advantage of the women to either allow them to pay less or nothing at all on goods, or to offer them a more comfortable seat—for instance, making sure they do not sit between passengers at the back of the vehicle. From what the informant says, poverty is one fundamental factor that has led women into trans-border trade, and the view of this FGD participant in Abidjan further buttresses the point:

> I join trans-border trade because of poverty. I was suffering and there was no money to take care of my children. I came to join my husband here in Abidjan, but things became worse. My aunty later raised money for me and introduced me to trans-border trade. Now that I am involved in trans-border trade, things have changed for the better. You can see me now, I was skinny before but now I thank God that things have changed. My plan now is how to send my daughter to a good school in Lagos (Nigeria) (FGD, Female, Trans-border trader, 38 years, Adjame-Abidjan, 18 May 2014).

In other words, the informal trans-border trade is a means of helping people, especially migrant women, to move out of poverty. It is common knowledge within the migrant community that the easiest way to assist a person who is in dire need of help is to introduce such an individual to trans-border trade (observation/interaction, Adjame-Abidjan, 26 May 2014).

Aside from the above, another opportunity which trans-border trade has created for women is the acquisition of financial capital which, as shown in the statements made below, has enhanced their decision-making role in their respective family settings:

> Before now, it is my husband who takes all the major decisions in the family, but now that I am involved in trans-border trade and I have money, we now take some decisions together. In fact, I decided the school our children attend. They were formerly attending a French school, where they only teach them English once in a week, but now they are attending a private school (owned by a Nigerian) where they can learn in the English language. After their secondary school education, I will bring them to Nigeria for further studies. The point is that I am working because of my children (IDI, Female, 40 years, Treichiville, Abidjan, 2 June 2014).

Migrant women engaged in trans-border trade are now gaining autonomy through financial independence as they become significant income-earners through their trade. Consequently, they have now shaped new gender norms and identities for themselves. To some extent, women are now utilising their new freedom of movement to expand their capital base. They now have aspirations that go beyond the need to take care of their immediate families to that of consolidating their stay in the host country. What we found is that improvement in information and communication technology had actually influenced female migration and trading along the Nigerian–Ivorian migratory corridor.

Discussion of Findings

This chapter attempts to fill the lacuna in female migration scholarly research by situating the experiences of a trans-border migrant trade group, the Ejigbo-Yoruba women traders, within gender and migratory perspectives. It has been hitherto assumed that most migratory process have focused more on male participation, while female migration was seen as merely 'associational'. Rather, this chapter has affirmed that there has been an unprecedented wave of female migration and that this has defied the status quo on gender norms and social construction. An increasing number of females in Africa, and particularly in West Africa, are entering into labour migration to better their lives. Ejigbo-Yoruba migrant women traders, in particular, are increasingly entering into the migration stream purposely to pursue their economic desires and aspirations.

Women's experience of migration is not limited to Africa but is global and has shaped new gender perspectives and identities. The findings from the study have also highlighted the fact that the need to earn income and support the family in Côte d'Ivoire, due to abandonment of family responsibilities by some men, have been the major reason for the involvement of Ejigbo-Yoruba women in migration and trans-border trade. This finding supports studies of researchers such as Mijere (2006), Little (2007) and Morris and Saul (2000), who noted that the main push factors towards informal cross-border trade are a lack of formal employment and low wages from the formal employment, pushing people to look for other ways to supplement their incomes. The study found that the need for socio-economic survival in the host society has been the impetus for engaging in internal and trans-border trade. The new trends in trade patterns have created the need to diversify in an emergent global market.

Concluding Comments

In summary, while an increasing number of women in Africa, particularly in West Africa, are entering into labour migration for better lives, this chapter depicted the history of women migration in West Africa and then narrowed it to Ejigbo-Yoruba women in Côte d'Ivoire. Given the need to earn income, alleviate poverty and create support platforms for family welfare, Ejigbo-Yoruba women from Nigeria have adopted migration as a coping mechanism. Also, given the immense growth of female migration and the role of informal trading activities in West African economies, attention should be given to the issue of the feminisation of migration in the mainstream migratory and trade policy decisions. Female migration should be accorded due attention and women's invaluable contributions to employment and wealth creation, poverty reduction and regional integration should be recognised by the ministries of trade and commerce in West African countries and by the Economic Community of West African States. Hence, there is a need for government and stakeholders in the West African sub-region to put in place appropriate policies that will take into account women who are entering the migration stream purposely to pursue their economic desires and aspirations and ultimately reduce poverty. The study specifically recommends that the functionality of inter-regional mobility, as practised by Ejigbo-Yoruba female migrants, should be utilised as a platform for reducing poverty and fostering growth and development within the continent.

References

Adedokun, O. A. (2003). "The Rights of Migrants Workers and Members of their Families" in Nigeria. *UNESCO*, SHS/2003/MC/7.

Adegbola, L. (1972). *The impact of migration on the rural economy of Osun division of Western Nigeria*. Unpublised Ph.D. thesis, University of Ibadan.

Adeniran, A. (2012). *Social networking and identity construction among Ejigbo-Yoruba Cote d'Iviore*. Unpublished Ph.D. thesis, University of Ibadan.

Adeniran, A., & Olutayo, O. (2015). Regional networking and identity integration in West Africa: Case study of Ejigbo-Yoruba in Cote d'Ivoire. In A. O. Olutayo & A. I. Adeniran (Eds.), *Regional economic communities: exploring the process of socio-economic integration in Africa*. CODESRIA: Dakar.

Adepoju, A. (1996). International migration in and from Africa: Dimensions, challenges and prospects. In A. Adepoju (Ed.), *Population, human, resources and development in Africa*. Dakar: PHRDA.

Adepoju, A. (2005). Migration in West Africa. A Paper prepared for the *Policy Analysis and Research Programme of the Global Commission on International Migration (GCIM)*.

Adepoju, A. (2010). *International migration within, to and from Africa in a globalized world*. Legon-Accra Ghana: Sub-Saharan Publishers.
Adepoju, A., & Wiel, A. (2010). *Seeking greener pastures abroad: A migration profile of Nigeria*. Ibadan: Safari Books Ltd.
Adesina, Y. R., & Adebayo, P. F. (2009). Yoruba traders in Cote D'Ivoire: A study of the role of migrant settlers in the process of economic relations in West Africa. *An International Multi-Disciplinary Journal, Ethiopia, 3*(2), 134–147.
Afolayan, A. A. (1991). Female migrants in Nigeria and development. In T. A. Olowu & J. A. Akinwumi (Eds.), *Development Strategies in 21st century Nigeria. African Economic Research Consortium Inc* (pp. 245–258). Pretoria, South Africa: African Economic Research Consortium.
Akanle, O. (2009). Immigration cultism and the Nigerian migrants: Tidal dynamism in the age of globalization. In A. G. Adebayo & O. C. Adesina (Eds.), *Globalisation and transnational migrations: African and Africa in the contemporary global system* (pp. 181–200). UK: Cambridge Scholars Publishing.
Akanmu, A. G., & Adesina, O. C. (2009). Globalisation and transnational migration: An overview. In A. G. Adebayo & O. C. Adesina (Eds.), *Globalisation and transnational migrations: African and Africa in the contemporary global system* (pp. 10–20). UK: Cambridge Scholars Publishing.
Akinjogbin, L. A. (1980). The economic foundation of Oyo Empire. In I. A. Akinjogbin & S. O. Osoba (Eds.), *Economic and social history*. Ile-Ife, Nigeria: University of Ife Press.
Anarfi, J., & Kwankye, S. (2003). Migration from and to Ghana: A background paper. Development Research Centre on Migration, Globalisation and Poverty, University of Sussex, Brighton.
Bocquier, P., & Traoré, S. (2000). Urbanisation et dynamique migratoire en Afrique de L' Quest La croissance urbaine en panne. L' Harmattan In Decaluwe, B., J-C, Dumont, S. Mesple-Somps, V. Robichaud. Union économique et mobilité des facteurs le cas de L' UEMOA. Retrieved from http://www.creta.ecnulavalca/cahuer/lislesl00.hlm
Bourguignon, F. J. (2006). *Women on the move: Magnitudes, trends and impacts of international migration of women*. New York: United Nations.
Cohen, B. P. (1968). *Modern social theory*. London: Collier Macmillan.
Cross, C., & Omoluabi, E. (2006). *Views on migration in Sub-Sahara Africa: Proceedings of an African migration alliance workshop*. Cape Town: HSRC Press.

De Haas, H. (2006). International Migration and National Development: Viewpoints and Policy Initiatives in Countries of Origin. The Case of Nigeria. Working Papers Migration and Development Series. *Report No. 6*, Radboud University, Nijmegen and Directorate General for International Cooperation (DGIS), Ministry of Foreign Affairs, The Netherlands.

Elwell, F. (2005). The Sociology Max Weber. Retrieved January 5, 2010, from http://www.faculty.rsu.edu/felwell/theorists/weber/whome.htm

Human Development Report. (2009). *United Nations Development Programme*. New York: UNDP.

Kivisto, P., & Faist, T. (2010). *Beyond a border: The causes and consequences of contemporary immigration*. Thousand Oaks: Sage.

Kymzer, I. (2009). *Female migration in Ukraine: Determinants and consequences*. IMPALLA/ESPANET Conference: The European Social Model in a Global Perspective.

Labinjoh, J. (2002). The sociological trio: An introduction to Marx, Weber and Durkheim. In U. C. Isiugo-Abanihe, A. N. Isamah, & J. O. Adesina (Eds.), *Currents and perspectives in sociology* (pp. 149–160). Ibadan: Malthouse Press Limited.

Little, P. D. (2007, March). Unofficial cross-border trade in Eastern Africa. Paper presented at the *FAO Workshop on staple Food Trade and Market Policy Options for Promoting Development in Eastern and Southern Africa*, FAO Headquarters, Rome.

Lutz, H. (2010). Gender in the migratory process. *Journal of Ethnic and Migration Studies, 36*(10), 1647–1663.

Mahadi, A. (1989). The roles of neighbouring countries in the Nigerian Civil War. In T. N. Tamuno & S. C. Ukpabi (Eds.), *Nigeria since independence: The first 25 years* (vol. VI, pp. 252–276).

Marphatia, A. A., Morris, G., & Saul, M. (2000). Literature review of women cross border traders in West Africa (since 1985). In *Women's business links: A preliminary assessment of women cross-border traders in West Africa*. Final report submitted to the West African Regional Programme, Development Alternatives, Inc. WIDTECH.

Meagher, K. (2007). Manufacturing disorder: Liberalisation, informal enterprise and economic 'ungovernance' in African small firm clusters. *Development and Change, 38*(3), 473–503.

Mijere, N. J. (2006). *Informal cross-border trade in the South African Development Community (SADC)*. Addis Ababa, Ethiopia: OSSREA Publications.

Morris, G. A., & Saul, M. (2000). Women's business links: A preliminary assessment of women cross-border traders in West Africa. Final report submitted to the West Africa regional programme, Development Alternatives, Inc. WIDTECH, August.

Muzenda, G. (2015). "Choiceless choices of women in migration" in moving stories women and migration in Africa. *Journal on Africa Women's Experiences.*, *6*, 6–11.

Ndaye, N. (1999). *Violence towards women: The case of the trade in women for Prostitution and Response by the International Organization for Migration.* A speech delivered by the Deputy Director General International Organisation for Migration.

Piper, N. (2005). Gender and migration. Paper presented for *Policy Analysis and Research Programme.* Geneva: Global Commission on International Migration.

United Nations Development Program. (2009). *Human Development Report.* Geneva: UNDP.

Timothy Abayomi Atoyebi is a Lecturer/Researcher in the Department of Sociology, Bowen University, Nigeria. He holds a PhD in Industrial Sociology and Sociology of Development from the University of Ibadan, Nigeria. His research interests lie in gender studies, culture, trans-national migration, border studies, rural livelihoods studies, regional/trade integration and research methods. In 2008, he benefited from the cultural and academic exchange programme opportunity at Baylor University, Waco, the University of Mary Hardin-Baylor, Belton, Dallas Baptist University, Dallas, Hardin Simon University and Howard Payne University in Texas, USA.

19

Conclusion

Adebusuyi Isaac Adeniran
and Lanre Olusegun Ikuteyijo

The African continent is currently faced with a plethora of challenges which make the maximisation of the continent's human and natural resources a mirage. This book has identified some of these challenges and alternative frameworks have been suggested to take the continent to a higher pedestal in terms of viable and sustainable human development.

In terms of governance, African nations need to awaken from the bane of maladministration which has resulted from years of political instability, a glaring lack of ideology, the absence of viable institutions and widespread institutional corruption. These have plagued the continent and served as an albatross to development over the years. The premium on power post-colonial Africa has been exceptionally high, and the institutional mechanisms for moderating political competition are absent (Ake 1996). For the continent to move forward, a general re-orientation of what leadership entails must be brought into play. Leaders must move

A.I. Adeniran (✉) • L.O. Ikuteyijo
Department of Sociology and Anthropology, Obafemi Awolowo University, Ile-Ife, Osun State, Nigeria

A. Adeniran, L. Ikuteyijo (eds.), *Africa Now!*,
https://doi.org/10.1007/978-3-319-62443-3_19

from rendering lip service to effective governance to the demonstration of positive examples. Moreover, at all levels of governance official positions must be seen as an opportunity to serve the African people, and not as an opportunity for personal aggrandisement. The public good must be the overriding principle informing public spending, especially on education and technology.

Most African countries have been ranked high in terms of corruption. For instance, in 2015 Transparency International reported that nearly 75 million people in sub-Saharan Africa were estimated to have paid a bribe in the previous year while many were forced to pay to get access to the basic services that they needed desperately (Transparency International 2015). More recently, in 2017, the United Nations Office on Drugs and Crime, in conjunction with the National Bureau of Statistics, revealed that about 400 billion naira had been offered as bribes in the public service sector in Nigeria.

Although the African Union has adopted a number of normative frameworks that enshrine the values of accountable and democratic governance, many African countries are still gripped by a sit-tight mentality where leaders are either holding on to power or planning to hand over power to their children or relatives. However, the Economic Community of West African States seems to be coming to terms with this challenge as the Gambian experience sends warning signals to prospective sit-tight leaders. Democracy needs to be firmly entrenched in Africa in order to guaranty sustainable development. Crucially, however, the emergence of a visionary leadership is needed if transformational development is to be attained in Africa.

Africa has vast natural resources such as gold, petroleum, diamond, copper and uranium, among many others. These resources, if well harnessed, could ensure sustainable and inclusive growth for the continent. However, the development of these resources has been hampered by a number of factors including the resource curse syndrome; weak institutional policies; widespread corruption; and poor natural resources management policies. Collier and Laroche (2015) have noted that natural resources in most post-colonial African history have fuelled conflict and filled the pockets of political leaders much more than they have benefited the population. As noted in this book, mismanagement of natural resources in Nigeria, Kenya, and Côte d'Ivoire has had telling consequences in these countries. Some of these resources have become means of division, strife and contention rather than means of enhancing development. There is therefore a need for governments of resource-endowed countries to get the natural resource management policy chain right and spend time building robust institutions, thereby ensuring resource revenues to boost sustainable growth and development.

Education in Africa is another area which has been examined in this book. Access to early childhood education in terms of primary and post-primary education, including vocational education and training, remains a key challenge for the continent (Hopgood 2012). With almost 50% of Africans reportedly under the age of 15, Africa is no doubt the most youthful continent in the world. However, African countries are lagging far behind other regions in early childhood development (State of Education Report 2015). In 2013 for example, only about 79% of primary school age children in sub-Saharan Africa were enrolled in school (UNESCO 2015). Over 59 million children were out of school, while gender disparity continues to characterize school enrolment to the detriment of girls (Sustainable Development Goals Knowledge Platform 2016). This situation has also been exacerbated by the growing conflicts in various parts of Africa. In Northern Nigeria for example, the Boko Haram insurgency has displaced millions of people, including schoolchildren. Similar tales abound in Sudan, Libya and other crisis-prone areas in the continent. Even in areas where there are no conflicts, discrimination against girls has led to decreasing enrolment of young girls, which is exacerbated by the continuing use of child labour, which denies children their right to basic education. In the coming years, concerted efforts are required on the part of all stakeholders to ensure that the education of young girls is considered a top priority in most African countries.

The nexus between migration and development has also been critically discussed in this book. Particular attention has been paid to the various dimensions of migration, namely regional integration, migration for higher education and the feminisation of migration. Across the continent, there have been attempts by regional authorities to facilitate the free flow of people, with a view to enhancing regional integration and economic prosperity. However, various impediments to these laudable goals, including xenophobic tendencies, poor policy implementation, lack of cooperation and flagrant corruption should be addressed in order to harness the benefits of coordinated human mobility. The need to incorporate international cooperation and integration into policy goals, objectives and programmes has also been reiterated.

Since the mid-1990s, the security of the African continent has taken centre stage in the media. Across the continent, conflicts of various dimensions have been reported, with telling consequences in terms of social, political and economic effects, from xenophobic attacks in South Africa to ethno-religious violence in Kenya, Nigeria and Sudan. The Arab Spring also marked a watershed in the security chronicle of the continent. The rising incidence of armed violence, kidnapping, insurgency and terrorism among other novel security challenges illustrates the precarious state of security in Africa. Scholars writing in this book have advocated for increased regional cooperation as well as the empowerment of the youthful population as some of the tools to address this situation. It is time Africa looked inwards for solutions to the crises rather than rely completely on external aid, which comes with inevitable consequences.

Overall, this book has singled out identifiable institutional inadequacies as the cause of the discouraging state of human emancipation in Africa. This notwithstanding, the following suggestions are deemed apt in addressing the situation:

- Non-dependent endogenous policy forecasting must be embraced for any tangible progress to be achieved in the largely re-ordered modernity;
- Development programming, when implemented in Africa, must incorporate the institutional and historico-cultural specificities of Africans, so as to avoid unnecessary social friction;
- Educational curricula should be designed to align with the reality of specific developmental requirements;
- Most importantly, the entire citizenry must always be seen as partners in progress and, as such, they must always be involved in all efforts targeted at societal change; and for the process of societal emancipation to be made sustainable in Africa, Africans must be seen as a people prepared for business.

References

Ake, C. (1996, November 13). For Africa, the way forward. *The Guardian* (Lagos), p. 29.

Collier, P., & Laroche, C. (2015). Harnessing natural resources of inclusive growth. Retrieved from https://www.theigc.org/wp-content/uploads/2015/03/IGCGrowthBriefNRMCollier20151.pdf assessed on 12/04/2017

Hopgood, S. (2012). Key issues on education in Africa. This is Africa Online. Retrieved from http://www.thisisafricaonline.com/News/Key-issues-on-education-in-Africa?ct=true

State of Education in Africa Report. (2015). A report card on the progress, opportunities and challenges confronting the African education sector. Published by African-American Institute. Retrieved from www.aaionline.org

Sustainable Development Goals Platform. (2016). *Sustainable Development Goal 4: Ensure inclusive and equitable quality education and promote lifelong learning opportunities for all.* Retrieved from www.sustainabledevelopment.un.org

Transparency International. (2015). *People and corruption in Africa: Africa survey 2015*. Retrieved from www.transparency.org

UNESCO. (2015). *Regional overview, Sub-Saharan Africa: Education for All Global Monitoring 2015 Report.* United Nations Educational Cultural and Scientific Organization. Retrieved from www.efareport.unesco.org

Adebusuyi Isaac Adeniran holds a PhD in Development Sociology. He is presently a Senior Lecturer/Researcher at Obafemi Awolowo University, Ile-Ife, Nigeria and a Visiting Researcher at The Harriet Tubman Institute for Research on Global Migrations of African Peoples, York University, Canada and The Nordic African Institute, Uppsala University, Sweden. He has been a recipient of an International Sociological Association doctoral scholarship (2011), an Africa Initiative doctoral research grant (2011) and a Comparative Research Network grant from the Council for the Development of Social Science Research in Africa (2012). He has published widely in renowned local and international outlets. Among his recent publications are 'Engendering an Endogenous Framework for Socio-economic Development in the ECOWAS Sub-region' (*Critical Sociology*, Vol. 40(2), March 2014) and *Regional Economic Communities: Exploring the Process of Socio-economic Integration in Africa* (a co-edited book)

(CODESRIA, 2014). He specialises in ECOWAS (the Economic Community of West African States) and international development, Chinese–African relations, transnationalism, migration, integration and identity studies.

Lanre Olusegun Ikuteyijo is a lecturer and researcher in the Department of Sociology and Anthropology at Obafemi Awolowo University, Nigeria. He obtained his PhD in Sociology and Anthropology from the same university, specialising in criminology, migration and urbanisation. His research interests include youth migration, urban crime and community policing. He has published in reputable journals, contributed to a number of books and written a number of encyclopaedia entries published by SAGE and Blackwell Publishing. He was a visiting researcher at the Department of History, McMaster University, Canada between 2011 and 2012 under the auspices of the Centre for International Governance Innovation, Waterloo, Canada. Ikuteyijo has received a number of academic grants and fellowships, including: a Qualitative Research Network Africa Grant for his PhD thesis; a Council for Development of Social Research in Africa research grant; and an Africa Initiative research grant for work at the Brown International Advanced Research Institute (Population and Development). His latest publications include: 'Illegal Migration and Policy Challenges in Nigeria' (*African Portal*); 'The Image of Nigeria Police: Lessons from History' (*Journal of Applied Security Research*, Volume 9, Taylor and Francis Group, 2014); and 'EU Migration Policies and the Criminalisation of the Senegalese Irregular Migration Flows' (in *Between Migrations to Europe and Returns: International Commentary*, edited by Michelle Gonneli: Vol. 10, no. 35, Scuola Superiore Sant'Anna, Pisa, Italy).

Index[1]

[1] Note: Page numbers followed by "n" refers to notes.

A. Adeniran, L. Ikuteyijo (eds.), *Africa Now!*,
https://doi.org/10.1007/978-3-319-62443-3

FSC
www.fsc.org
MIX
Papier | Fördert gute Waldnutzung
FSC® C083411

Kolibri 360 GmbH
Ferdinand-Jühlke-Straße 7
99095 Erfurt, Deutschland
produktsicherheit@kolibri360.de